CARTILAGE:

MOLECULAR ASPECTS

BRIAN HALL

STUART NEWMAN

CRC Press
Boca Raton Ann Arbor Boston London

Library of Congress Cataloging-in-Publication Data

Cartilage: molecular aspects/[editors] Brian K. Hall, Stuart A. Newman.
p. cm.
Includes bibliographical references and index.
ISBN 0-8493-8817-1
1. Cartilage--Growth--Molecular aspects. 2. Cartilage--Aging--Molecular aspects. I. Hall, Brian Keith, 1941- . II. Newman, Stuart A.
[DNLM: 1. Cartilage--growth & development. 2. Cartilage--metabolism. WE 300 M718]
QP88.23M65 1991
611'.0183--dc20
DNLM/DLC
for Library of Congress 90-15176
CIP

Developed by Telford Press

This book represents information obtained from authentic and highly regarded sources. Reprinted material is quoted with permission, and sources are indicated. A wide variety of references are listed. Every reasonable effort has been made to give reliable data and information, but the author and the publisher cannot assume responsibility for the validity of all materials or for the consequences of their use.

Direct all inquiries to CRC Press, Inc., 2000 Corporate Blvd., N.W., Boca Raton, Florida 33431.

International Standard Book Number 0-8493-8817-1

Library of Congress Card Number 90-15176
Printed in the United States

Preface

Cartilage is a tissue of interest to a diversity of biological, biochemical, molecular and biomedical researchers and students. One of us having edited a three-volume series on cartilage in 1983, we felt that the time had come for a volume devoted to molecular aspects of cartilage. This is that volume. The rationale for this book is that since the chapters for the previous series were written (mostly in late 1981, early 1982), our knowledge of the molecular biology of cartilage has increased dramatically. Techniques for isolation and cloning of genes for important cartilage products; molecular aspects of cartilage mineralization; biophysical studies on cartilage organization; characterization of factors that induce cartilage formation; and molecular aspects of cartilage degradation have all advanced rapidly over the intervening decade.

Cartilage plays diverse roles in the body; a definitive supporting tissue in some organs; the basis of low friction surfaces in joints; and as a transient morphogenetic template during embryogenesis and skeletal repair. The versatility of cartilage derives from the remarkable properties of its extracellar matrix, the wide spectrum of regulatory systems that influence the synthesis and degradation of the unique molecular constituents of this matrix, and the development and growth of the cells that produce it. This book treats each of these topics, providing the first comprehensive reviews of the molecular biology of the genes specifying the collagenous and noncollagenous proteins of the cartilage matrix (Chapter 1), and of the biophysical-chemical properties of this matrix that suit it to the varying functions it performs (Chapter 2). Up-to-date critical reviews and evaluations of the unique requirements on the energy metabolism of cartilage as a nonvascular tissue (Chapter 3), the role of polypeptide growth-inducing factors and ions in the regulation of cartilage growth and development (Chapters 4 and 5), and the mineralization and degradation of cartilage matrix (Chapters 6 and 7) make this volume a valuable source for results, concepts and mechanisms from the multiple disciplines that bear on modern research into the biology of cartilage.

All the chapters are written by acknowledged experts, active in the diverse fields of cartilage research. The chapters are up-to-date, authoritative, comprehensive and synthetic, affording insights into the molecular mechanisms underlying the development, differentiation, maintenance, metabolism, function and/or degradation of cartilage. This volume will be valuable for all those interested in cartilage, whether from biochemical, molecular, biophysical, biological, biomedical, function or clinical points-of-views, and whether the interest is in developing cartilage in the embryo, differentiation of cartilage *in vitro,* function of cartilage in the adult, maintenance and/or degradation of cartilage at joint surfaces, cartilage in normality or in disease, or in function or dysfunction.

Brian K. Hall
Stuart A. Newman
April 1991

The Editors

Brian K. Hall, Ph.D., D.Sc. is Izaak Walton Killam Research Professor in the Department of Biology, Faculty of Science, Dalhousie University, Halifax, N.S., Canada. He also holds an appointment as Professor of Physiotherapy, Faculty of Health Professions, Dalhousie University.

Professor Hall received his B.Sc. (Hons.) degree in 1965, his Ph.D in Zoology in 1969, and a D.Sc. in Biological Sciences in 1978 from the University of New England, Armidale, NSW, Australia. He was appointed a Teaching Fellow in the Department of Zoology, UNE in 1965, Assistant Professor in the Department of Biology, Dalhousie University in 1968, Associate Professor in 1972, Professor in 1975 and Killam Research Professor in 1990. He served as Chair of the Biology Department, Dalhousie University from 1978-85.

Professor Hall is a Fellow of the Royal Society of Canada, and member of the American Society of Zoologists, International Society for Differentiation, British Society for Developmental Biology, International Society for Developmental Biology, Society of Vertebrate Palaeontology, The Bone and Tooth Society and the British Connective Tissue Society. He is an editor of *Anatomy and Embryology,* member of the editorial board of the *Journal of Craniofacial Genetics and Developmental Biology,* member of the international advisory board of the *Croatian Medical Journal,* past associate editor of the *Canadian Journal of Zoology* and past member of the advisory editorial board of *Bone*.

Professor Hall has worked on the development and differentiation of cartilage and bone for 25 years. He has presented 60 invited lectures at international meetings and symposia and 90 guest lectures at universities and institutes. He is the author or coauthor of 5 books, 125 papers in refereed scientific journals, and author of 33 chapters in edited works or conference proceedings. He edited a three-volume series on Cartilage (1983) and is editor of the seven-volume series Bone being published by CRC Press. The current focus of his research is on the development and evolution of neural crest-derived craniofacial cartilage and bone.

Stuart A. Newman, Ph.D. is Professor of Cell Biology and Anatomy at New York Medical College, Valhalla, New York.

Dr. Newman received his A.B. in chemistry in 1965 from Columbia University, and his Ph.D. in chemical physics in 1970 from the University of Chicago. He did postdoctoral work in biophysics at the University of Chicago, and in developmental biology at the University of Sussex, U.K. He was an Instructor in Anatomy at the University of Pennsylvania and an Assistant Professor of Biological Sciences at the State University of New York at Albany before joining New York Medical College in 1979.

Dr. Newman has been an INSERM Fellow at the Pasteur Institute, Paris (1984) and a Fogarty Senior International Fellow at Monash University, Melbourne, Australia (1989). His research interests include mechanisms of skeletal pattern formation in the vertebrate limb, and regulation of chondrogenesis at the nuclear level.

Contents

Chapter 1
The Active Genes of Cartilage 1
W. B. Upholt and B. R. Olsen

Chapter 2
Physicochemical Aspects of Cartilage Extracellular Matrix 59
Wayne D. Comper

Chapter 3
Energy Metabolism in Cartilage 97
Irving M. Shapiro, Toshizaku Tokuoka, and Susan F. Silverton

Chapter 4
Cartilage Growth and Differentiation Factors 131
Saeid M. Seyedin and David M. Rosen

Chapter 5
Ionic Regulation of Chondrogenesis 153
Rocky S. Tuan

Chapter 6
The Growth Plate: Cellular Physiology, Cartilage Assembly and Mineralization 179
A. Robin Poole

Chapter 7
Cartilage Degradation 213
Jenny A. Tyler

Index 257

1
The Active Genes of Cartilage

W. B. UPHOLT
Department of Biostructure and Function
School of Dental Medicine
University of Connecticut Health Center
Farmington, Connecticut

B. R. OLSEN
Department of Anatomy and Cellular Biology
Harvard Medical School
Boston, Massachusetts

Introduction

In this chapter, the structure and molecular biology of the major cartilage components is described. We review first the structure and regulation of the genes encoding the collagen components of cartilage fibrils. This is followed by a discussion of the genes coding for the cartilage proteoglycan core protein, link protein, and cartilage matrix protein. Finally, we review the molecular biology and speculate on the function of type X collagen, a short collagenous molecule synthesized by hypertrophic cartilage.

The mixture of collagen fibrils and proteoglycan aggregates gives hyaline cartilage its unique physiological properties. The collagen fibrils provide elements of high tensile strength, while the proteoglycan aggregates exert an internal swelling pressure due to their hydrophilic character. The collagen fibrils contain type II collagen as the major component and smaller amounts of types IX and XI collagen (Mendler *et al.*, 1989). The collagen fibrils of normal cartilage are thin with type IX collagen molecules located at or close to the surface (Vaughan *et al.*, 1988). The precise location of type XI collagen molecules within the fibrils has not yet been definitively determined, but they may form a central core (Mendler *et al.*, 1989). It has been suggested that type XI collagen within cartilage fibrils somehow plays a role in the regulation of fibrillar diameters, and that type IX collagen molecules on the fibrillar surface provide a molecular link between the fibrils and other matrix components, such as proteoglycans (Ninomiya *et al.*, 1989).

The proteoglycan aggregate is composed of a variety of molecules including hyaluronic acid, the major proteoglycan of cartilage and link protein (Sandell, 1987). In addition to these macromolecules, a number of other proteins have been described in cartilage including cartilage matrix proteins, and a variety of others which may be cartilage-specific, many of which have been characterized only partially.

Several components that are not unique to cartilage have been described in cartilage. These include types V and VI collagens in articular cartilage (Eyre and Wu, 1987; Mayne, 1989), and anchorin CII (Mollenhauer and von der Mark, 1983). Since the functional significance of the presence of these components in cartilage is unclear, they are not discussed here, even if extensive data on the structure of their genes are available.

The Type II Collagen Gene

Type II Collagen

Type II collagen is the major component of cartilage extracellular matrix fibrils (Mendler *et al.*, 1989). The molecules are composed of three identical polypeptide subunits, named α1(II) chains (Bornstein and Sage, 1980; Eyre, 1980). The protein and gene structures of type II collagen are similar to those of types I, III, V, and XI collagens. In addition to its localization to cartilage, type II collagen is also present in the vitreous of the eye, in the nucleus pulposus of intervertebral disks, and in the embryonic chick primary corneal stroma (Bornstein and Sage, 1980; Eyre, 1980; von der Mark *et al.*, 1977). In addition, it has been shown to be present under several embryonic epithelia (Fitch *et al.*, 1989; Kosher and Solursh, 1989; Thorogood *et al.*, 1986; Thorogood, 1988). The presence of type II collagen mRNA in developing chick (McDonald and Tuan, 1989; Kravis and Upholt, 1985) and mouse (Lovell-Badge *et al.*, 1987) calvaria has also been reported.

Molecular Cloning of the α1(II) Collagen Gene

The structure and expression of the type II collagen gene (COL2A1) has been reviewed recently (Upholt, 1989). Thus the discussion here will be limited to the major points previously reported together with a discussion of more recent information regarding this gene.

Complementary DNA clones for type II collagen have been isolated from a variety of species (Figure 1) including human (Elima *et al.*, 1985; Baldwin *et al.*, 1989a), chicken (Lukens *et al.*, 1983; Ninomiya *et al.*, 1984; Vuorio *et al.*, 1982; Young *et al.*, 1984; Deak *et al.*, 1985), rat (Kohno *et al.*, 1984), and cow (Sangiorgi *et al.*, 1985b). Most of the cDNA clones consist of relatively short sequences (500 to 1500 bp) near the 3′ ends of the genes with three exceptions. A cDNA clone corresponding to 500 nucleotides at the 5′ end of the rat mRNA has been cloned by primer extension using rat chondrosarcoma mRNA (Kohno *et al.*, 1984). Also, a series of cDNA clones

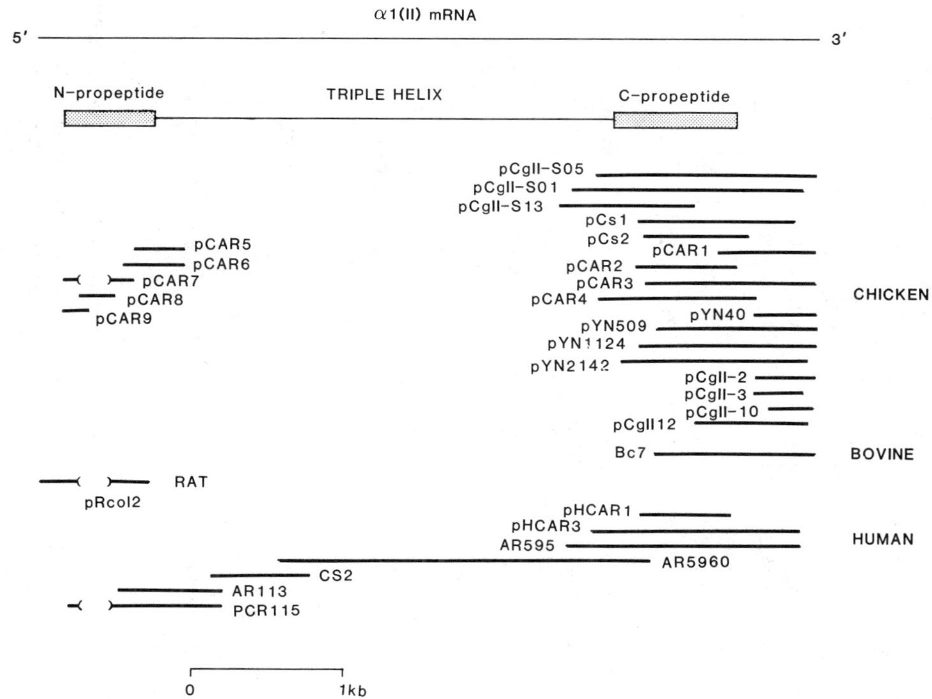

Figure 1. Diagram showing the relative size and location of chicken, bovine, rat, and human cDNAs encoding the α1(II) collagen chain. The mRNA and the domain structure of the translation product are shown at the top. The parentheses indicate the absence of the alternatively spliced second exon. References: pCGII-S01, 05 and 13, Deak *et al*. (1985); pCsl and 2, Young *et al*. (1984); pCAR1 and 2, Vuorio *et al*. (1982); pCAR3 and 4, Barembaum and Upholt (unpublished); pCAR5-9, Nah and Upholt (unpublished); pYN40, 509, 1124, and 2142, Ninomiya *et al*. (1984); pCGII-2, 3, 10, and 12, Lukens *et al*. (1983); Bc7, Sangiorgi *et al*. (1985b); pRco12, Kohno *et al*. (1985); pHCAR1 and 3, Elima *et al*. (1985); AR595, 5960, 113, CS2, and pPCR115, Baldwin *et al*. (1989b).

encoding the amino propeptide of chicken type II collagen have been constructed using polymerase chain reaction techniques (Nah and Upholt, unpublished), and a series of cDNA clones covering nearly the full length of the human mRNA have been constructed by a combination of primer extension and polymerase chain reaction techniques using RNA isolated from knee condyles, femoral heads, and tibiae from a spontaneous 24-week human abortion (Baldwin *et al.*, 1989a). Su *et al.* (1989b) have also reported the nucleotide sequence of the entire mature human α1(II) chain. The clones corresponding to the human type II collagen mRNA have allowed the assembly of the complete nucleotide sequence for a type II collagen mRNA and the complete amino acid sequence of the translation product of the human type II collagen gene. Comparison of hydropathy profiles of the type II procollagen amino acid sequence deduced from the cDNA sequence with hydropathy profiles for other human collagens (Baldwin *et al.*, 1989a) extends and confirms previous observations that COL2A1 and COL1A1 are more closely related than any other pair of known genes for fibrillar collagens. These findings suggest either that the genes for type II collagen and the α1 chain of type I collagen have resulted from the most recent gene duplication event in the family of fiber forming collagens or that these two genes may have been subject to more similar selective pressures than other collagen genes.

Part or all of the human COL2A1 gene has been isolated by several groups (Sangiorgi *et al.*, 1985a; Strom and Upholt, 1984b; Nunez *et al.*, 1986; Cheah *et al.*, 1985; Sandell, 1985; Weiss *et al.*, 1982; Vikkula and Peltonen, 1989) and approximately 50% of the exons have been mapped and sequenced. The chicken gene has also been isolated (Sandell *et al.*, 1983, 1984; Young *et al.*, 1984) and, although the region encoding the 8 to 9 5′-most exons of the chicken type II collagen gene has not yet been cloned, all 46 exons in the cloned portion have been mapped and each exon has been at least partially sequenced (Upholt and Sandell, 1986). A genomic clone corresponding to the 5′ end of the rat type II collagen gene has been isolated and the region of the promoter-1st exon-1st intron has been characterized by DNA sequencing (Kohno *et al.*, 1985). Two overlapping genomic clones containing approximately 17 kb of the bovine proα1(II) gene have also been isolated and the 3′ most exon has been defined within the sequences. Cosmid clones containing the entire mouse type II collagen gene have also been isolated and characterized (Vuorio and de Crombrugghe, personal communication).

The Exon Organization of the α1(II) Collagen Gene is Similar to that of Other Fibrillar Collagen Genes

The studies of the exon/intron organization of the type II collagen gene (Figure 2) show a high degree of similarity with the genes for α1(I), α2(I), α2(III), and α2(V) collagens (Upholt and Sandell, 1986; Ramirez *et al.*, 1985; Ramirez, 1989; Upholt *et al.*, 1989). Within the triple-helical coding region there is only one position where a difference in the exon/intron organization

A

	1	2	2B	3	3B	4	5	6
COL2A1	85-91	(210)	17	33	33	54	102-105	81 (9)
COL1A1	82-103	195		35		36	102	72 (9)
COL1A2	70	—	11	15-18		36	84-93	54-57 (9)
COL3A1	79-85	200-203		51		114		81 (27)

B

	SP	Cys-rich	J1	TH	NTH	TH	J2
COL2A1	28-30	(70)	3	12	4	63	21-22
COL1A1	27-34	65	9	—	—	51	19
COL1A2	23	—	7-9	—	—	45	13-14
COL3A1	26-28	67-68	8	—	—	39	24
COL5A2	5 30	5 76	5 3	12	4	63*	24

Figure 2. Comparison of the gene and protein structure of the amino prepropeptide of type II collagen with other collagens. (A) Exon/intron organization. (B) Protein domain organization. Data are based on information available for the human, chicken, mouse and rat genes. Variations in sizes represent interspecies variations. Numbers in parentheses for exon 2 and Cys-rich domain indicate an alternatively spliced region. Numbers in parentheses for exon 6 correspond to the number of bp in exon 6 which encode part of the major triple helical domain. The COL1A2 gene is indicated as lacking exon 2 because the small 11 bp exon does not encode a large Cys-rich domain. In part B, the protein has been divided into 7 domains: the signal peptide domain (SP) which has been defined by the size of exon 1 and includes a few amino acids between the cleavage site and the Cys-rich domain; the Cys-rich domain defined by the boundaries of exon 2; a junction domain (J1) between the Cys-rich domain and the amino-propeptide triple helical domain (TH); a 4 amino acid interruption (NTH) in the amino-propeptide triple helical domain; and the junction domain (J2) which encodes the N-proteinase cleavage site and which is defined by the ends of the amino-propeptide triple helical and the mature triple helical regions. References: Ramirez (1989); COL2A1, Nunez *et al.* (1986), Nah and Upholt, unpublished; COL1A1, Tromp *et al.* (1988), Finer *et al.* (1987); COL1A2, Boedtker *et al.* (1985), Kuivaniemi *et al.* (1988), Schmidt *et al.* (1984); COL3A1, Benson-Chanda *et al.* (1989), Wood *et al.* (1987), Liau *et al.* (1985), Yamada *et al.* (1984); COL5A2, Woodbury *et al.* (1989).

has been observed; the genes for type II and type III collagens and for the α2(I) chain of type I collagen have two 54 bp exons with an intervening intron whereas the α1(I) gene has a single exon containing the corresponding 108 bp.

In the carboxyl-terminal propeptide, the positions of the intron interruptions have been conserved between the α1(I), α2(I), α1(II), α1(III), and α2(V) genes, but there are several positions where codons have been deleted or inserted, particularly in the region encoding the C-terminal protease cleavage site.

At the end encoding the amino-terminal propeptide there are considerably greater differences between the various proteins and the structures of their genes, making it difficult to define corresponding regions.

The human (Baldwin *et al.*, 1989a) and rat (Kohno *et al.*, 1984) type II collagen cDNA sequences lack the region corresponding to the 2nd exons of the α1(I) and α1(III) collagen genes. This exon encodes a large globular domain with many disulfide-bonded cysteines. The amino propeptide of chicken type II collagen as analyzed by Curran and Prockop (1982) also lacks this domain. However, Ryan *et al.* (1990) and Vuorio and deCrombrugghe (personal communication) have recently identified sequences in the first intron of the human and mouse type II collagen genes, respectively, which encode a potentially functional exon corresponding to the large second exon of the α1(I) and α1(III) collagen genes. Using oligonucleotide probes that bridge the junctions between the two ends of this exon and the two flanking exons and an oligonucleotide probe bridging the first and third exons, Ryan and Sandell (1990) found hybridization patterns which suggest that forms of type II collagen mRNA both with and without the exon are present in human cartilage. These results suggest that this exon may be alternatively spliced in some situations.

The type II collagen gene also may encode the α3 chain of type XI collagen. This hypothesis is based on experiments showing very similar gel electrophoresis patterns and properties of CNBr and V8 protease peptides for the triple-helical regions of these two chains (von der Mark *et al.*, 1982; Clark and Richards, 1985; Sussman *et al.*, 1984; Furuto and Miller; 1983) and the lack of detection of a second gene in chicken closely related to the type II collagen gene (Sheffield and Upholt, 1985). However, there are no published data regarding the structure of the N-terminal propeptides of the α3(XI) chain other than those suggesting that the N-terminal propeptide is not removed from the mature α3(XI) chain (Eyre and Wu, 1987).

Type II collagen as characterized by the rat, human, and chicken (Nah and Upholt, unpublished) cDNA sequences contains an N-propeptide triple-helical domain with many similarities to the α2 chain of type V collagen (Woodbury *et al.*, 1989) but different from either type I or type III collagen. The amino-terminal propeptide triple-helical domain in types II and α2(V) chains differs from that in α1(I) or type III chains in that it is considerably longer (22 to 25 Gly-X-Y triplets in types II and V collagens vs. 13 to 17 in type I or III collagens) and in that types II and V contain a nonhelical interruption of four amino acid residues after the first four triple-helical repeats. In the α1(I) collagen gene the triple-helical region of the amino-terminal propeptide is encoded by exons 3 to 5 from the 5′ end of the gene (D'Alessio *et al.*, 1988). The longer N-propeptide triple-helical domain of type II collagen is in part encoded by a longer exon (54 vs. 36 bp) corresponding to exon 4 of the α1(I) gene and in part by two additional exons preceding the 54 bp exon (Su *et al.*, 1989a; Upholt, Makela, and Mannisto, personal communication). These two exons either do not have counterparts in the α1(I) collagen gene or have been fused in the α1(I) gene to form the third exon of the type I gene (or alternatively interrupted by introns in the type II gene). In fact, there are three exons in the type II gene (an additional one of 17 bp as well as the two 33 bp exons) which appear to encode the region of the type II N-propeptide corresponding to that encoded by exon 3 (35 bp) of the α1(I) gene. There is insufficient nucleotide or amino acid sequence similarity in the amino-terminal end of this triple-helical region to come to any definite conclusions regarding whether exon 3 of the α1(I) gene corresponds to the fusion of the three type II exons (or alternatively the three exons resulting from the subdividing of exon 3 of the type I gene) or whether the type II collagen gene (and possibly the α2(V) gene) contains two (or three) exons which have no counterpart in the α1(I) gene.

The presence of the large apparently alternatively spliced exon in the type II gene corresponding to exon 2 of the α1(I) gene suggests that the 17 and 33 bp exons (second and third exons not counting the large alternatively spliced exons) are not vestigial remnants of the cysteine-rich domain exons as was previously thought (Ramirez, 1989). The small second (11 bp) exon of the α2(I) gene also might correspond to the 17 bp exon of the type II collagen gene rather than being a vestigial remnant exon corresponding to the 195 bp exon 2 of the α1(I) gene (Boedtker *et al.*, 1985; Ramirez, 1989).

These extensive differences in the organization of the protein and the gene in the regions encoding the N-terminus of the types II and I collagen amino-terminal propeptides contrast with the high degree of similarity in the remainder of the proteins and genes as discussed above. The regions of similarity between the type II collagen sequence and the α2 (V) sequence are intriguing since in the remainder of the type II collagen gene (the triple-helical and carboxyl-terminal propeptide) type II and α1(I) are more closely related than any other known pair of fibrillar collagen chains. The similarity of the type II amino propeptide region to that of type V is also interesting in that: (1) types V and XI collagens are considered to be close relatives constituting a specialized subclass of the fibrillar collagen family (Eyre and Wu, 1987); (2) α1(XI) chains have been found in association with type V collagen in bone (Niyibizi and Eyre, 1989); and (3) the α1(II) gene may encode the α3(XI) chain.

Regulation of the α1(II) Collagen Gene

The major mRNA for type II collagen is approximately 5000 nucleotides in length, including approximately 160 nucleotides of 5′ nontranslated sequence (human and rat) and 439, 513, and 434 nucleotides of 3′ nontranslated sequence for the human (Elima *et al.*, 1987), chick (Sandell *et al.*, 1984; Ninomiya *et al.*, 1984), and bovine (Sangiorgi *et al.*, 1985b) RNAs, repectively. A second minor RNA species of approximately 7000 nucleotides has been observed in some chicken tissues, particularly in those tissues expressing moderate amounts of type II collagen mRNA (Vuorio *et al.*, 1982). The origin of the additional 2000 nucleotides has not been identified.

Most studies on expression of the gene have used chick tissues, although some studies used rat tissues. A variety of differentiating systems have been studied, including chicken limb chondrogenesis (Kravis and Upholt, 1985; Kosher *et al.*, 1986b), chick mandibular chondrogenesis (Mina *et al.*, in press), the developing chick cornea (Hayashi *et al.*, 1988), and redifferentiating tibial chondrocytes from chick limb (Castagnola *et al.*, 1988). In both the developing limb and mandible, low levels of type II collagen mRNA are detected in the mesenchyme (or ectomesenchyme) well prior to overt chondrogenesis and at times when type II collagen protein has not been reported to be present in the tissue (Kravis and Upholt, 1985; Kosher *et al.*, 1986b; Mina *et al.*, in press). Upon differentiation there is a large increase (approximately 100-fold) in the amount of type II collagen mRNA per cell. Some of these studies have been done using culture systems in which limb mesenchyme differentiates into cartilage in culture paralleling its normal differentiation *in vivo* (Kravis and Upholt, 1985; Kosher *et al.*, 1986b). These systems are particularly useful in that they permit experimental manipulation.

The cells containing the low level of type II collagen mRNA are not limited to those which subsequently differentiate into cartilage but appear to include a larger population of undifferentiated mesenchymal cells in the limb or neural crest-derived undifferentiated ectomesenchymal cells in the developing mandible (Nah *et al.*, 1988; Kosher and Upholt, unpublished; Mina *et al.*, 1991). There have been several reports of collagen located at epithelial-mesenchymal interfaces throughout the early developing chick embryo which reacts with antibodies against type II collagen, although it is unclear which cells are synthesizing the immunoreactive material (Thorogood *et al.*, 1986; Thorogood, 1988; Kosher and Solursh, 1989; Fitch *et al.*, 1989). As of the present such studies have been reported only in chicken tissues.

These studies of low levels of type II collagen mRNA or immunoreactive protein should be interpreted with caution, since it is possible that the probes presently in use may not distinguish between the α1 chain of type II collagen and the α3 chain of type XI collagen as the mRNAs for two chains are presumed to be transcribed from a single gene. It is presently unclear whether the mRNAs for these two chains are identical or whether they may be differently transcribed or processed from the same gene. The two pepsin treated protein chains presently can be distinguished only by the higher level of hydroxylation and glycosylation of lysine residues in the α3(XI) chain compared to the α1(II) chain (Furuto and Miller, 1983). Resolution of these questions awaits both further characterization of the α3(XI) procollagen chain and its mRNA as well as more complete studies of the collagens being synthesized in these tissues.

Gerstenfeld *et al.* (1989) have published a careful quantitative analysis of collagen expression in cultured chondrocytes from the caudal half of 16-day-old chicken embryonic sterna and the ventral half of 12-day-old chicken embryo vertebra. These studies, together with previous studies from other laboratories (Allebach *et al.*, 1985; Finer *et al.*, 1985), suggest that type II collagen gene expression is regulated at a variety of levels in differentiated chondrocytes, including regulation of steady state mRNA levels, of efficiency of translation, and of efficiency of processing, and that the precise mechanisms of regulation and the level of type II collagen mRNA vary depending upon the type of cell (Castagnola *et al.*, 1988; Gerstenfeld *et al.*, 1989; Horton *et al.*, 1988; Allebach *et al.*, 1985). The results which have been obtained with the various systems suggest that caution should be used in generalizing results obtained with any one system to all types of cartilage.

Several laboratories have studied the regulation of expression of the type II collagen gene in "dedifferentiated" chondrocytes obtained by treatment of cultured chondrocytes with various agents including bromodeoxyuridine (Saxe *et al.*, 1985), phorbol myristate acetate (Finer *et al.*, 1985), Rous sarcoma virus (Allebach *et al.*, 1985), and retinoic acid (Horton *et al.*, 1987b). Each of these agents causes a partially reversible change in the chondrocyte phenotype toward a more fibroblast-like cell with reduced levels of type II collagen mRNA. Treatment of human costal chondrocytes in culture with interleukin 1 results in decreased levels of type II collagen mRNA (Goldring *et al.*, 1988). Horton *et al.* (1989) have recently reported that transforming growth factor-β and fibroblast growth factor act to inhibit type II collagen synthesis in cultured chick sternal chondrocytes and that the two growth factors act synergistically. This synergistic effect is most likely mediated through transcriptional control, as chondrocytes transfected with a plasmid carrying the type II collagen promoter and first intron ligated to the CAT gene have dramatically reduced CAT activity following treatment with the two growth factors. In contrast, TGF-β promotes cartilage differentiation and type II collagen gene expression when applied to differentiating limb mesenchyme in culture (Kulyk *et al.*, 1989), fibroblast growth factor alone inhibits the process, and its antichondrogenic effect is potentiated by TGF-β.

In situ hybridization studies (Figure 3) using probes for type II collagen mRNA have been reported in the developing chick (Nah *et al.*, 1988; Swalla *et al.*, 1988; Devlin *et al.*, 1988) and human (Sandberg *et al.*, 1987) limb, chick mandible (Mina *et al.*, unpublished), cornea (Hayashi *et al.*, 1986, 1988), 8-week-old chick epiphyseal cartilage (Oshima *et al.*, 1989), chick calvarium (McDonald and Tuan, 1989) and sternum (Hayashi *et al.*, 1986; LuValle *et al.*, 1989), and micromass cultures of chick limb bud mesenchyme (Gregg *et al.*, 1989). These results are consistent with the previous biochemical studies showing that there are large increases in the level of type II collagen mRNA at the time of chondrogenesis. Studies on the developing cornea (Figure 4) that include immunohistochemistry at the electron microscope level are particularly interesting in that they clearly show that epithelial cells can synthesize type II collagen (Hayashi *et al.*, 1988).

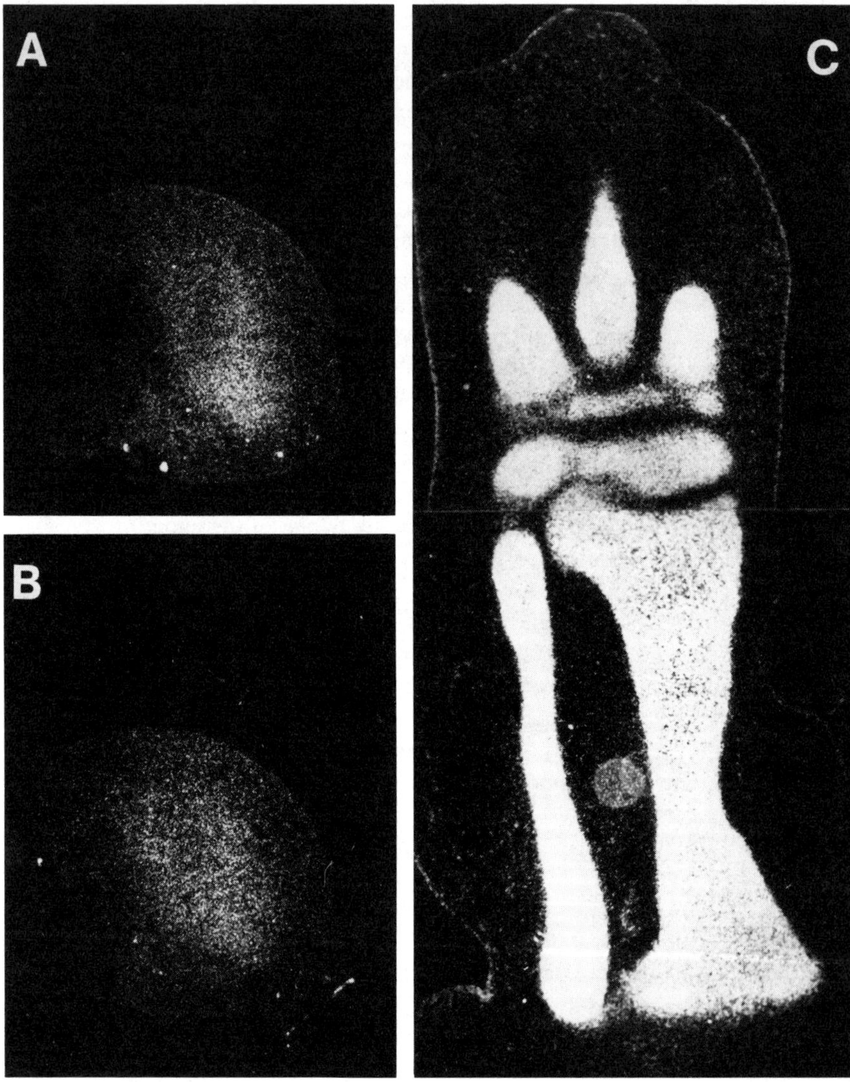

Figure 3. Hybridization of type II RNA probes to stage 24 and stage 32 chick embryo limbs. (A) dark field autoradiograph of a frontal section through a state 24 wing bud that was hybridized with ^{35}S-labeled single-stranded type II collagen RNA probe illustrating a hybridization signal in the condensing central core of the limb at the onset of chondrogenesis; (B) section adjacent to the above that was prehybridized with an approximately 200-fold excess of unlabeled $\alpha1(I)$ cRNA before being hybridized with the labeled type II collagen probe; (C) frontal section through a stage 32 (7-day) forelimb that was hybridized with the same ^{35}S-labeled type II collagen probe used in (A) and (B). $\times 40$ (Reproduced from Nah *et al.*, with permission.)

The 5′ end of the type II collagen gene has not yet been isolated from the chicken genome but has from the human and rat genes. Sequences of the 5′ flanking region, the first exon, and part or all of the first intron of the rat and human genes have been determined (Su *et al.*, 1989a; Kohno *et al.*, 1985; Nunez *et al.*, 1986; Ryan *et al.*, 1990). The genes from these two species show considerable similarity in the 5′ flanking sequences and in parts of the first intron, but show little similarity to corresponding regions of the α(I) collagen gene. Both the rat and the human genes contain ATATAA sequences at approximately −25 bp from the transcription start site, but neither contains an obvious "CCAAT" element on either strand in the expected location. The genes contain the GC-rich hexanucleotide GGGCGG on both strands as well as other sequences (inverted complements and consensus sequences corresponding to known regulatory elements in other genes) which are putative control sequences, although there have been no published mutagenesis experiments correlating control functions with these sequences.

Horton *et al.* (1987a, 1989) have reported transfection experiments in chicken cells using potential regulatory sequences of the rat type II collagen gene coupled to the gene for chloramphenicol acetyltransferase. The 5′-flanking sequences acted as a weak promoter with little difference in activity in chondrocytes compared with other mesenchymally derived cells. However, when an 800-bp region of the first intron was included in the plasmid together with the 5′-flanking promoter sequence, expression of chloramphenicol acetyltransferase was increased 13- to 18-fold in cartilage cells compared to noncartilage cells.

Lovell-Badge *et al.* (1985, 1987) transformed mouse embryonic stem cells with a human cosmid clone containing the entire gene including approximately 4.5 kb of 5′ and 2.2 kb of 3′ flanking DNA and examined mouse chimeras containing the gene. Expression of the gene was restricted to those tissues transcribing the endogenous gene, and there was a low level of inappropriate expression of the transformed human gene in the cultured mouse embryonic stem cells following transformation which was not detected in the chimeric animals.

Studies of the methylation of cytosine residues of the type II collagen gene in chicks and humans provide little evidence consistent with methylation as a primary regulatory mechanism for this gene. In chickens, the type II collagen gene was found to be hypomethylated in cartilage compared to other tissues (Fernandez *et al.*, 1985; Sheffield and Upholt, unpublished). However, there was no change in the methylation pattern following "dedifferentiation" of chondrocytes treated with bromodeoxyuridine or retinoic acid (Fernandez *et al.*, 1985), and hypomethylation of the gene followed, rather than preceded, or accompanied differentiation of chick limb mesenchyme into cartilage (Sheffield and Upholt, unpublished). Vuorio *et al.* (1987) examined 12 *Sma*I sites throughout the human gene and found a complex pattern of variation of the level of methylation of the gene which depended both on position in the gene

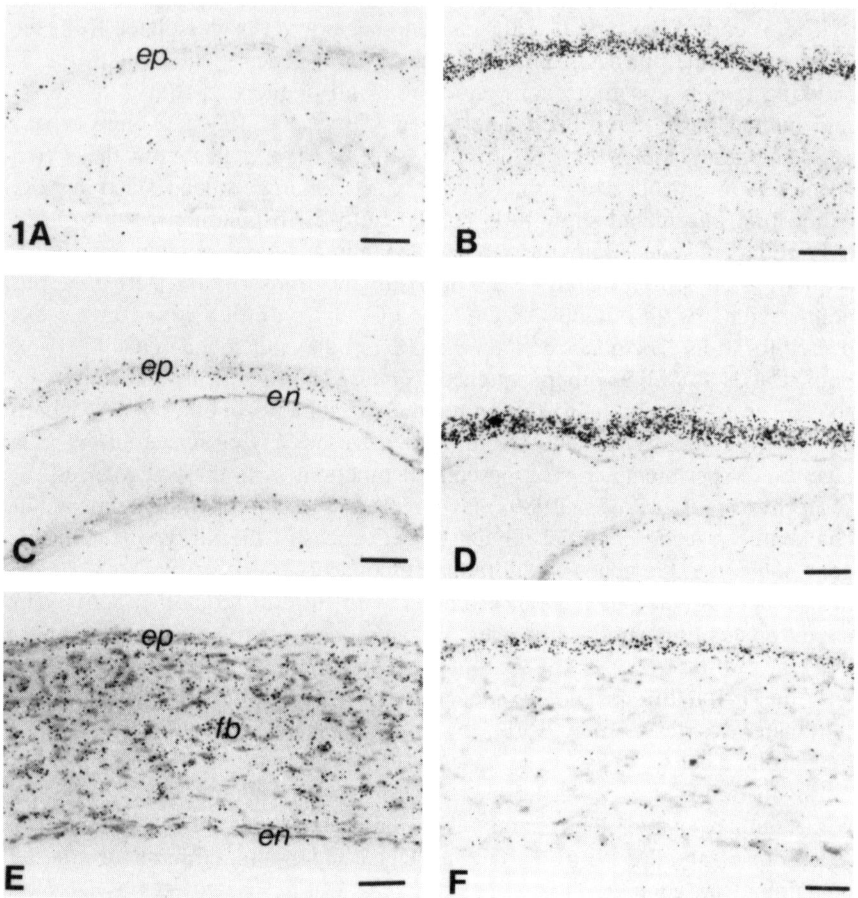

Figure 4. Sections of chick cornea processed for *in situ* hybridization with ^{3}H-labeled cDNA probes for type I (A, C, E, G) and type II (B, D, F, H) collagen mRNAs. The epithelial layers (ep) of panels A to D are sectioned obliquely and therefore appear thicker than in other panels. At stage 18 (A, B) few silver grains are noted over the epithelium with the type I collagen probe, while there is heavy labeling with the type II probe. At stage 26 (C, D) the density of grains is increased over the epithelium with both probes, a small number of grains are recognized over the endothelium (en) with both probes. At stage 31 (E, F) the type I probe labels the stromal fibroblasts (fb), while the type II probe still shows significant labeling of the epithelium. At stage 39 (G, H) there is heavy labeling of the stromal fibroblasts with the type I collagen probe, and moderate labeling of the epithelium and endothelium with the type II probe. Bar, 20 μm. (Reproduced from Hayashi *et al.*, 1988, with permission.)

and on the tissue source of the DNA. In all tissues examined, the sites at the 5′ end of the gene were unmethylated; only in cartilage and placental DNA were all sites unmethylated.

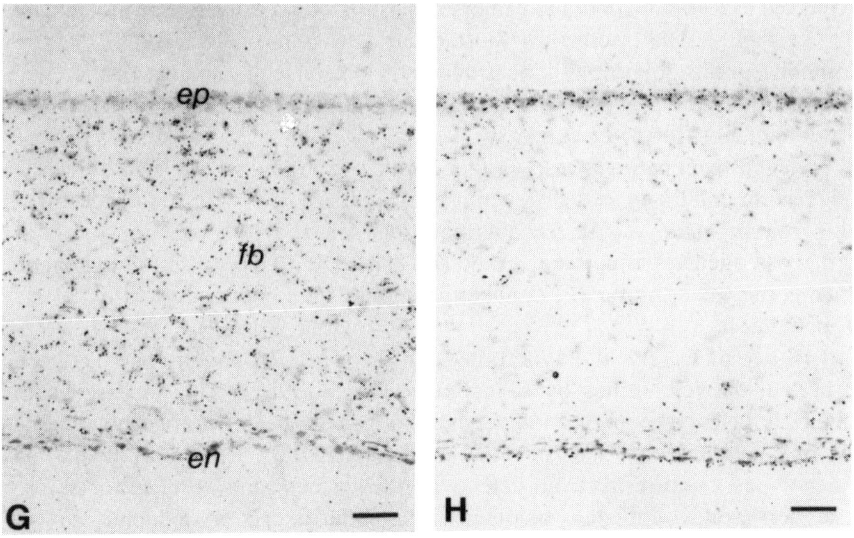

FIGURE 4 (continued).

Chromosomal Location

The human type II collagen has been localized to chromosome 12 in the region of 12q13 (Solomon *et al.*, 1985; Strom *et al.*, 1984a; Huerre-Jeanpierre *et al.*, 1986a; Law *et al.*, 1986; Arheden *et al., 1989)*. Law *et al.* (1986) have reported the gene to be in the region q14.3 by comparison of probe hybridization to somatic cell hybrids with two different terminal deletions of chromosome 12. More recently, Arheden *et al.* (1989) have noted that the chromosome segment 12q13–15 has been consistently rearranged in several human tumors and suggest the 12q13 is the correct position. A number of polymorphic restriction sites (PvuII, HindIII, BamHI, HinfI) (Weaver and Knowlton, 1989; Sykes, 1983; Driesel *et al.*, 1982; Nunez *et al.*, 1985; Sykes *et al.*, 1985a; Wordsworth *et al.*, 1988; Eng and Strom, 1985; Strom, 1988b; Vaisanen *et al.*, 1988; Vikkula and Peltonen, 1989) have been mapped throughout the gene as well as a highly polymorphic (hypervariable) region downstream from the 3′ end of the mRNA (Sykes *et al.*, 1985b; Pope *et al.*, 1984; Stoker *et al.*, 1985). These polymorphic sites have been used to investigate possible linkage of the type II collagen gene with a number of different hereditary connective tissue disorders with potential for involvement of the type II collagen gene.

Lee *et al.* (1989) have recently described a family with spondyloepiphyseal dysplasia in which one of the alleles of the type II collagen gene has a deletion of 390 bp that removes an entire exon of 108-bp encoding acid residues 964 to 999 near the carboxyl terminus of the triple-helical domain. Since type II collagen is composed of three identical alpha chains, this deletion would be expected to result in 1/8 of the trimers consisting of normal trimeric molecules, 1/8 of triple-helical molecules shortened by 36 amino acids, and 3/4 of overmodified, unstable triple-helices due to the presence of one or two shortened chains in the trimer. These unstable trimers would most likely not be secreted from the cell. Murray *et al.* (1989) have reported biochemical studies of type II collagen from patients with various forms of spondyloepiphyseal dysplasia suggesting that the type II collagen is overmodified. They further suggest that this may be analogous to the overmodifications of type I collagen found in lethal osteogenesis imperfecta. Anderson *et al.* (1989) have also found genetic linkage between the type II collagen gene and spondyloepiphyseal dysplasia congenita.

Linkage of the Stickler syndrome with the type II collagen gene has been reported in two families by Francomano *et al.* (1987). This disease has a clinically heterogeneous phenotype and is also genetically heterogeneous based on linkage studies (Watkins *et al.*, unpublished data). It is likely that some phenotypes could result from defects in other genes such as those for chains of type IX or XI collagens or other extracellular matrix components.

Palotie *et al.* (1989) have recently suggested that a predisposition for familial osteoarthrosis is linked to the type II collagen gene.

Defects in type II collagen have been reported associated with type II achondrogenesis (Langer-Saldino) (Eyre, 1988; Eyre *et al.*, 1986; Godfrey and Hollister, 1988). Vissing *et al.* (1989) have identified a glycine to serine substitution near the carboxyl terminus of the triple helical domain in a relatively mild case of type II achondrogenesis-hypochondrogenesis. Defects in type II collagen have also been reported associated with Kniest dysplasia (Poole *et al.*, 1988), although there have been no reports regarding genetic linkage between the type II collagen gene and this disease. A number of laboratories have examined families with achondroplasia, and in all cases no linkage has been found with the type II collagen gene (Wordsworth *et al.*, 1988a; Strom, 1988; Francomano and Pyeritz, 1988). Francomano *et al.* (1988) have also published data excluding linkage of Marfan syndrome to the type II collagen gene in three families.

The Type XI Collagen Genes

Type XI Collagen

Type XI collagen was first discovered as three minor collagenous polypeptides of cartilage, provisionally named 1α, 2α, and 3α (Burgeson and Hollister, 1979). Because of a lack of definitive data on the subunit composition of trimeric molecules, a type designation was initially avoided. Instead, names such as K collagen (Miller, 1985) or 1α, 2α, 3α collagen have been used. Recently, however, Morris and Bachinger (1987) have demonstrated that the three chains are assembled into one heterotrimeric molecule. The three chains have been renamed α1(XI), α2(XI), and α3(XI) collagen.

The $\alpha1$(XI) and $\alpha2$(XI) chains are products of distinct genes, and both cDNA and genomic clones are available for the two chains in humans. However, as discussed above, the identity of $\alpha3$(XI) remains unclear. As mentioned earlier, peptide mapping data suggest, in fact, that the $\alpha3$(XI) is very similar, perhaps identical, to $\alpha1$(II) (Burgeson *et al.*, 1982). It is possible, therefore, that $\alpha3$(XI) transcripts are derived from the $\alpha1$(II) collagen gene.

Type XI collagen has been found as a minor constituent of cartilage in a number of different locations, including epiphyseal growth cartilage and articular cartilage (Eyre and Wu, 1987). Isolation of cDNA clones encoding $\alpha1$(XI) chains from libraries made with RNA isolated from transformed fibroblasts and human placenta (Bernard *et al.*, 1988) suggests that the expression of at least the $\alpha1$(XI) chain may not be restricted to cartilage. In fact, analysis of type V collagen from bone indicates that the $\alpha1$(XI) collagen chains may form heterotrimeric molecules with $\alpha1$(V) and $\alpha2$(V) collagen chains (Niyibizi and Eyre, 1989). This finding points to the close similarity between type XI and type V collagens. The two collagens are clearly related, not only structurally but functionally as well. Co-expression of types II and XI closely resembles that of types I and V in non-cartilage matrices, and type XI molecules are components of cartilage collagen fibrils, together with type II and IX collagens (Mendler *et al.*, 1989). It has therefore been suggested that type XI collagen helps to regulate the diameter of type II-containing fibrils, as type V collagen is thought to regulate the diameter of type I-containing fibrils (Eyre and Wu, 1987).

Molecular Cloning of the $\alpha1$(XI) and $\alpha2$(XI) Collagen Genes

Both cDNA and genomic clones encoding portions of the human $\alpha1$(XI) and $\alpha2$(XI) collagen chains have been isolated (Bernard *et al.*, 1988; Kimura *et al.*, 1989a). In addition, a cDNA specific for rat $\alpha1$(XI) has been characterized (Bernard *et al.*, 1988). The first $\alpha1$(XI) cDNA, OK4, was isolated from a library made with RNA from an SV40-transformed fibroblast cell line, GM637. Its identification was initially unclear, and the clone was therefore referred to as a type V collagen-related cDNA (Weil *et al.*, 1987). Isolation of cDNAs with similar or identical sequences from rat and human chrondrocyte libraries, respectively, suggested that the clones represented type XI collagen sequences. Final proof of identity came from the demonstration that amino acid sequences of three cyanogen bromide peptides derived from $\alpha1$(XI) collagen were contained within the sequence defined by OK4 (Bernard *et al.*, 1988).

Several cDNA clones encoding human $\alpha1$(XI) have now been characterized in addition to OK4 (Bernard *et al.*, 1988). Together they cover about 5 kb of DNA, encoding the 3′-nontranslated region, the C-propeptide, and about 99% of the triple-helical domain (Figure 5). Within the triple-helical domain, there is a preference for T and A in the wobble position of codons. This is similar to the $\alpha1$(III) and $\alpha2$(V) genes, but different from the $\alpha1$(I), $\alpha2$(I),

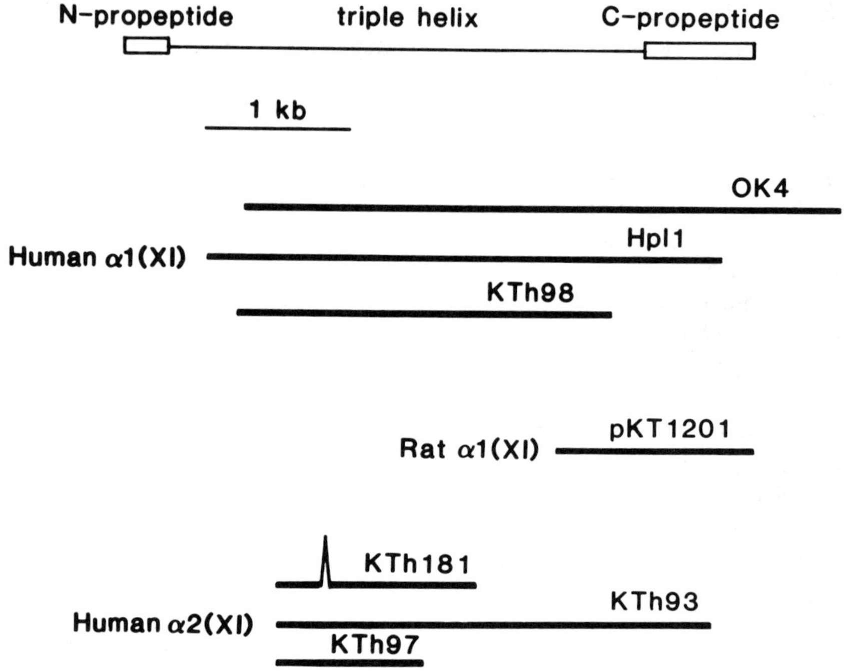

Figure 5. A diagram showing the domain structure of α1(XI) and α2(XI) collagen chains and the lengths and relative locations of cDNAs encoding the two chains. The triple-helical domain is indicated by a line; the propeptide regions are indicated by boxes. cDNAs encoding the α1(XI) chain are available from both human and rat (Bernard *et al.*, 1988), while cDNAs for the α2(XI) chain have been reported only from human (Kimura *et al.*, 1989a).)

and α1(II) genes which show a preference for T and C in this position. This suggests that the α1(XI) gene (COL11A1) is more closely related to the types III and V than to the types I and II collagen gene. This conclusion is supported by calculations of both amino acid and nucleotide sequence divergence, which indicate that COL11A1 is most divergent from COL1A1 and least from COL5A2.

Three overlapping cDNAs encoding a portion of the α2(XI) chain have been isolated (Kimura *et al.*, 1989a). Together, they code for about 2/3 of the C-propeptide and about 80% of the triple-helical domain (Figure 5). Identification is based on the complete identity between the DNA-derived sequence and the amino acid sequences of three tryptic peptides derived from α2(XI) collagen chains. A pairwise sequence comparison between α2(XI) and other fibrillar collagen chains shows that α2(XI) is closely related to α1(XI); in the triple-helical sequence region there is almost 80% similarity between the two chains.

Triple-Helical Domain

The amino acid sequences derived from the α1(XI) and α2(XI) cDNAs explain some of the previously reported properties of type XI molecules. For example, it has been found that bovine type XI molecules are not cleaved by synovial or gastric mucosal collagenase under conditions that generate 3/4 and 1/4 fragments of types I, II, and III collagens (Eyre *et al.*, 1984). The collagenase cleavage site in types I, II, and III is located between residues 775 (Gly) and 776 (Ile) (Kuhn, 1987). A collagenase cleavage sequence is absent from both α1(XI) and α2(XI) chains in this region. A potential cleavage sequence is found in the α1(XI) chain at residues 175 and 176, but since this region of the triple-helix has not yet been covered by cDNA sequences in the α2(XI) chain, it is not known whether this site is present in both chains.

In fibrillar collagens, lysyl residues involved in intermolecular cross-links are located in the amino and carboxyl telopeptides and at positions 87 and 930 of the triple-helix (Kuhn, 1987). In the α1(XI) and α2(XI) sequences, there are candidate cross-link lysines in similar positions. In α1(XI) the triple-helical domain contains such a residue at position 87, and both α1(XI) and α2(XI) chains have a second cross-link residue in position 927. Both chains also contain two consecutive lysyl residues in the carboxyl telopeptide region which could participate in cross-link formation.

Carboxyl Propeptide Region

While the complete sequence of the C-propeptide region of human α1(XI) chains has been deduced from cDNA sequencing (Bernard *et al.*, 1988), only about 75% of the C-propeptide region of the α2(XI) chain have been sequenced (Kimura *et al.*, 1989a). Examination of the available sequence shows, however, that the structure of the C-propeptide of type XI collagen is similar to that of the fibrillar collagens. Thus, both α1(XI) and α2(XI) chains contain seven or eight cysteinyl residues in the same relative locations within the C-propeptides as other fibrillar procollagen chains. The absolute distance (in amino acid residues) between the cysteinyl residues is somewhat shorter in α2(XI) than in other fibrillar procollagen chains (Kimura *et al.*, 1989a).

Chromosomal Location

Isolation of probes for human α1(XI) and α2(XI) collagen chains has permitted identification of the chromosomes on which COL11A1 and COL11A2 reside. *In situ* hybridization of metaphase chromosomes shows that COL11A1 is located in region p21 of chromosome 1 (Henry *et al.*, 1988), while COL11A2 is located on chromosome 6 in region p21 (Kimura *et al.*, 1989a). This localization on different chromosomes adds to the list of dispersed chromosomal sites for fibrillar collagen genes. With the exception of COL3A1 and COL5A2, which are both on chromosome 2, the fibrillar collagen genes are all on different chromosomes, with COL1A1 on 17, COL1A2 on 7, and COL2A1 on chromosome 12. This contrasts with the head-to-tail arrangement

Gene	Chromosomal Location	References
COL1A1	17q21-17q22	Huerre *et al.* (1982) Solomon *et al.* (1984) Retief *et al.* (1985)
COL1A2	7q21-7q22	Junien *et al.* (1982) Henderson *et al.* (1983) Solomon *et al.* (1984) Retief *et al.* (1985)
COL2A1	12q12-12q14	Strom *et al.* (1984a) Solomon *et al.* (1985) Huerre-Jeanpierre *et al.* (1986a) Law *et al.* (1986) Arheden *et al.* (1989)
COL3A1	2q24-2q33	Emanuel *et al.* (1985) Solomon *et al.* (1985) Huerre-Jeanpierre *et al.* (1986a)
COL4A1	13q33-13q34	Pihlajaniemi *et al.* (1985) Solomon *et al.* (1985; 1987) Emanuel *et al.* (1986) Bowcock *et al.* (1987) Griffin *et al.* (1987) Boyd *et al.* (1988)
COL4A2	13q33-13q34	Griffin *et al.* (1987) Killen *et al.* (1987) Solomon *et al.* (1987) Boyd *et al.* (1988)
COL4A5	Xq22	Hostikka *et al.* (1990)
COL5A1	?	
COL5A2	2q24-2q32	Emanuel *et al.* (1985) Huerre-Jeanpierre *et al.* (1986b)
COL6A1	21q223	Weil *et al.* (1988)
COL6A2	21q223	Weil *et al.* (1988)
COL6A3	2q37	Weil *et al.* (1988)
COL9A1	6q12-6q14	Kimura *et al.* (1989b)
COL9A2	?	
COL9A3	?	
COL11A1	1p21	Henry *et al.* (1988)
COL11A2	6p212	Kimura *et al.* (1989a)
COL13A1	10q22	Shows *et al.* (1989)

Figure 6. The chromosomal location of the human collagen genes.

of the α1(IV) and α2(IV) genes on chromosome 13, and the α2(VI) and α2(VI) genes on chromosome 21 (Figure 6).

The Type IX Collagen Genes

Type IX Collagen

Type IX collagen molecules consist of three genetically distinct polypeptide subunits, named α1(IX), α2(IX), and α3(IX) (van der Rest *et al.*, 1985).

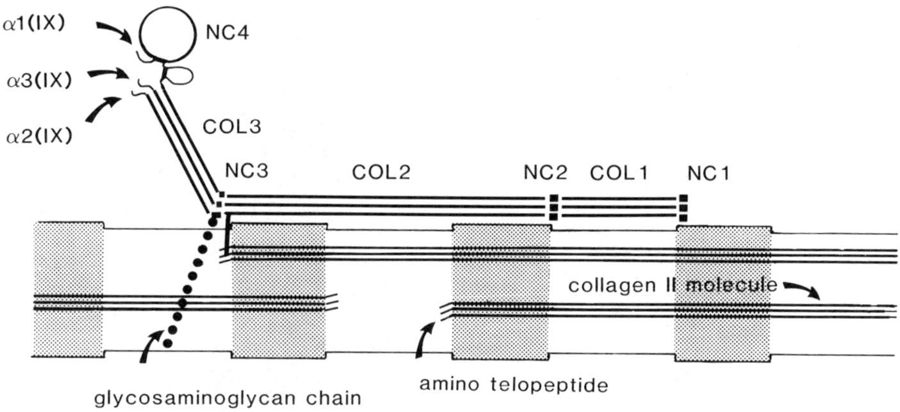

Figure 7. A diagram showing a type IX collagen molecule on the surface of a collagen type II-containing cartilage fibril. The three genetically distinct subunits of the type IX molecule form three triple-helical domains (COL1, COL2, COL3) separated and flanked by non-triple-helical sequences (NC1, NC2, NC3, NC4). A glycosaminoglycan chain is attached to the α2(IX) chain in the NC3 region. Collagen type IX molecules are covalently cross-linked to collagen type II. One such cross-link is shown between the α2(IX) chain and the amino telopeptide of a type II collagen molecule. (Reproduced from Nishimura *et al.*, 1989, with permission.)

Each subunit contains three domains of Gly-X-Y sequence repeats separated by non-triple-helical sequences. Thus, the triple-stranded molecule contains three triple-helical domains (Figure 7). The non-triple-helical region that separates the amino-terminal and central triple-helical domains represents a flexible hinge, allowing type IX molecules, located in a periodic manner along cartilage collagen fibrils, to project their amino-terminal triple-helical domain as an arm into the perifibrillar matrix (Irwin and Mayne, 1986; Vaughan *et al.*, 1988) (Figure 8). In cartilage, this arm continues into a globular domain of about 250 amino acid residues (Vasios *et al.*, 1988). This domain may interact with cartilage proteoglycans and is formed by the amino-terminal portion of the α1(IX) collagen chain. A second chain, α2(IX), carries a glycosaminoglycan chain which is covalently attached to a seryl residue within the flexible hinge-region of the molecule (see above) (Vaughan *et al.*, 1985; Bruckner *et al.*, 1985; Huber *et al.*, 1986; Konomi *et al.*, 1986; McCormick *et al.*, 1987). Type IX collagen is therefore also a proteoglycan, and is, in fact, identical to the proteoglycan PG-Lt first described by Noro *et al.* (1983).

Type IX collagen is a major component of cartilage collagen fibrils, together with the fibrillar collagens II and XI (Mendler *et al.*, 1989). However, the expression of type IX collagen genes is not limited to cartilage. Type IX collagen is also present in the vitreous of the eye and the central portion (nucleus pulposus) of intervertebral disks (Ayad *et al.*, 1982; Wright and Mayne, 1988). It is also present under several embryonic epithelia, including the embryonic chick corneal epithelium (Fitch *et al.*, 1988; Svoboda *et al.*,

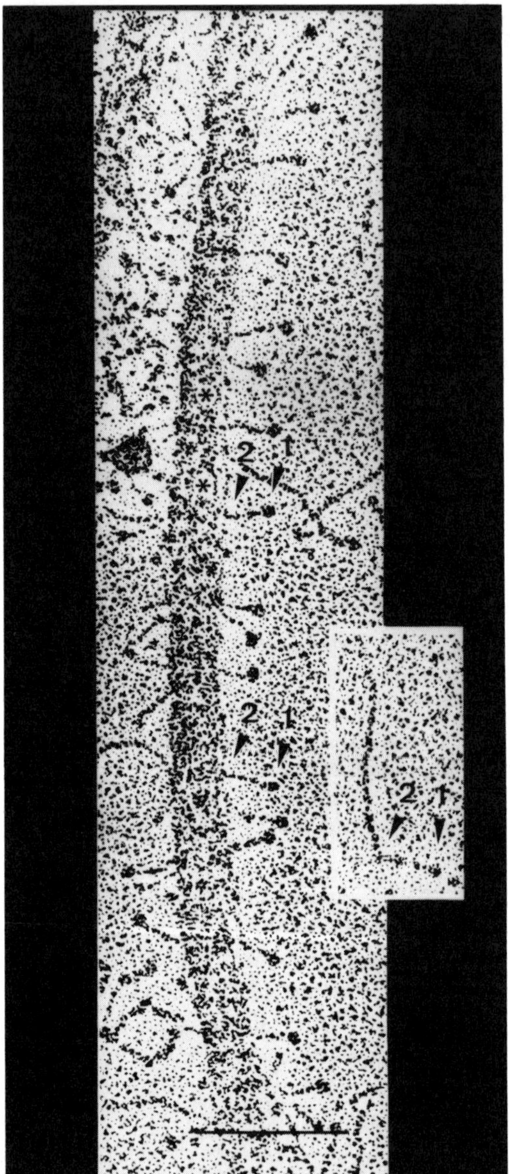

Figure 8. Rotary shadowing image of a cartilage collagen fibril showing the rod-like COL3 domains projected as arms from the fibrillar surface. The globular domain at the tip of the arms corresponds to the amino-terminal non-triple-helical region (NC4) of the $\alpha 1(IX)$ chain. The insert shows the rotary shadowing image of a single type IX collagen molecule. (Reproduced from Vaughan *et al.*, 1988, with permission.)

1988). The presence of type IX collagen in the embryonic chick primary corneal stroma is particularly interesting because of the transient nature of the expression and because the corneal form of the molecule is different from the cartilage form in that most of the amino-terminal globular domain of α1(IX) is absent from the cornea (Svoboda *et al.*, 1988; Nishimura *et al.*, 1989). Using a monoclonal antibody against type IX collagen, both Svoboda *et al.* (1988) and Fitch *et al.* (1988) found that type IX collagen is transiently expressed in 5-day-old chick embryonic corneas. In addition, Svoboda *et al.* (1988) found that both α1(IX) and α2(IX) mRNAs are present in 6- and 8-day-old corneas but almost undetectable in extracts of 11-day-old corneas. Surprisingly, while the α2(IX) mRNA is the same size in cornea and cartilage, Northern-blot analysis demonstrated that the α1(IX) mRNA is about 700 nucleotides shorter in cornea than in cartilage. Sequencing of cDNA encoding the 5′ region of the cornea α1(IX) transcript, combined with sequence analysis of the α1(IX) gene, has provided an explanation for the size difference by demonstrating that the α1(IX) gene has two alternative promoters and transcription start sites (Nishimura *et al.*, 1989). The two promoters are located 20 kb apart, flanking a series of exons that encode the globular domain at the amino end of type IX molecules in cartilage. In chondrocytes, the most 5′ promoter is used and the domain is expressed; in embryonic corneal epithelial cells, the most 3′ transcription start site is used, the resulting mRNA is about 700 nucleotides shorter than in cartilage, and most of the amino-terminal globular sequence is not expressed. This finding raises the possibility that modulation of the expression of the amino-terminal globular domain of type IX molecules represents a mechanism by which cells in different tissues can modulate the interaction of collagen fibrils with their surrounding matrix. Definitive evidence that a shortened form of type IX collagen exists as a protein in cornea is not yet available. However, the existence of such a form, probably identical to the corneal form, has been demonstrated in the vitreous of the eye by Wright and Mayne (1988).

Molecular Cloning of the α(IX) and α2(IX) Collagen Genes

Overlapping genomic clones encoding the α1(IX) and α2(IX) collagen genes (COL9A1 and COL9A2) have been isolated from chicken libraries (Figure 9) (Lozano *et al.*, 1985; Vasios *et al.*, 1988; Ninomiya *et al.*, 1989). Analysis of these clones demonstrates that COL9A2 is relatively small, only about 10 kb in size, and contains 32 exons. The precise transcription start site of the chicken α2(IX) gene has not been precisely mapped. Therefore, the size of the most 5′ exon, 315 nucleotides, must be considered tentative. Several hundred nucleotides of sequence upstream of this exon have been determined, but it is not known whether this sequence contains upstream promoter elements.

The α1(IX) gene is much larger than the α2(IX) gene, spanning almost 100 kb (Figure 9). The central portion of the gene has not yet been cloned.

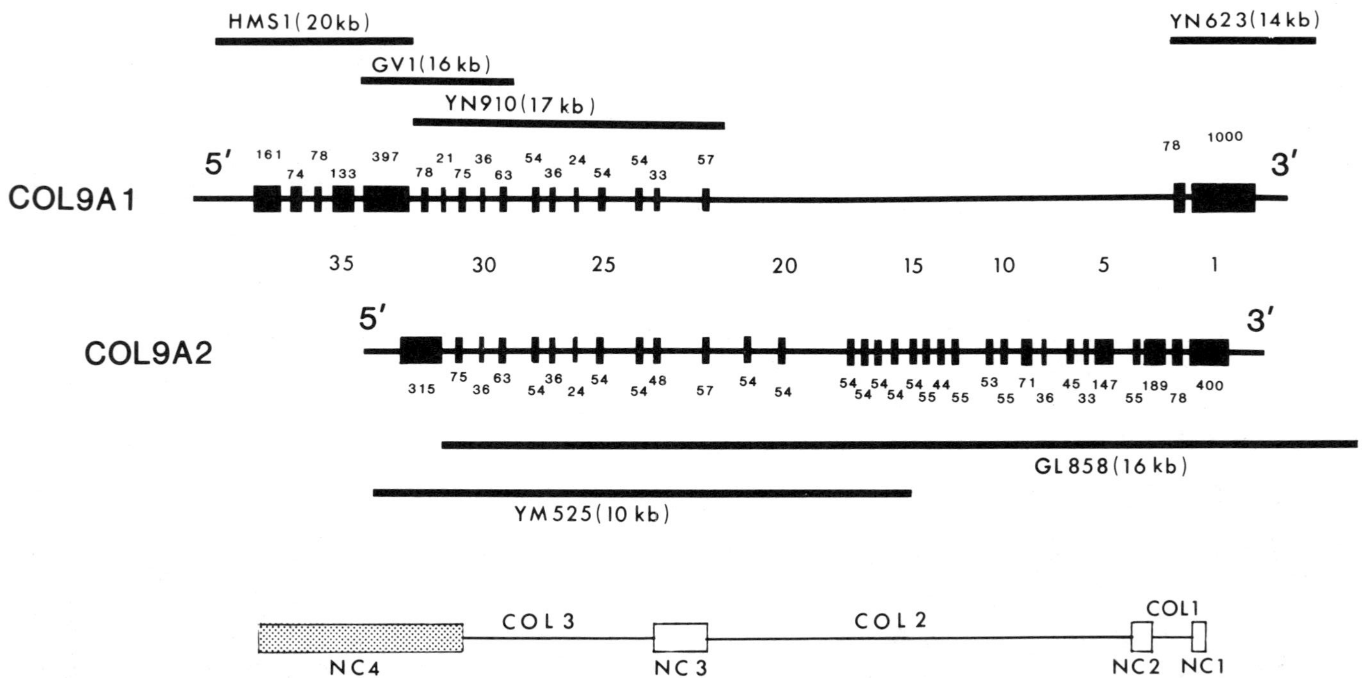

Figure 9. Diagram showing the exon structure of the chicken α1(IX) and α2(IX) collagen genes (COL9A1 and COL9A2) together with the genomic clones that have been isolated for the two genes. The exons are numbered (between the two genes) from the 3′ end, and their sizes (in bp) are given above (COL9A1) and below (COL9A2) the diagrams. The diagram at the bottom of the figure indicates the positions (relative to exons) of the various domains of the type IX collagen chains. Triple-helical domains (lines) are COL1, COL2, and COL3; non-triple-helical domains are NC1, NC2, NC3, and NC4. A large NC4 domain is only present in the cartilage form of α1(IX). Note that genomic clones spanning the gap between YN910 and YN623 have not yet been isolated. Assuming that the exon structure is the same in this region of the α1(IX) gene as it is in the α2(IX) gene, this gap corresponds to exons 3 to 21. (Reproduced from Ninomiya *et al.*, 1989, with permission.)

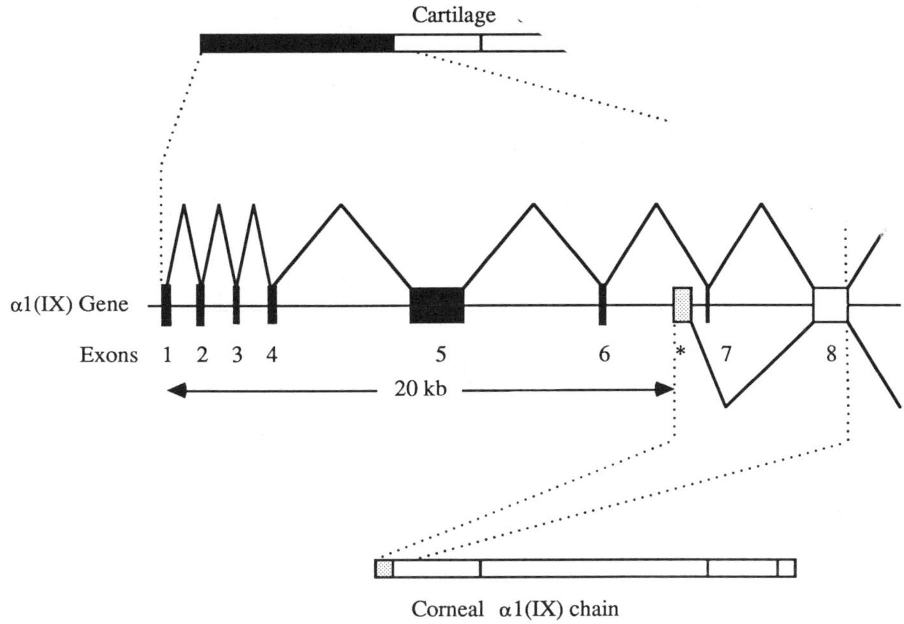

Figure 10. Diagram of the 5′ portion of the α1(IX) collagen gene with exons 1 to 8. An exon utilized in embryonic chick cornea (*) is located within the intron between exons 6 and 7 and about 20 kb downstream of exon 1. The cartilage α1(IX) collagen chain contains a large amino-terminal globular domain encoded by exons 1 to 7, whereas the short corneal α1(IX) chain contains instead a small peptide encoded by exon (*) connected to the sequence encoded by exon 8. (Reproduced from Nishimura *et al.*, 1989, with permission.)

However, assuming that the number of exons are the same in COL9A1 and COL9A2 within this gap, the missing part of the gene would include exons 3 to 21 (as counted from the 3′ end of the gene). Both S1 nuclease and primer extension studies have been done to map the transcription start sites in chondrocytes and corneal epithelial cells (see above), demonstrating that COL9A1 has two major transcription start sites and promoters (Figure 10) (Nishimura *et al.*, 1989) spaced about 20 kb apart. The 5′-flanking sequences in the two regions both contain a TATA-box in the appropriate position upstream of the transcription start, but they otherwise are quite different. For example, the downstream promoter, utilized in cornea, contains two CAAT-boxes not contained in the upstream promoter. This promoter contains instead GC-rich elements. The upstream transcription start site utilized in cartilage defines the 5′ border of exon 1. The size of this exon, 161 nucleotides (Figure 9), must be considered an average number because there is a cluster of start sites in this region of the gene. Exon 1 encodes a 150-nucleotide-long 5′-nontranslated region and 4 2/3 amino acid codons of the signal peptide while the rest of the signal peptide is encoded by exon 2 (Figure 9). The amino-terminal amino acid sequence of the secreted form of cartilage α1(IX) has not yet been determined. Therefore, the precise cleavage site for the signal peptidase is not known. However, it is believed that the most likely cleavage site is the bond between residues 23 (alanine) and 24 (threonine). If this is correct, exon 2 would encode 18 1/3 codons of the signal peptide and 6 1/3 codons of the 243-residue-long non-triple-helical domain NC4 of cartilage α1(IX) chains. The remaining portion of NC4 is encoded by exons 3 to 7, plus the first codon of exon 8 which encodes an arginyl residue that links NC4 to the amino-terminal triple-helical domain COL3 (Vasios *et al.*, 1988).

The exon sizes in this region of the α1(IX) gene vary considerably in length, ranging from 21 bp (exon 7) to 397 bp (exon 5) (Figure 9). The introns are also quite variable in length. The longest are the two introns that flank exon 5; they are both almost 7 kb long. The shortest are those between exons 1 and 2 and exons 6 and 7, which are about 350 and 650 bp, respectively. Of special interest is the intron between exons 6 and 7, in that it contains the alternative transcription start and exon 1 utilized in cornea (Figure 10). This alternative exon encodes a 5′-nontranslated sequence of about 90 nucleotides and 24 amino acid residues of translated sequence. This amino acid sequence has the hydrophobic characteristics of a signal peptide; it is believed that the cleavage site for signal peptidase is between residues 23 (alanine) and 24 (glutamine). During processing of the α1(IX) transcript in cornea, the corneal exon 1 is spliced to exon 8 (Figure 10). Thus, the sequence of exon 7 is not contained within cornea α1(IX) mRNA and the amino-terminal non-triple-helical sequence of the cornea α1(IX) chain consists of only two amino acid residues; one residue is contributed by the alternative exon 1 and one residue is contributed by exon 8 (Nishimura *et al.*, 1989).

Since the 5′-nontranslated sequences of the chicken α1(IX) transcripts in cartilage and cornea are encoded by two different exons in the gene, it is not surprising that they are entirely different. The 5′-nontranslated region of the human cartilage α1(IX) transcript, about as long as the chicken sequence, is also quite different, showing only 38% similarity with the chicken sequence.

The Multiple Domain Structure of α1(IX) and α2(IX) Collagen Chains

The nucleotide sequence of the cDNA pYN1738 provided the first evidence for the presence of seven distinct sequence domains in α1(IX) collagen, three triple-helical domains separated by non-triple-helical sequences and flanked by non-triple-helical sequences at the amino and carboxyl ends of the polypeptide chain (Figure 11) (Ninomiya and Olsen, 1984). The α2(IX) chain shows the same structural organization, not surprisingly, since α1(IX) and α2(IX) chains participate in the formation of a heterotrimeric type IX collagen molecule. However, there are differences between the two chains in the precise length of some of the domains. One difference is in the size of the amino-terminal non-triple-helical sequence. This domain is 266 amino acid residues

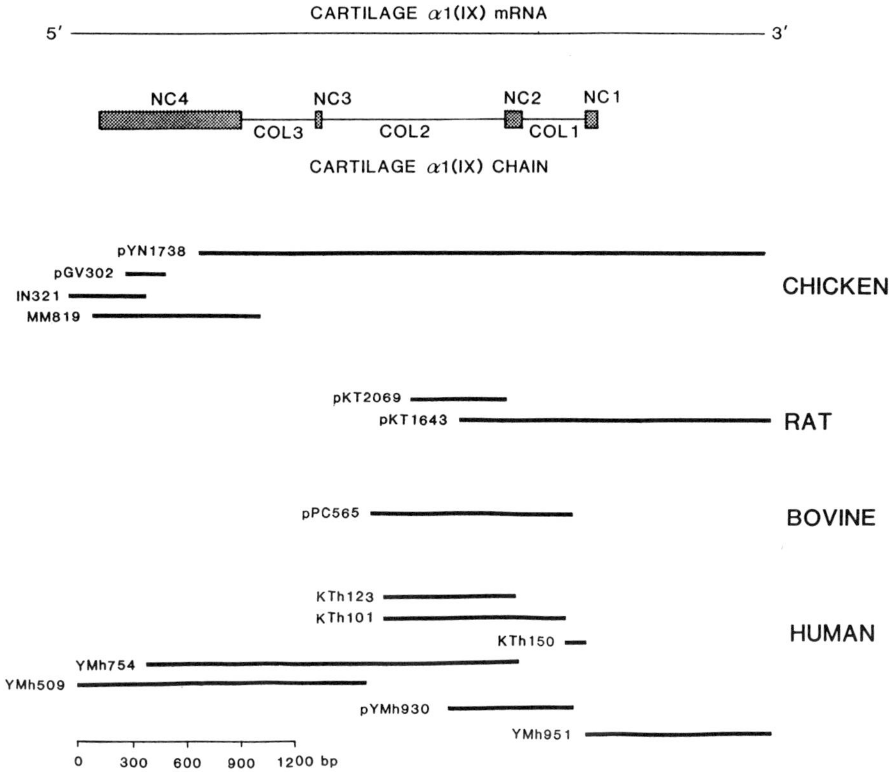

Figure 11A. Diagram showing the relative size and location of chicken, rat, bovine, and human cDNAs encoding the cartilage α1(IX) collagen chain. The mRNA and the amino acid sequence domains of the translation product are shown at the top. Triple-helical domains (COL1, COL2, COL3) are indicated by lines; non-triple-helical domains (NC1, NC2, NC3, NC4) are indicated by boxed-in areas. (Modified from Ninomiya *et al.*, 1990).

in chicken cartilage α1(IX), including the signal peptide, but only 24 residues in α2(IX). It is interesting to note that in the corneal form of α1(IX) the domain is also short, 25 residues (Nishimura *et al.*, 1989). A second difference is the size of the non-triple-helical sequence NC3 that joins the amino-terminal and central triple-helical domains. This sequence is 12 amino acid residues in the α1(IX) chain but 17 residues in α2(IX) (McCormick *et al.*, 1987). The extra five residues in α2(IX), VEGSA, contain the seryl residue that represents the attachment site for a glycosaminoglycan side chain (McCormick *et al.*, 1987). A third difference is the size of the carboxy-terminal non-triple-helical domain. This is 21 amino acid residues in α1(IX), but only 15 residues in α2(IX) (Ninomiya and Olsen, 1984; Ninomiya *et al.*, 1985). Nevertheless, the triple-helical sequence domains in α1(IX) and α2(IX) chains are the same lengths and have similar sequence characteristics. For example, the amino-

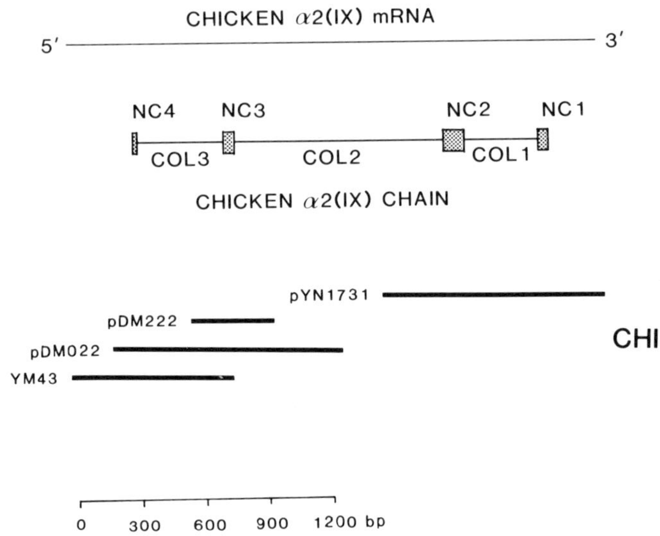

Figure 11B. Diagram showing the relative size and location of chicken cDNAs encoding the cartilage α2(IX) collagen chain. The mRNA and the amino acid sequence domains of the translation product are shown at the top. Triple-helical domains (COL1, COL2, COL3) are indicated by lines; non-triple-helical domains (NC1, NC2, NC3, NC4) are indicated by boxed-in areas. (Reproduced from Ninomiya *et al.*, 1990, with permission.)

and carboxy-terminal triple-helical domains in both chains contain imperfections in the Gly-X-Y repeat structure in the same locations.

Extensive data allowing firm conclusions about the extent of conservation of this multidomain structure of type IX collagen during evolution are not yet available. However, analysis of partial sequences of rat, bovine, and human α1(IX) chains (Figure 11) suggests that, at least for this chain, the lengths of the different domains are highly conserved. Thus, the only notable difference between chicken and human cartilage α1(IX) is that the amino-terminal non-triple-helical domain contains two more amino acid residues in the human than in the chicken chain (Ninomiya *et al.*, 1990).

The domains of triple-helical and non-triple-helical sequences in type IX chains are encoded by multiple exons whose sizes depend on the nature of the domains they encode (Figure 9). All the 32 exons in the chicken α2(IX) gene have been sequenced, and this gene serves as a prototype for the type IX collagen genes. As shown in Figure 9, six exons in the gene are junction exons in that they encode both triple-helical and non-triple-helical sequences. The carboxy-terminal triple-helical domain COL1 and the carboxyl portion of the central COL2 domain are encoded by exons that appear related to a 36 bp unit. This is also the case for the amino-terminal portion of the COL3 domain. However, the carboxyl end of COL3 and the amino-terminal half of

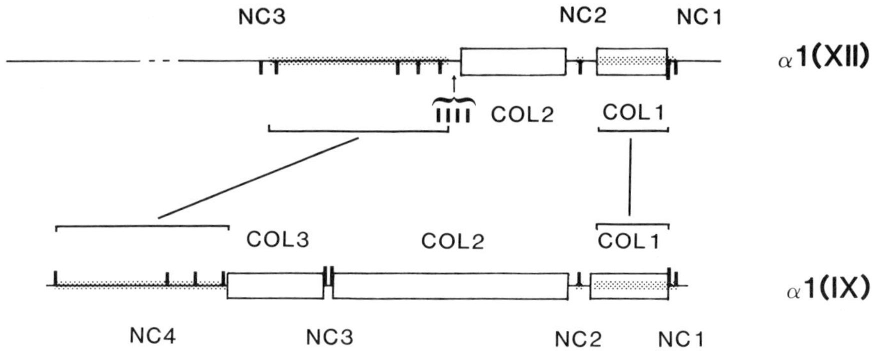

Figure 12. Diagram showing comparison between domains of α1(XII) (top) and α1(IX) (bottom) collagen chains. Triple-helical domains are indicated by boxes; non-triple-helical domains are shown as straight lines. Cysteinyl residues are indicated by short vertical lines. Regions of sequence similarity are indicated by shaded areas. (Reproduced from Gordon *et al.*, 1989, with permission.)

COL2 are encoded by 54 bp exons or a modification of that size (53 or 55 due to split codons). This exon organization is very different from that of fibrillar collagen genes, and leads to the conclusion that the type IX collagen genes have followed an evolutionary pathway different from that of the genes that encode the family of fibrillar collagens (Ninomiya *et al.*, 1990).

Type IX Collagen is a Member of a Family of Nonfibrillar Collagens

The use of cDNAs encoding chicken α1(IX) and α2(IX) collagen chains to probe Northern-blots of RNA isolated from a variety of sources has demonstrated that tissues such as tendon or bone do not express type IX genes. However, embryonic chick tendon fibroblasts do contain mRNAs capable of synthesizing bacterial collagenase-sensitive polypeptides in cell-free systems that are similar in size to type IX chains (Gordon *et al.*, 1987). To explore the nature of such RNAs/polypeptides in detail, Gordon *et al.* (1987) screened a tendon fibroblast cDNA library for the presence of DNAs that encode Gly-X-Y repeat sequences, and discovered a novel collagen chain designated α1(XII) collagen. Of particular interest was the finding that this chain contains a carboxy-terminal triple-helical sequence domain with a high degree of sequence similarity to the COL1 domain of α1(IX) and α2(IX) collagen chains (Figure 12). Sequencing of additional cDNA clones demonstrates also that a portion of the amino-terminal non-triple-helical sequence of α1(XII) is similar to the corresponding domain in the cartilage α1(IX) chain (Gordon *et al.*, 1989). These similarities, as well as similarities in the exon structure in the 3′ region of the α1(XII) and type IX genes, have led to the conclusion that α1(XII) and type IX chains are homologous proteins (Gordon *et al.*, 1989). However, this homology is only partial; despite striking similarities between the chains in certain regions, there are also clear differences between types

IX and XII collagens. The use of a monoclonal antibody against α1(XII) as probe for Western-blots of embryonic tendon extracts or fibroblast culture media shows that α1(XII) chains are 220 kDa in size, as compared with only about 100 kDa for cartilage α1(IX) (Dublet *et al.*, 1989; Sugrue *et al.*, 1989). This size difference is due to a much larger amino-terminal non-triple-helical domain in α1(XII) than in α1(IX); about 190 kDa in α1(XII), but only 25 kDa in cartilage α1(IX) (Dublet *et al.*, 1989). A second difference is that type XII collagen appears to be, at least in tendon, a homotrimer of three α1(XII) chains, compared with the heterotrimeric nature of type IX collagen (Dublet *et al.*, 1989). Finally, α(XII) chains contain only two, relatively small, triple-helical domains, while type IX chains contain three.

The large size of α1(XII) makes it very unlikely that α1(XII) mRNA could be responsible for the cell-free synthesis of type IX-sized collagenous polypeptides with tendon RNA preparations (see above.) Therefore, it is quite possible that collagen chains with the size of type IX chains are synthesized by tendon fibroblasts in addition to type XII collagen.

Chromosomal Location

DNA probes encoding human type IX collagen chains are only available for the α1(IX) chain (Kimura *et al.*, 1989b; Muragaki *et al.*, 1990). COL9A1 is therefore the only type IX gene for which the chromosomal location has been determined. By *in situ* hybridization of metaphase chromosome spreads, Kimura *et al.* (1989b) have found that COL9A1 is located on the long arm of chromosome 6 within the region 6q12–6q14. Another cartilage collagen gene, COL11A2, has also been localized to this chromosome, but to the short arm, in the region 6p21.

Cartilage Proteoglycan Core Protein

Core Protein Structure

The major large proteoglycan of cartilage is a predominant component of the cartilage extracellular matrix (for recent reviews see Wight and Mecham, 1987; Sandell, 1987; Ruoslahti, 1988; Hassell *et al.*, 1986; Doege *et al.*, in press). In spite of intense study of this molecule, the characterization of the core protein has been hampered until recently due to combination of the large amount of carbohydrate (80 to 90% of the total mass of this molecule is carbohydrate with at least three to four different types of carbohydrate side chains), the variation in the amount and size of the side chains, and the relatively large size of the core protein. Based on biochemical, electron microscopical, and immunological studies, the protein has been described as being heterogeneous in size and containing a minimum of four to five domains, two connected globular domains at the amino terminus, one of which contains lectin activity and interacts with hyaluronic acid and link protein in the protein aggregate, another (adjacent) region enriched in keratan sulfate side chains, a region toward the carboxyl terminus enriched in chondroitin sulfate side chains, and a globular region at the carboxyl terminus. Initial studies of the molecular biology of the proteoglycan core protein used antibodies and cell-free translation to identify the translation product as a protein with an apparent relative molecular weight of 340,000, to aid in the partial characterization of its mRNA, and in studies of the biosynthesis and subsequent processing of the proteoglycan (Upholt *et al.*, 1976; Vertel *et al.*, 1984; Vuorio *et al.*, 1982). Doege *et al.* (1990b) have recently suggested that the molecule be called aggrecan.

Molecular Cloning of the Core Protein Gene

Complementary DNA clones corresponding to the mRNAs for the chicken (Sai *et al.*, 1986), rat (Doege *et al.*, 1986a), bovine (Oldberg *et al.*, 1987; Antonsson *et al.*, 1989), human (Dudhia and Hardingham, 1989a; Baldwin *et al.*, 1989b; Doege *et al.*, 1991) and porcine (Dudhia and Hardingham, 1989a) core proteins have been isolated (Figure 13). Initially, relatively short cDNA clones (400 to 1229 bp) were identified from expression libraries by screening with antibodies. The initial clones were then used to rescreen the libraries and to identify longer clones (Doege *et al.*, 1987; Stirpe *et al.*, 1987). The first cDNA clones were confirmed as corresponding to the core protein by the fact that the deduced amino acid sequence contained a 20 to 21-residue-long sequence previously identified in a cyanogen bromide peptide derived from bovine nasal cartilage proteoglycan (Perin *et al.*, 1984). These cDNAs hybridized specifically to poly A+ RNAs from the respective cartilages. In the case of the rat (Doege *et al.*, 1987), a series of overlapping clones were constructed by primer extension. The sequence of these clones includes a 2124-residue open reading frame beginning with a methionine followed by a typical signal peptide sequence suggesting that this represents the coding sequence for an entire core protein. The size of the initial translation product of the rat core protein as deduced from the cDNA sequence is 220,952 Da, considerably smaller than the size estimated by SDS-polyacrylamide gel electrophoresis of the cell-free translation product.

The rat and chick cDNA clones have been used either as probes (Sasaki *et al.*, 1989) or to design oligonucleotide probes (Baldwin *et al.*, 1989b) for the isolation of cDNA (Sasaki *et al.*, 1989; Baldwin *et al.*, 1989b) and genomic (Doege *et al.*, 1990a, 1991) clones corresponding to the human proteoglycan core protein mRNA.

The increasing amount of data concerning the amino acid sequence, exon/intron structure, and biochemical properties of various regions of the core protein show this protein and its gene to be very interesting from evolutionary and structural points of view. At least five of its domains have been related to domains present in other proteins consistent with evolutionary hypotheses suggesting that proteins evolve in part by exon shuffling (see below). These five domains are also present in the large fibroblast proteoglycan, versican,

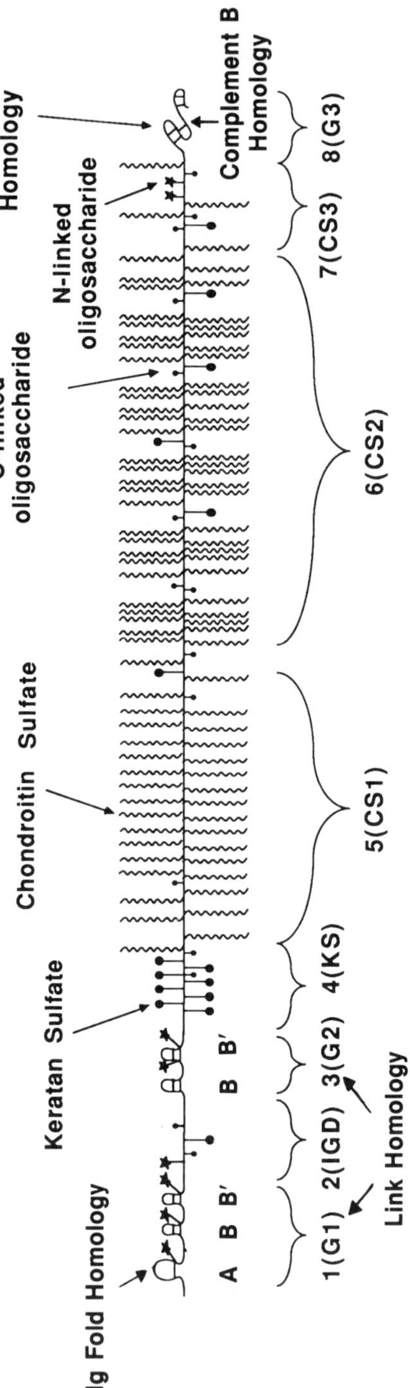

Figure 13A. Diagram showing the domain structure of cartilage proteoglycan core protein. (Modified from Doege *et al.*, 1987, with permission.)

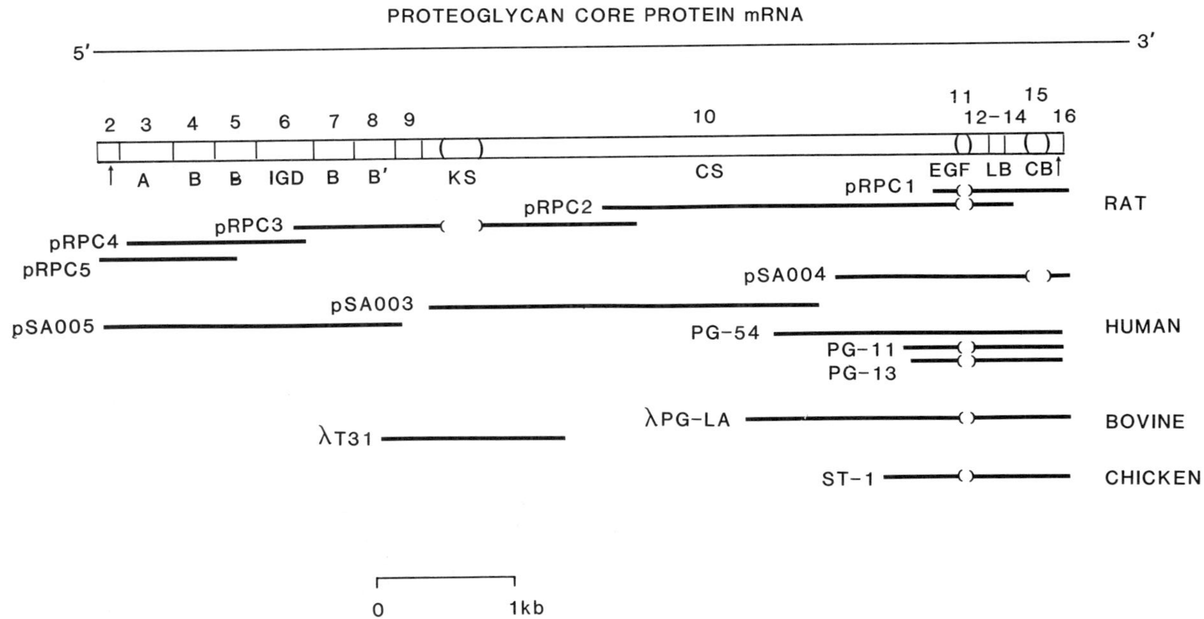

Figure 13B. Diagram showing the relative size and location of rat, human, bovine, and chicken cDNAs encoding the proteoglycan core protein. The mRNA and the domain structure of the translation product are shown at the top. A, B, and B′, regions with homology with link protein domains; IGD, interglobular domain; KS, keratan sulfate region; CS, chondroitin sulfate region; EGF, epidermal growth factor homology; LB, lectin binding protein homology domain; CB, complement B homology domain; arrows indicate the positions of the AUG and terminator codons. Parentheses indicate either the KS domain present in the human and bovine genes but absent in the rat gene or the absence of alternatively spliced domains encoding the EGF homology domain or the complement B homology domain. References: ST-1, Sai *et al.* (1986); pRPCl-5, Doege *et al.* (1986a, 1987; λPG-LA, Oldberg *et al.* (1987); λT31, Antonsson *et al.* (1989); PG-11, 13, and 54, Baldwin *et al.* (1989b); pSA003, 004, and 005, Sasaki *et al.* (1989).

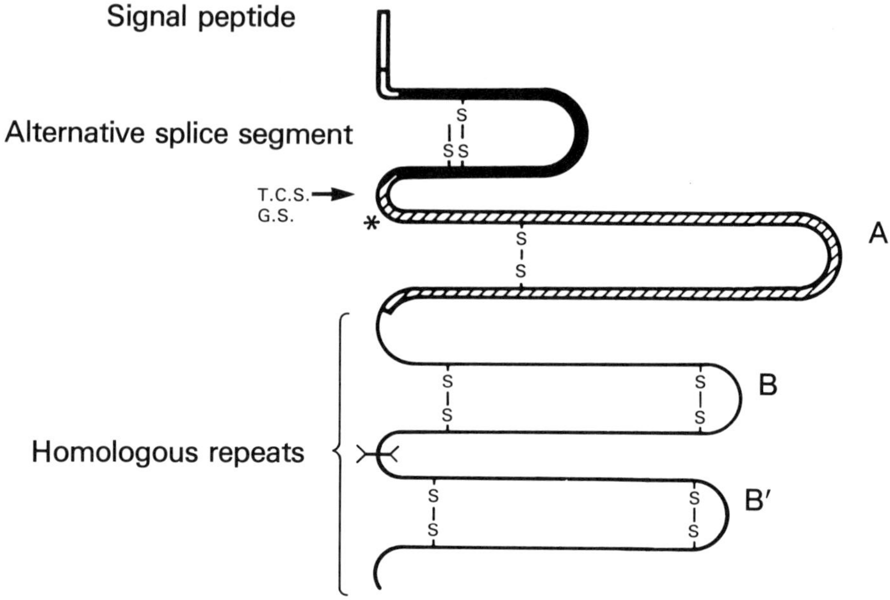

Figure 14A. Diagram showing the domain organization of the link protein. T.C.S., amino trypsin cleavage site; G.S., potential glycosylation site; A, B, and B′, domains also present in the proteoglycan core protein. (Modified from Figure 7 of Rhodes *et al.* (1988)).

in the same order (Zimmerman and Ruoslahti, 1989), suggesting a possible evolutionary relationship between the two genes.

Doege *et al.* (1987) have analyzed the resulting coding sequence for the rat core protein and divided it into eight structural domains (Figure 13). Five of these correspond to the previously suggested domains (see above). The additional domains consist of an interglobular domain between the two globular domains at the amino terminus of the protein and a division of the chondroitin sulfate enriched domain into three separate Ser-Gly-rich domains with somewhat different characteristics. They have also found that two of the Ser-Gly-rich domains may be divided into further subdomains, suggesting that the larger domains have evolved by tandem repetitions of smaller domains.

Domains 1 and 3 (GD1 and GD2) of the core protein have homology with themselves, with link protein, with versican, and with a human lymphocyte homing molecule (Goldstein *et al.*, 1989; Stamenkovic *et al.*, 1989). Core protein domain 1 contains regions homologous to domains A, B, and B′ of link protein (Figures 13 and 14) and is responsible for the interaction of the proteoglycan with hyaluronate and link protein. Core protein domain 3 also contains regions homologous to domains B and B′ of link protein, but contains no region corresponding to link protein domain A and does not interact with

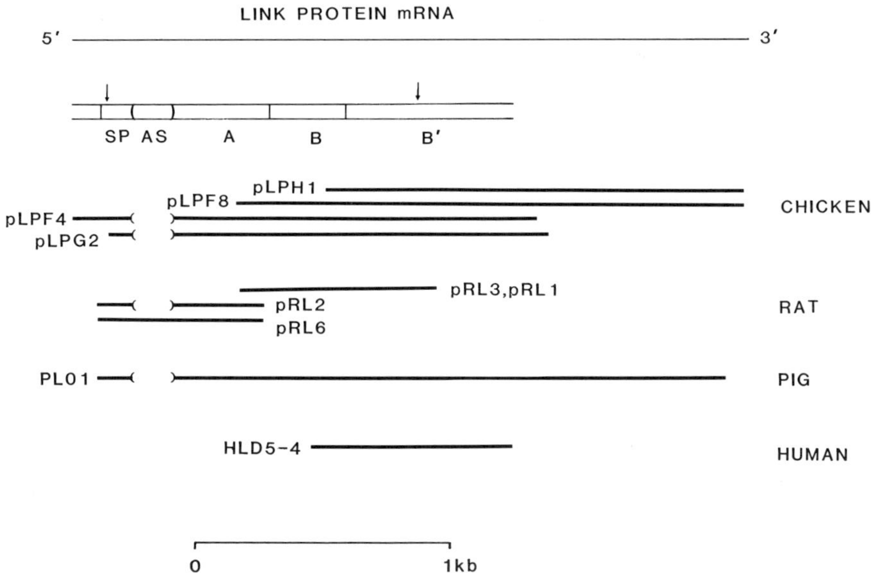

Figure 14B. Diagram showing the relative size and location of chicken, rat, pig and human cDNAs encoding link protein. The mRNA and domain structure of the translation product is shown at the top. SP, signal peptide; AS, alternatively spliced domain; A, B, and B′, conserved domains defined in text. References: pLPG2, F4, F8, and H1, Deak *et al.* (1986); pRL1, 2, 3, and 6, Rhodes *et al.* (1988); PL01 and HLD5-4, Dudhia and Hardingham (1989b).

hyaluronate and link protein (Fosang and Hardingham, 1989). The region corresponding to A is replaced in GD2 by core protein domain 2 or the interglobular domain (IGD). The regions of the core protein and link protein which are homologous with the lymphocyte molecule are the link protein repeated units B and B′ which are each present twice in the core protein. Perkins *et al.* (1989) have shown that the globular region of the link protein at the amino terminus (domain A, also present at the amino terminus of domain 1 of the core protein) corresponds to an immunoglobulin fold-like sequence.

Following the G2 domain (Figure 13) there is a region of 108 amino acid residues encoded in the bovine and human core protein genes but absent in the rat gene (Antonsson *et al.,* 1989). This region consists of a highly conserved hexamer peptide motif repeated consecutively 23 times. Remnants of five of these repeats are present in the rat gene. This region has a glutamic acid, serine, and proline composition consistent with the amino acid composition determined for a fragment of the keratan sulfate enriched region of the bovine proteoglycan. The absence of a corresponding region in the rat gene is consistent with the finding that the rat chondrosarcoma proteoglycan does not contain keratan sulfate.

Additional information may be gained by a comparison of rat, bovine, human, and chicken sequences. The G2 region is not highly conserved, and only 66% of the amino acids are identical between the rat and cow (Antonsson *et al.*, 1989). In contrast, the first chondroitin sulfate region (CS1) is 88% identical between the rat and cow. The more 3′ bovine cDNA of Oldberg *et al.* (1987) begins in the second Gly-Ser-rich domain (CS2), and these authors have also noted the repeats present in this domain. Doege *et al.* (1990a) have reported that there is a single large exon of approximately 4 kb which encodes the entire chondroitin sulfate domain in the human gene. The chicken sequence begins near the amino terminus of the seventh domain (the third Gly-Ser-rich domain) and extends beyond the end of the coding sequence. Genomic clones corresponding to the chicken gene have been described, thus defining the positions of intron interruptions for the eighth or carboxyl terminal globular domain.

As noted by Doege *et al.* (1987) and Dudhia and Hardingham (1989a), the third Ser-Gly-rich domain is less highly conserved compared to the second such domain. This is particularly evident when the chicken sequences are compared with the rat and bovine sequences. This region is highly variable between species in the number of Ser-Gly sequences (from 4 in the chicken to 11 or 12 in the cow, with only 2 of the sites conserved in all three species). In addition there are three positions of insertions or deletions of coding sequence. The rat sequence contains two potential asparagine glycosylation sites in this region, whereas neither the chicken nor bovine sequences contain such sites in this domain.

Baldwin *et al.* (1989b) have shown that there is an alternatively spliced region between the third Ser-Gly-rich region (domain 7) and the lectin binding-region (domain 8, see below) which encodes an EGF-like domain. Using polymerase chain reaction techniques they have shown that RNAs both containing and not containing this region are present in total RNA extracted from freshly isolated cartilage cells and that the relative amount of the RNA containing the region encoding the EGF-like domain is increased following culturing the same cells for 4 days. The similar tandemly repeated EGF-like domains are found at a corresponding position in the human fibroblast proteoglycan, versican (Zimmerman and Ruoslahti, 1989).

Following the alternatively spliced EGF-like region there is a carboxyl terminal globular domain with homology to vertebrate lectin sequences including five conserved cysteine residues (Tanaka *et al.*, 1988; Oldberg *et al.*, 1987; Doege *et al.*, 1987). This region is in addition to the region with lectin activity reacting with hyaluronic acid at the opposite end of the protein. Of all of the sequences available for comparison between the three species, this is the most highly conserved. The region of the lectin protein which contains the homology with the core protein is believed to have binding activity for a terminal N-acetylglucosamine, and the finding suggests that the proteoglycan might have a sugar binding region in the carboxyl terminal globular domain.

Halberg *et al.* (1988) have expressed the carboxyl terminal globular domain of the protein containing the putative lectin-like sequences in an *in vitro* transcription and translation system using a portion of the cDNA clone. Affinity chromatography on immobilized sugars suggests that there is sugar specificity of binding, although no binding constants have been determined, and the affinity of the expressed domain is lower than that of other known binding domains.

The boundary between the third Ser-Gly-rich region (or the EGF-like region) and the beginning of the region with lectin homology corresponds to an intron interruption in the chicken gene suggesting that, as has been found in other genes, the exons define structural units of proteins (Tanaka et al., 1988). In this case the lectin homology region is encoded by three exons and the end of the third exon occurs at the carboxyl end of the lectin homology region. A comparison of the exon/intron organization of the chicken core protein gene with that of the rat hepatic lectin gene (RHL-1) and the human pulmonary surfactant apoprotein gene (HPSAP) shows some significant similarities in the positions of exon boundaries.

The lectin homology region is followed in the chicken gene by an exon encoding a 60 amino acid region with four cysteines conserved between all three species. Doege *et al.* (1987) have suggested that this and the subsequent region might comprise a globular domain functionally separate from the lectin domain. Patthy (1987) has reported that precisely the region encoded by this exon is homologous with complement B, further supporting the idea that this is an evolutionarily separate functional domain. Based on a report by Doege *et al.* (1991), it appears that this exon is alternatively spliced in humans.

The final exon encodes 26 to 27 amino acids, contains no cysteines, and is generally less conserved particularly when the chicken sequence is compared to the rat and bovine sequences.

In the various species, the cDNA probes hybridize to mRNAs of 7.5 to 12 kb. The rat probe hybridizes to two mRNA species of 8.2 and 8.9 kb (Doege *et al.,* 1987), the chicken probes to an RNA of 8 to 9 kb (Sai *et al.,* 1986), the bovine probe to an RNA of 12 kb (Oldberg *et al.,* (1987), and the human probe to an RNA of 7.5 kb (Doege *et al.,* 1991). The work of Doege *et al.* (1987) suggests that there is about 230 bp of 5′-untranslated sequence (based on two independent cDNA clones with common 5′ ends). This 230 bp, together with the approximately 6400 bp of coding sequence, add up to about 6.6 kb, substantially shorter than the 8 to 9 kb size of the RNA. Based on their results they suggest that the additional sequence is an extensive 3′ noncoding region. None of the cDNA clones that have been described contain either a poly A region or a polyadenylation signal (probably due to the fact that in the original cDNA clones were screened for expression thus selecting for sequence extending into the coding region). Tanaka *et al.* (1988) have found that the chicken genomic clone contains a $(dA)_{14}$ sequence 17 bp downstream from the 3′ end of their cDNA which could have acted as a template for an oligo dT primer. They have also found some potential polyadenylation sequences 160 and 313 bp downstream from the termination codon but no further experiments have been done to define the 3′ end of the mRNA.

Regulation of the Core Protein Gene

A limited number of studies have been done to examine regulation of expression of the core protein gene. Its expression appears to be limited to cartilaginous tissues, although weak signals were detected in dot blots using RNA from embryonic chicken liver, brain, heart, and breast muscle (Sai *et al.*, 1987). However, the weak signals have not been further studied by more definitive techniques such as Northern blot hybridization or by protection experiments. Unlike type II collagen mRNA, no RNA was detected in pre-cartilage limb mesenchyme. A detailed study of changes in steady-state levels of the mRNA in chick limb mesenchyme differentiating into cartilage *in vitro* has been published (Kosher *et al.*, 1986a) showing that the mRNA begins to accumulate at the onset of overt chondrogenesis coincident with condensation of cells and that the rate of accumulation is about half the rate of accumulation of type II collagen mRNA. Expression of the core protein mRNA also appears to be more sensitive to cAMP than is expression of the type II collagen gene. The expression of the gene in the developing limb has also been examined by *in situ* hybridization (Mallein-Gerin *et al.*, 1988; Stirpe and Goetinck, 1989).

Tsonis and Goetinck (1988) have investigated core protein mRNA in the developing chick eye and found it to be present at relatively constant levels between days 8 and 11 at about 20% of the level present in 14-d sternal cartilage. In the eye of chicks homozygous for the cartilage proteoglycan-deficient mutant, nanomelia, the level was reduced approximately tenfold from the normal level. Stirpe *et al.* (1987) and O'Donnell *et al.* (1988) have investigated the expression of the RNA in the cartilage of nanomelic chicks, using the 1.2 kb chicken cDNA, ST-1, as well as a slightly longer 1.65 kb clone, pPG525 isolated by screening a lambda gt11 cartilage cDNA library with ST-1. These experiments (Stirpe *et al.*, 1987) showed a single mRNA band of normal size on Northern-blots but at a level reduced approximately 20 times, suggesting that nanomelia is a regulatory mutation. Southern-blot analysis of genomic restriction fragments using the same probes also showed no change in the band pattern, suggesting that there is not a major deletion in the gene in the region detected by these probes.

Following the findings of Stirpe *et al.* (1987), O'Donnell *et al.* (1988) reported that the homozygous nanomelia mutants do synthesize an immunologically related, but slightly smaller, glycoprotein which does not undergo the posttranslational processing characteristic of the normal cartilage core protein, and that this smaller glycoprotein is not detected in biosynthetic studies with chondrocytes from normal chickens. These findings create a dilemma regarding the nature of the nanomelia mutation. The most likely possibility is that a mutation in the coding sequence results in the translation of a truncated protein from mRNA of normal size, and that the mutation additionally leads to either less efficient processing of the transcript or decreased stability of the mRNA. These conclusions are limited by the fact that the cDNA probes used in this study only detect about 20% of the coding sequence of the RNA near the 3′ end of the translated sequences and by the fact that the immunoprecipitated protein has not been definitively shown to correspond to a truncated version of the normal core protein.

Link Protein

Link protein of cartilage proteoglycan aggregates is involved in stabilizing the interactions between hyaluronic acid and proteoglycan subunits in cartilage extracellular matrix. In most cartilages at least two forms (and sometimes three) of this molecule have been described and show much similarity by protein and immunological criteria. A part of the difference between the various forms appears to result from differences in glycosylation.

Cloning of the Link Protein Gene

Following publication of partial amino acid sequences for the bovine nasal cartilage link protein (Perin *et al.*, 1980; Neame *et al.*, 1985) and the rat chondrosarcoma link protein (Baker *et al.*, 1983), Deak *et al.* (1986) used a synthetic oligonucleotide probe deduced from a region of the amino acid sequence which was identical in bovine cartilage and rat chondrosarcoma link proteins to isolate a clone from a chicken cDNA library (Figure 14). The resulting clone was used to isolate nine other clones which together cover the entire protein coding sequence of the mRNA and contain approximately 2.5 kb of nucleotides. Probes from all regions of the clones hybridized to two major RNA species in sternal cartilage of 6.0 and 5.8 kb and a minor species of 3.0 kb.

In parallel with the work on the chicken link protein cDNA, Doege *et al.* (1986b) used the available amino acid sequence data to confirm the identity of a rat cDNA clone of 778 bp which they isolated by screening an expression library with polyclonal and monoclonal antibodies against the rat chondrosarcoma link protein prepared by Caterson *et al.* (1985) (Figure 14). In the rat chondrosarcoma the cDNA hybridized to four RNA species of 1.5, 2.1, 2.5, and 5.5 kb.

More recently Dudhia and Hardingham (1989b) used antibodies against porcine link protein to isolate cDNA clones from laryngeal cartilage cDNA expression library. They also isolated human link protein cDNAs from a human articular chondrocyte cDNA library using one of the porcine clones as a probe (Figure 14) (Dudhia and Hardingham, 1990).

Using the previously isolated cDNA clone as probes, genomic clones for the rat (Rhodes *et al.*, 1988) and chicken (Kiss *et al.*, 1987) genes have been isolated. The link protein gene appears to contain at least five or six exons, an uncertainty in part due to the fact that in neither case has the 5′ end of gene been described.

The rat gene has been shown to contain one internal exon of 159 bp which is alternatively spliced (Rhodes *et al.*, 1988). Comparative hybridization analysis suggests that this exon is present in only about 10% of the RNA. The amino acids encoded in this exon have not been identified by amino acid sequencing of the protein, although if only a fraction of the protein contained this region it might not be clearly observed, and no systematic search for its existence in the protein has been reported. The presence of the region encoded by this exon would substantially change the properties of the protein as the region includes ten positively charged amino acids as well as three cysteines, and it would increase the molecular mass of the protein from 39 to 45 kDa.

At present the 3′ and 5′ ends of the chicken and rat link protein genes are poorly defined, and the available data are not easily interpreted by any simple hypothesis.

The chicken link protein gene (Kiss *et al.*, 1987) contains at least one exon at the 5′ end which contains no protein coding sequences. R-loop analysis suggests that this exon is not present in the existing genomic clones. Introns in the chicken gene are very large, one being >39 kb, so that the extent of the gene is >80 kb. The alternatively spliced exon of the rat link protein gene has not been reported in the chick gene, but only two cDNA clones which span the region of this exon have been characterized, and no specific report of an effort to locate this exon has been published.

As mentioned above, multiple poly A+ RNA species have been detected in both species. Hybridization analysis using a variety of probes suggests that the different RNAs detected by Northern blot analysis result from the use of alternative polyadenylation sites, although the possibility of some alternative splicing events in the 3′ nontranslated regions of the mRNAs has not been excluded. The probe for the alternatively spliced exon of the rat gene shows no difference in Northern blot hybridization compared to probes for the adjacent exons, presumably because the Northern blot analysis performed does not adequately resolve differences of this amount in RNAs of this size.

In the characterization of the chicken cDNA clones, clones with two different 3′ ends were detected and poly A tails were observed at the 3′ ends of each. Somewhat surprisingly, probes unique to the 3′ end of the clones extending approximately 900 bp further in the 3′ direction hybridized to all of the RNA bands, giving a pattern identical with that seen using probes from cDNAs with the shorter 3′ ends. The position of the end of the clones with the longer 3′ nontranslated region is consistent with the polyadenylation site in these clones corresponding to the shorter 3.0 kb RNA detected on the Northern blot analysis. Possible explanations for the absence of mRNA species detected only by the longer clones but not by the shorter clones could be the existence of an internal poly A tract encoded in the RNA on which the oligo dT might prime or the existence of a very rare mRNA terminating at an earlier position. The region of the possible poly A tract has not been sequenced in the genomic DNA or in the longer cDNA clones.

It is intriguing that rat cartilage has a shorter RNA species of about 1.5 kb whose 3′ end appears to correspond to the 3′ end of the chicken cDNA clones with the shorter 3′ nontranslated sequences. This finding might suggest that the same site is also used as a polyadenylation site in the chicken gene, but that the amount of RNA terminating at that site is difficult to detect. Furthermore, the cDNA cloning procedure might preferentially select those shorter clones with a 3′ end at a minor polyadenylation site.

Analysis of the amino acid sequence both from direct amino acid sequencing of the protein (Neame *et al.*, 1985, 1986) as well as by analysis of the cDNA clones (Doege *et al.*, 1986b; Kiss *et al.*, 1987; Rhodes *et al.*, 1988) suggests that link protein can be divided into three or four domains (not including the domain encoded by the alternatively spliced exon), an amino terminal domain with signal peptide sequence, and three looped domains (Figure 14). The last two of these looped domains (B and B′) are structurally related, suggesting a tandem repeat of similar domains. Goetinck *et al.* (1987) have used synthetic peptides derived from the chicken link protein cDNA sequence to show that each of the two tandemly repeated sequences contains two potential sites for interaction with hyaluronic acid. Each of the two domains is encoded by a single separate exon. Comparisons of the deduced amino acid sequences of the porcine and human link proteins with those for the chicken and rat proteins show that the first 120 N-terminal residues contain the majority of amino acid substitutions and that the remaining sequence towards the C-terminus which encodes the two repeated domains is highly conserved (Dudhia and Hardingham, 1989b). Interestingly, sequences homologous to the link protein repeated domains (B and B′) have been found in a human lymphocyte homing molecule (Goldstein *et al.*, 1989; Stamenkovic *et al.*, 1989). The three structural domains of the link protein (excluding the alternatively spliced region) also seem to be present in the core protein of versican (Zimmerman and Ruoslahti, 1989) and of the large cartilage proteoglycan (see below), and Perkins *et al.* (1989) have suggested that the loop domain (region A) at the amino terminus of both link protein and core protein corresponds to an immunoglobulin fold-like sequence.

Regulation of the Expression of the Link Protein Gene

A limited number of experiments have been done to examine the expression of the link protein genes showing that the rat link protein mRNAs are expressed in rat chondrosarcoma and rat sternal chondrocytes and not in rat embryo fibroblasts (Doege *et al.*, 1986b) and that the chick mRNA is expressed in chick sterna and not in calvaria (Kiss *et al.*, 1987). Tsonis and Goetinck (1988) examined the expression of the gene in the embryonic chick eye, showing that it is expressed in cornea, sclera, choroid, retina, lens capsule, lens epithelium, and lens fibers, but not in pigment epithelium. These studies also showed that the same pattern of mRNA species is expressed as in 14-day sterna, but that it is expressed (in the whole eye) at about 20% of the level in sterna, and that there is little change in the relative level of expression between days 8 and 11 of development. In contrast to the proteoglycan core protein and cartilage matrix protein mRNA, there was no reduction in the level of link protein RNA in 11-day eyes from nanomelic chicks compared with normal 11-day eyes. The expression of the gene in the developing chick has been examined by *in situ* hybridization showing that link protein mRNA transcripts appear in the cartilage forming regions of the limb together with proteoglycan core protein transcripts somewhat later than the transcripts of the type II collagen gene. Transcripts appear earlier in the notochord some time prior to stage 21.

Cartilage Matrix Protein

Cartilage matrix protein is a major cartilage constituent, consisting of homotrimeric molecules of 54 kDa subunits (Paulsson and Heinegard, 1979, 1981, 1982, 1984). It appears to bind to collagen (Argraves *et al.*, 1987) and cartilage proteoglycan (Paulsson and Heinegard, 1979, 1984), but the precise function in cartilage is not yet determined.

Cloning of the Matrix Protein Gene

cDNA clones covering most of the chicken cartilage matrix protein mRNA have been isolated by immunological screening of a pUC8/pUC9 cDNA library (Argraves *et al.*, 1984). The cDNAs have subsequently been used to isolate the entire gene for the chicken protein (Kiss *et al.*, 1989), and nucleotide sequence analysis has established the complete coding sequence of the chicken matrix protein translation product.

The translated form of matrix protein contains 493 amino acid residues. Peptide sequencing of the protein isolated from chicken sternal cartilage shows that the amino-terminal residue of the mature protein is residue 24 of the translation product, and the first 23 residues are therefore likely to represent the signal peptide (Kiss *et al.*, 1989). The identity between the protein sequence and the DNA-derived sequence also provides definitive proof that the DNA sequence is indeed that of cartilage matrix protein.

Analysis of the sequence has revealed a number of interesting structural features for the matrix protein. First, the protein is composed of two homologous repeats separated by an EGF-like domain. Within each repeat, two cysteinyl residues, one at the beginning of the repeat and one at the end, are connected by an intrachain disulfide bridge. The degree of sequence similarity between the two repeat domains is quite high, with 40% of the amino acid residues being identical. Second, the sequences of the two repeats show significant similarity to sequence domains within von Willebrand factor, complement factors B and C2, the α chains of the integrins Mac-1, p 150, 95, and LFA-1, and type VI collagen (Kiss *et al.*, 1989). Finally, the sequence domain structure of the matrix protein is also reflected in the exon structure of the gene. The central EGF-like domain is encoded by a single exon, and each of the two repeat domains is encoded by two exons. However, the two exons are not the same size for the two repeat domains. Therefore, it has been proposed that the two exons have evolved by insertion of introns in two larger duplicated exons (Kiss *et al.*, 1989).

Regulation and Expression of the Matrix Protein Gene

Matrix protein RNA has been shown to be present in cartilage but not in bone (Argraves *et al.*, 1987). In developing chick embryo limb bud tissue, the appearance of the RNA coincides with the histological detection of cartilage. Also, in high density limb bud cultures only differentiated cartilage nodules react with an antiserum against the polypeptide encoded by matrix protein cDNA (Argraves *et al.*, 1987), suggesting that matrix protein expression is turned on during chondrocyte differentiation.

The Type X Collagen Gene

Type X Collagen

Type X collagen, independently discovered as a short bacterial collagenase-sensitive polypeptide in chondrocyte cultures by at least three laboratories (Schmid, 1980; Gibson *et al.*, 1981; Capasso *et al.*, 1982), contains homotrimeric molecules of subunits that are only about one-half the size of the polypeptides of fibrillar collagens. Each subunit contains three distinct sequence domains; a relatively short (52 residues including a signal peptide of 18 residues) amino-terminal non-triple-helical domain; a triple-helical domain of 460 amino acid residues, and a carboxy-terminal, non-triple-helical domain of 162 amino acid residues (Figure 15). The size of the carboxyl domain was initially reported as 170 residues in length (Ninomiya *et al.*, 1986); however, resequencing of the type X cDNA and gene shows that one nucleotide was incorrectly left out of the initial sequence. This has now been corrected, and the resulting shift in reading frame makes the domain 162 instead of 170 residues (Yamaguchi *et al.*, 1989).

As discussed below, the structure of the type X collagen gene is fundamentally different from that of fibrillar collagens or type IX. Most striking is the absence in the type X gene of multiple short exons encoding the triple-helical domain while triple-helical domains in fibrillar collagens (types I, II, II, V, and XI) and fibril-associated collagens (types IX and XII) are coded for by short exons. The evolutionary relationship between type X collagen and fibrillar and fibril-associated collagens is therefore not clear. Type X collagen may, in fact, represent a case of convergent evolution rather than a divergent one. For this reason, and also because type X collagen does not

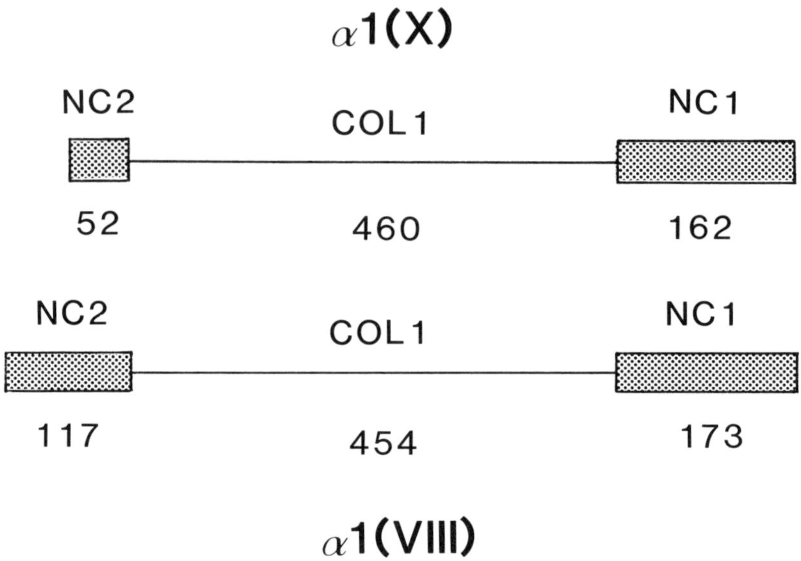

Figure 15. Diagram comparing the domain structure of chicken α1(X) and rabbit α1(VIII) collagen chains. Triple-helical domains (COL1) are shown as lines; non-triple-helical domains (NC1, NC2) are indicated by boxes. The sizes of the domains (in amino acid residues) are shown below each domain. (Reproduced from Ninomiya *et al.*, 1990, with permission.)

type IX collagen, we have chosen to discuss type X collagen at the end of this chapter.

Molecular Cloning of the α1(X) Collagen Gene

Ninomiya *et al.* (1986) isolated the first type X collagen cDNA from a cDNA library made with chick hypertrophic chondrocyte mRNA, using a differential hybridization screen. The positive probe was ^{32}P-labeled RNA from long-term cultured tibial hypertrophic chondrocytes, and the negative probe was ^{32}P-labeled RNA from 17-day-old-chick sternal cartilage. Nucleotide sequence analysis of the cDNA pYN3116 showed that its insert encoded only a small portion of type X mRNA. It was therefore used as a probe to isolate a large fragment of the chicken type X collagen gene (Ninomiya *et al.*, 1986). In addition, a second cDNA, encoding the 5′ portion of type X mRNA, was isolated by primer extension cloning and used to isolate a genomic fragment covering the 5′ region of the type X gene (LuValle *et al.*, 1988).

Nucleotide sequence analysis of three genomic clones that encode most of the chicken α1(X) gene (COL10A1) has demonstrated that this gene is unique among vertebrate collagens in that it is relatively small, only 5 kb, and contains only three exons (Figure 16). This is in stark contrast to the fibrillar collagen genes that contain more than 50 exons and the type IX genes with over 30 appear to be an integral part of the cartilage collagen fibril structure, as is

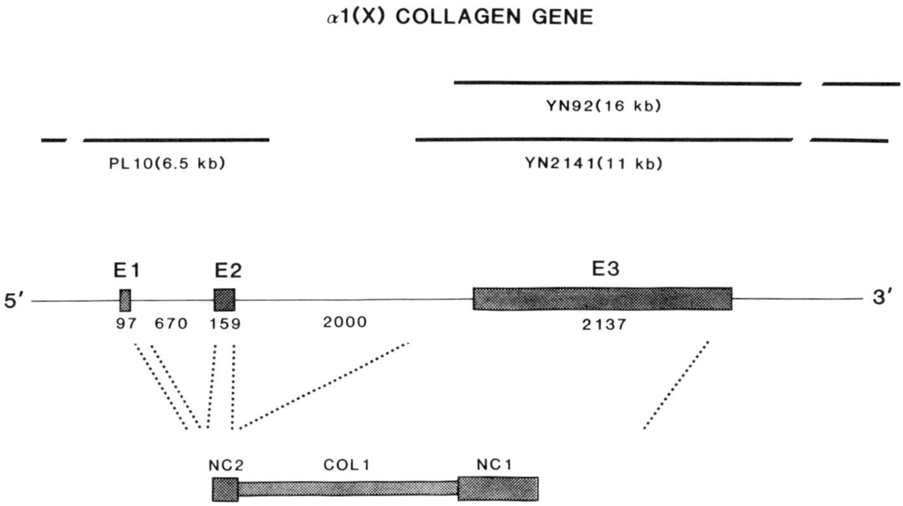

Figure 16. Diagram showing the exon structure (center) of the chicken α1(X) collagen gene, the genomic clones that have been isolated (top) and the domain structure of the α1(X) chain (bottom). Exon and intron sizes (in bp) are given below the diagram. Exon 1 encodes 5′ untranslated sequences, exon 2 codes for most of the amino-terminal non-triple-helical domain NC2, while exon 3 codes for the COL1 and NC1 domains. (Reproduced from Ninomiya *et al.*, 1990, with permission.)

exons. Particularly striking is the fact that whereas the triple-helical domains of fibrillar collagens are encoded by multiple, small exons, the type X triple-helix is coded for by one large exon of 2137 nucleotides. This exon (exon 3 as counted from the 5′ end of the gene) also encodes the carboxy-terminal non-triple-helical domain and 4 2/3 codons of the amino-terminal non-triple-helical sequence. The rest of this sequence as well as the signal peptide is encoded by the 159 nucleotide-long exon 2. Exon 1, 97 nucleotides in length, encodes the 5′ untranslated portion of α1(X) mRNA (Figure 16).

Regulation of the α1(X) Collagen Gene

Biochemical (Schmid and Linsenmayer, 1983; Gibson and Flint, 1985; Grant *et al.*, 1985) and immunofluorescence (Schmid and Linsenmayer, 1985a, 1985b; Linsenmayer *et al.*, 1986) data as well as *in situ* hybridization with type X-specific probes (LuValle *et al.*, 1989) suggest that expression of type X collagen is restricted to cartilage undergoing endochondral ossification. Thus, in the embryonic chick tibiotarsus, immunoreactivity for type X collagen with the monoclonal antibody X-AC9 is first seen at 7.5 days of development in the area corresponding to the primary center of ossification (Schmid and Linsenmayer, 1985a). The area of reactivity then extends from the primary center, in parallel with the extension of the cartilage that contains hypertrophic chondrocytes. At later stages, the expression of X clearly follows the development of chondrocyte hypertrophy, with appearance of immunoreactive material lagging slightly behind cellular hypertrophy. In embryonic chick sternal cartilage, type X immunoreactive protein and mRNA appears in the cephalic half at day 17 or 18 of development when this portion of the sternum mineralizes and ossifies, while the caudal portion remains cartilaginous and negative with type X probes (Gibson *et al.*, 1984). Nuclear run-off assays with nuclei isolated from the cephalic and caudal portions of 18-day-old sterna show that the type X collagen gene is transcribed at a rate that is at least 20-fold higher in nuclei from the cephalic portion than in nuclei from the caudal part (LuValle *et al.*, 1989). This correlates well with the increase in type X collagen mRNA levels in the cephalic portion and the undetectable level of RNA in the caudal region. It also agrees well with studies of cultured chick tibial chondrocytes showing that increased synthesis of type X collagen is due to an increase in the rate of transcription of the gene (Castagnola *et al.*, 1988). It is therefore likely that the cell-specific expression of type X collagen in hypertrophic chondrocytes is due primarily to a transcriptional regulatory mechanism.

The Function of Type X Collagen in Hypertrophic Cartilage

The function of type X collagen in cartilage is not known. However, several possible roles have been suggested. One possibility is that type X collagen may facilitate the removal of cartilage during endochondral ossification (Schmid and Linsenmayer, 1987). In support of this speculation is that since type X collagen molecules contain two cleavage sites for vertebrate collagenase as compared to the single site in type II collagen, collagenase activity within hypertrophic cartilage may degrade type X more rapidly and open up the matrix (Schmid *et al.*, 1986; Schmid and Linsenmayer, 1987).

A second speculation is that type X collagen may play a role in calcification (Schmid and Linsenmayer, 1987). During the ossification of the cephalic region of chick sternal cartilage, the matrix becomes calcified after the appearance of type X collagen. The presence of a hydrophobic segment in the carboxy-terminal non-triple-helical domain of the α1(X) chain as initially deduced from the DNA-derived sequence has been considered of relevance to this hypothetical role in mineralization (Ninomiya *et al.*, 1986; Schmid and Linsenmayer, 1987). In fact, it has been speculated that type X collagen may be associated with matrix vesicles through a transmembrane domain (Ninomiya *et al.*, 1986). This possibility can now be eliminated for two reasons. First, the association between matrix vesicles and type X collagen appears to be an artifact due to nonspecific binding of the collagen to the filters used to isolate the vesicles (Schmid, personal communication). Second, the revision of the sequence of chicken α1(X) prompted by its similarity to that of α1(VIII) (see below) effectively eliminates the hydrophobic sequence segment towards the carboxyl end of the molecule (Yamaguchi *et al.*, 1989).

The similarity between α1(X) and α1(VIII) is striking and unexpected (Figure 15). The complete sequence of the rabbit α1(VIII) chain has been determined from two overlapping cDNAs isolated from a rabbit cornea endothelial library, and a comparison of that sequence with the chicken α1(X) shows that the triple-helical domains are almost identical in length in the two chains; 460 amino acid residues in α1(X) vs. 454 residues in α1(VIII) (Yamaguchi *et al.*, 1989). The triple-helical domains in both chains contain eight imperfections in the Gly-X-Y repeat structure in almost the same relative locations. This structural similarity raises the question of whether there is a functional similarity between types X and VIII collagens as well.

It is believed that type VIII collagen is a major component of the hexagonal lattice structure of Descemet's membrane (Sawada, 1982; Sawada *et al.*, 1984). Such a lattice may create a matrix that resists compression, yet maintains an open, porous structure. Perhaps type X collagen plays a similar role in hypertrophic cartilage by functioning as a scaffold to prevent collapse of the cartilage matrix as the proteoglycans and type II-containing fibrils are degraded. In support of this speculation is the remarkable thermal stability of type X triple-helices, even after cleavage with vertebrate collagenase (Schmid *et al.*, 1986). In fact, the *Tm* of the central triple-helical 32 kDa fragment released by vertebrate collagenase is higher than 40°C, and this fragment must therefore be stable as a triple-helix at body temperature.

References

E.S. Allebach, D. Boettiger, M. Pacifici, and S.L. Adams. Control of types I and II collagen and fibronectin gene expression in chondrocytes delineated by viral transformation. *Mol. Cell. Biol.* **5**: 1002–1008 (1985).

I.J. Anderson, R.B. Goldberg, R.W. Marion, W.B. Upholt, and P. Tsipouras. Spondyloepiphyseal dysplasia congenita: genetic linkage to type II collagen (COL2A1), *Am. J. Hum. Genet.* **46**: 896–901 (1990).

P. Antonsson, D. Heinegard, and A. Oldberg. The keratan sulfate-enriched region of bovine cartilage proteoglycan consists of a consecutively repeated hexapeptide motif. *J. Biol. Chem.* **264**: 16170–16173 (1989).

W.S. Argraves, F. Deak, K.J. Sparks, I. Kiss, and P. Goetinck. Structural features of cartilage matrix protein deduced from cDNA. *Proc. Natl. Acad. Sci. U.S.A.* **84**: 464–468 (1987).

K. Arheden, N. Mandahl, S. Heim, and F. Mitelman. In situ hybridization localizes the human type II α1 collagen gene (COL2A1) to 12q13. *Hereditas* **110**: 165–167 (1989).

S. Ayad, M.Z. Abedin, J.B. Weiss, and S.M. Grundy. Characterization of another short-chain disulfide-bonded collagen from cartilage, vitreous and intervertebral disc. *FEBS Lett.* **139**: 300–304 (1982).

J.R. Baker, B. Caterson, and J.E. Christner. The link proteins: some aspects of their structure. *Prog. Clin. Biol. Res.* **110B**: 17–24 (1983).

C.T. Baldwin, A.M. Reginato, C. Smith, S.A. Jimenez, and D. J. Prockop. Structure of cDNA clones coding for human type II procollagen. The α1(II) chain is more similar to the α1(I) chain than two other alpha chains of fibrillar collagens. *Biochem. J.* **262**: 521–528 (1989a).

C.T. Baldwin, A.M. Reginato, and D.J. Prockop. A new epidermal growth factor-like domain in the human core protein for the large cartilage-specific proteoglycan. *J. Biol. Chem.* **264**: 15747–15750 (1989b).

V.D. Bennett, I.M. Weiss, and S.L. Adams. Cartilage-specific 5′ end of chick α2(I) collagen mRNAs. *J. Biol. Chem.* **264**: 8402–8409 (1989).

V. Benson-Chanda, S.M.-W. Weil, M.-L. Chu, and F. Ramirez. Cloning and analysis of the 5′ portion of the human type-III procollagen gene (COL3Al). *Gene* **78**: 255–265 (1989).

M. Bernard, H. Yoshioka, E. Rodriguez, M. van der Rest, T. Kimura, Y. Ninomiya, B.R. Olsen, and F. Ramirez. Cloning and sequencing of pro-α1(XI) collagen cDNA demonstrates that type XI belongs to the fibrillar class of collagens and reveals that the expression of the gene is not restricted to cartilagenous tissue. *J. Biol. Chem.* **263**: 17159–17166 (1988).

H. Boedtker, M. Finer, and S. Aho. The structure of the chicken α2 collagen gene. *Ann. N.Y. Acad. Sci.* **460**: 85–116 (1985).

P. Bornstein and H. Sage. Structurally distinct collagen types. *Annu. Rev. Biochem.* **49**: 957–1003 (1980).

A. M. Bowcock, J.M. Hebert, A.M. Christiano, E. Wijsman, L.L. Cavalli-Sforza, and C.D. Boyd. The proαl(IV) collagen gene is linked to the D13S3 locus at the distal end of human chromosome 13q. *Cytogenet. Cell Genet.* **45**: 234–236 (1987).

C.D. Boyd, S.E. Toth-Fejel, I.K. Gadi, M. Litt, M.R. Condon, M. Kolbe, I. K. Hagen, M. Kurkinen, J.W. Mackenzie, and E. Magenis. The genes coding for human proαl(IV) collagen and proα2(IV) collagen are both located at the end of the long arm of chromosome 13. *Am. J. Hum. Genet.* **42**: 309–314 (1988).

R.E. Burgeson, P.A. Hebda, N.P. Morris, and D.W. Hollister. Human cartilage collagens. Comparison of cartilage collagens with human type V collagen. *J. Biol. Chem.* **257**: 7852–7856 (1982).

R.E. Burgeson and D.W. Hollister. Collagen heterogeneity in human cartilage: identification of several new collagen chains. *Biochem. Biophys. Res. Commun.* **87**: 1124–1131 (1979).

P. Bruckner, L. Vaughan, and K.H. Winterhalter. Type IX collagen from sternal cartilage of chicken embryo contains covalently bound glycosaminoglycans. *Proc. Natl. Acad. Sci. U.S.A.* **82**: 2608–2612 (1985).

O. Capasso, E. Gionti, G. Pontarelli, F.S. Ambesi-Impiombato, L. Nitsoh, G. Tajana, and R. Cancedda. The culture of chick embryo chondrocytes and the control of their differentiated functions in vitro. *Exp. Cell Res.* **142**: 197–206 (1982).

P. Castagnola, B. Dozin, G. Moro, and R. Cancedda. Changes in the expression of collagen genes show two stages in chondrocyte differentiation in vitro. *J. Cell Biol.* **106**: 461–467 (1988).

B. Caterson, J.R. Baker, J.E. Christner, Y. Lee, and M. Lentz. Monoclonal antibodies as probes for determining the microheterogeneity of the link proteins of cartilage proteoglycan. *J. Biol. Chem.* **260**: 11348–11356 (1985).

K.S. Cheah, N.G. Stoker, J.R. Griffin, F.G. Grosveld, and E. Solomon. Identification and characterization of the human type II collagen gene (COL2A1). *Proc. Natl. Acad. Sci. U.S.A.* **82**: 2555–2559 (1985).

C.C. Clark and C.F. Richards. Isolation and partial characterization of precursors to minor cartilage collagens. *Collagen Relat. Res.* **5**: 205–223 (1985).

S. Curran and D.J. Prockop. Isolation and partial characterization of the amino-terminal propeptide of type II procollagen from chick embryos. *Biochemistry* **21**: 1482–1487 (1982).

M. D'Alessio, M. Bernard, P.J. Pretorious, W. de Wet, and F. Ramirez. Complete nucleotide sequence of the region encompassing the first twenty-five exons of the human pro α1 (I) collagen gene (COL1A1). *Gene* **67**: 105–115 (1988).

F. Deak, W.S. Argraves, I. Kiss, K.J. Sparks, and P.F. Goetinck. Primary structure of the telopeptide and a portion of the helical domain of chicken type II procollagen as determined by DNA sequence analysis. *Biochem. J.* **229**: 189–196 (1985).

F. Deak, I. Kiss, K.J. Sparks, W.S. Argraves, G. Hampikian, and P.F. Goetinck. Complete amino acid sequence of chicken cartilage link protein deduced from cDNA clones. *Proc. Natl. Acad. Sci. U.S.A.* **83**: 3766–3770 (1986).

C.J. Devlin, P.M. Brickell, E.R. Taylor, A. Hornbruch, R.K. Craig, and L. Wolpert. In situ hybridization reveals differential spatial distribution of mRNAs for type I and type II collagen in the chick limb bud. *Development* **103**: 111–118 (1988).

K. Doege, P. Fernandez, J.R. Hassell, M. Sasaki, and Y. Yamada. Partial cDNA sequence encoding a globular domain at the C terminus of the rat cartilage proteoglycan. *J. Biol. Chem.* **261**: 8108–8111 (1986a).

K. Doege, J.R. Hassell, B. Caterson, and Y. Yamada. Link protein cDNA sequence reveals a tandemly repeated protein structure. *Proc. Natl. Acad. Sci. U.S.A.* **83**: 3761–3765 (1986b).

K. Doege, C. Rhodes, M. Sasaki, J.R. Hassell, and Y. Yamada. Molecular biology of cartilage proteoglycan (aggrecan) and link protein. In: *Extracellular Matrix Genes*, Eds L.J. Sandell and C.D. Boyd. Academic Press, Orlando, pp. 137–155, 1990b.

K. Doege, M. Sasaki, E. Horigan, J.R. Hassell, and Y. Yamada. Complete primary structure of the rat cartilage proteoglycan core protein deduced from cDNA clones. *J. Biol. Chem.* **262**: 17757–17767 (1987).

K. Doege, M. Sasaki, and Y. Yamada. Rat and human cartilage proteoglycan (aggrecan) gene structure. *Biochem. Soc. Trans.* **18**: 200–202 (1990a).

K.J. Doege, M. Sasaki, T. Kimura, and Y. Yamada. Complete coding sequence and deduced primary structure of the human cartilage large aggregating proteoglycan aggrecan. Human specific repeats, and additional alternatively spliced forms. *J. Biol. Chem.* **266**: 894–902 (1991).

A.J. Driesel, A.M. Schumacher, and R.A. Flavell, A Hind III restriction site polymorphism in the human collagen α1(I)-like gene on chromosome no. 7. *Hum. Genet.* **62**: 175–176 (1982).

B. Dublet, S. Oh, S.P. Sugrue, M.K. Gordon, D.R. Gerecke, B.R. Olsen, and M. van der Rest, The structure of avian type XII collagen. *J. Biol. Chem.* **264**: 13150–13156 (1989).

J. Dudhia and T. Hardingham. The primary structure of human cartilage link protein. *Nucl. Acids Res.* **18**: 2214 (1990).

J. Dudhia and T.E. Hardingham. cDNA sequences to human and porcine cartilage proteoglycan and link protein. *Trans. Orthopaed. Res. Soc.* **14**: 8 (1989a).

J. Dudhia and T.E. Hardingham. Isolation and sequence of cDNA clones for pig and human cartilage link protein. *J. Mol. Biol.* **207**: 749–753 (1989b).

K. Elima, J.K. Makela, T. Vuorio, S. Kauppinen, J. Knowles, and E. Vuorio. Construction and identification of a cDNA clone for human type II procollagen mRNA. *Biochem. J.* **229**: 183–188 (1985).

K. Elima, T. Vuorio, and E. Vuorio. Determination of the single polyadenylation site of the human pro α1(II) collagen gene. *Nucleic Acids Res.* **15**: 9499–9504 (1987).

B.S. Emanuel, L.A. Cannizzro, J.M. Seyer, and J.C. Myers. Human α1(III) and α2(V) procollagen genes are located on the long arm of chromosome 2. *Proc. Natl. Acad. Sci. U.S.A.* **82**: 3385–3389 (1985).

B.S. Emanuel, B.T. Sellinger, L.J. Gudas, and J.C. Myers. Localization of the human procollagen α1(IV) gene to chromosome 13q34 by *in situ* hybridization. *Am. J. Hum. Genet.* **38**: 38–44 (1986).

C.E.L. Eng and C.M. Strom. Analysis of three restriction fragment length polymorphisms in the human type II procollagen gene. *Am. J. Hum. Genet.* **37**: 719–732 (1985).

D.R. Eyre. Type II collagen deficiency in achondrogenesis (Langer-Saldino). *Pathol. Immunopathol. Res.* **7**: 90–94 (1988).

D.R. Eyre. Collagen: molecular diversity in the body's protein scaffold. *Science* **207**: 1315–1322 (1980).

D.R. Eyre, M.P. Upton, F.D. Shapiro, R.H. Wilkinson, and G.F. Vawter. Nonexpression of cartilage type II collagen in a case of Langer-Saldino achondrogenesis. *Am. J. Hum. Genet.* **39**: 52–67 (1986).

D. Eyre and J.-J. Wu. Type XI or 1α2α3α collagen. In: *Structure and Function of Collagen Types*, Eds R. Mayne and R.E. Burgeson. Academic Press, Orlando. pp. 261–281, 1987.

D.R. Eyre, J.J. Wu, and D.E. Woolley, All three chains of 1α2α3α collagen from hyaline cartilage resist human collagenase. *Biochem. Biophys. Res. Commun.* **118**: 724–729 (1984).

M.P. Fernandez, M.F. Young, and M.E. Sobel. Methylation of type II and type I collagen genes in differentiated and dedifferentiated chondrocytes. *J. Biol. Chem.* **260**: 2374–2378 (1985).

M.H. Finer, L.C. Gerstenfeld, D. Young, P. Doty, and H. Boedtker. Collagen expression in embryonic chicken chondrocytes treated with phorbol myristate acetate. *Mol. Cell. Biol.* **5**: 1415–1424 (1985).

J.M. Fitch, A. Mentzer, R. Mayne, and T.F. Linsenmayer. Acquisition of type IX collagen by the developing avian primary corneal stroma and vitreous. *Dev. Biol.* **128**: 396–405 (1988).

J.M. Fitch, A.M.R. Mentzer, and T.F. Linsenmayer. Independent deposition of collagen types II and IX at epithelial-mesenchymal interfaces. *Development* **105**: 85–96 (1989).

R.J. Focht and S.L. Adams. Tissue specificity of type I collagen gene expression is determined at both transcriptional and post-transcriptional levels. *Mol. Cell. Biol.* **4**: 1843–1852 (1984).

A.J. Fosang and T.E. Hardingham. Isolation of the N-terminal globular protein domains from cartilage proteoglycans. Identification of G2 domain and its lack of interaction with hyaluronate and link protein. *Biochem. J.* **261**: 801–809 (1989).

C.A. Francomano and R.E. Pyeritz. Achondroplasia is not caused by mutation in the gene for type II collagen. *Am. J. Med. Genet.* **29**: 955–961 (1988).

C.A. Francomano, R.M. Liberfarb, T. Hirose, I.H. Maumenee, E.A. Streeten, D.A. Meyers, and R.E. Pyeritz. The Stickler syndrome: evidence for close linkage to the structural gene for type II collagen. *Genomics* **1**: 293–296 (1987).

C.A. Francomano, E.A. Streeten, D.A. Meyers, and R.E. Pyeritz. Marfan syndrome: exclusion of genetic linkage to three major collagen genes. *Am. J. Med. Genet.* **29**: 457–462 (1988).

D.K. Furuto and E.J. Miller. Different levels of glycosylation contribute to the heterogeneity of α1(II) collagen chains derived from a transplantable rat chondrosarcoma. *Arch. Biochem. Biophys.* **226**: 604–611 (1983).

L.C. Gerstenfeld, M.H. Finer, and H. Boedtker. Quantitative analysis of collagen expression in embryonic chick chondrocytes having different developmental fates. *J. Biol. Chem.* **264**: 5112–5120 (1989).

G.J. Gibson, S.L. Schor, and M.E. Grant. Partial characterization of a low-molecular-weight collagen synthesized by chondrocytes cultured in collagen gels. *Biochem. Soc. Trans.* **9**: 550–551 (1981).

G.J. Gibson and M.H. Flint. Type X collagen synthesis by chick sternal cartilage and its relationship to endochondral development. *J. Cell Biol.* **101**: 277–284 (1985).

G.J. Gibson, B.W. Beaumont, and M.H. Flint. Synthesis of a low molecular weight collagen by chondrocytes from the presumptive calcification region of the embryonic chick sterna. The influence of culture with collagen gels. *J. Cell Biol.* **99**: 208–216 (1984).

M. Godfrey and D.W. Hollister. Type II achondrogenesis-hypochondrogenesis: identification of abnormal type II collagen. *Am. J. Hum. Genet.* **43**: 904–913 (1988).

P.F. Goetinck, N.S. Stirpe, P.A. Tsonis, and D. Carlone. The tandemly repeated sequences of cartilage link protein contain the sites for interaction with hyaluronic acid. *J. Cell Biol.* **105**: 2403–2408 (1987).

M.B. Goldring, J. Birkhead, L.J. Sandell, T. Kimura, and S.M. Krane. Interleukin 1 suppresses expression of cartilage-specific types II and IX collagens and increases types I and III collagens in human chondrocytes. *J. Clin. Invest.* **82**: 2026–2037 (1988).

L.A. Goldstein, D.F. Zhou, L.J. Picker, C.N. Minty, R.F. Bargatze, J.F. Ding, and E.C. Butcher. A human lymphocyte homing receptor, the hermes antigen, is related to cartilage proteoglycan core and link proteins. *Cell* **56**: 1063–1072 (1989).

M.K. Gordon, D.R. Gerecke, and B.R. Olsen. Type XII collagen: distinct extracellular matrix component discovered by cDNA cloning. *Proc. Natl. Acad. Sci. U.S.A.* **84**: 6040–6044 (1987).

M.K. Gordon, D.R. Gerecke, B. Dublet, M. van der Rest, and B.R. Olsen. Type XII collagen: a large multidomain molecule with partial homology to type IX collagen. *J. Biol. Chem.* **264**: 19772–19778 (1989).

W.T. Grant, M.D. Sussman, and G. Balian. A disulfide-bonded short chain collagen synthesized by degenerative and calcifying zones of bovine growth plate cartilage. *J. Biol. Chem.* **260**: 3798–3803 (1985).

B.C. Gregg, A. Rowe, P.M. Brickell, and L. Wolpert. Ectodermal inhibition of cartilage differentiation in micromass culture of chick limb bud mesenchyme in relation to gene expression and cell shape. *Development* **105**: 769–777 (1989).

C.A. Griffin, B.S. Emanuel, J.R. Hansen, W.K. Cavanee, and J.C. Myers. Human collagen genes encoding basement membrane α1(IV) and α2(IV) chains map to the distal long arm of chromosome 13. *Proc. Natl. Acad. Sci. U.S.A.* **84**: 512-516 (1987).

D.F. Halberg, G. Proulx, K. Doege, Y. Yamada, and K. Drickamer. A segment of the cartilage proteoglycan core protein has lectin-like activity. *J. Biol. Chem.* **263**: 9486–9490 (1988).

J.R. Hassell, J.H. Kimura, and V.C. Hascall. Proteoglycan core protein families. *Annu. Rev. Biochem.* **55**: 539–567 (1986).

M. Hayashi, Y. Ninomiya, K. Hayashi, T.F. Linsenmayer, B.R. Olsen, and R.L. Trelstad. Secretion of collagen types I and II by epithelial and endothelial cells in the developing chick cornea demonstrated by in situ hybridization and immunohistochemistry. *Development* **103**: 27–36 (1988).

M. Hayashi, Y. Ninomiya, J. Parsons, K. Hayashi, B.R. Olsen, and R.L. Trelstad. Differential localization of mRNAs of collagen types I and II in chick fibroblasts, chondrocytes, and corneal cells by in situ hybridization using cDNA probes. *J. Cell Biol.* **102**: 2302–2309 (1986).

A.S. Henderson, J.C. Myers, and F. Ramirez. Localization of the human α2(I) collagen gene (COL1A2) to chromosome 7q22. *Cytogenet. Cell Genet.* **36**: 586–587 (1983).

I. Henry, A. Bernheim, M. Bernard, M. van der Rest, T. Kimura, C. Jeanpierre, F. Barichard, R. Berger, B.R. Olsen, F. Ramirez, and C. Junien. Mapping of a human fibrillar collagen gene, pro α1(XI) (COL11A1), to the p21 region of chromosome 1. *Genomics* **3**: 87–90 (1988).

W.E. Horton, J. Cleveland, U. Rapp, G. Nemuth, M. Bolander, K. Doege, Y. Yamada, and J.R. Hassell. An established rat cell line expressing chondrocyte properties. *Exp. Cell Res.* **178**: 457–468 (1988).

W.E. Horton, J.D. Higginbotham, and S. Chandrasekhar. Transforming growth factor-beta and fibroblast growth factor act synergistically to inhibit collagen II synthesis through a mechanism involving regulatory DNA sequences. *J. Cell Physiol.* **141**: 8–15 (1989).

W.E. Horton, Y. Yamada, and J.R. Hassell. Retinoic acid rapidly reduces cartilage matrix synthesis by altering gene transcription in chondrocytes. *Dev. Biol.* **123**: 508–516 (1987b).

W. Horton, T. Miyashita, K. Kohno, J.R. Hassell, and Y. Yamada. Identification of a phenotype-specific enhancer in the first intron of the rat collagen II gene. *Proc. Natl. Acad. Sci. U.S.A.* **84**: 8864–8868 (1987a).

S.L. Hostikka, R.L. Eddy, M.G. Byers, M. Höyhtyä, T.B. Shows, and K. Tryggvason. Identification of a distinct type IV collagen α chain with restricted kidney distribution and assignment of its gene to the locusts of X chromosome-linked Alport syndrome. *Proc. Natl. Acad. Sci. U.S.A.* **87**: 1606-1610 (1990).

S. Huber, M. van der Rest, P. Bruckner, E. Rodriguez, K.H. Winterhalter, and L. Vaughan. Identification of the type IX collagen polypeptide chains. The α2(IX) polypeptide carries the chondroitin sulfate chain(s). *J. Biol. Chem.* **261**: 5965–5968 (1986).

C. Huerre, C. Junien, D. Weil, M.-L. Chu, M. Morabito, N. Van Cong, J.C. Myers, C. Foubert, M.S. Gross, D.J. Prockop, A. Boue, J.C. Kaplan, A. de la Chapelle, and F. Ramirez. Human type I procollagen genes are located on different chromosomes. *Proc. Natl. Acad. Sci. U.S.A.* **79**: 6627–6630 (1982).

C. Huerre-Jeanpierre, M.G. Mattei, D. Weil, K.H. Grzeschik, M.L. Chu, F.O. Sangiorgi, M.E. Sobel, F. Ramirez, and C. Junien. Further evidence for the dispersion of the human fibrillar collagen genes. *Am. J. Hum. Genet.* **38**: 26–37 (1986a).

C. Huerre-Jeanpierre, I. Henry, M. Bernard, P. Gallano, D. Weil, K.H. Grzeschik, F. Ramirez, and C. Junien, The proα2(V) collagen gene (COL5A2) maps to 2q14-2q-32, syntenic to the proα1(III) collagen locus (COL3A1). *Hum. Genet.* **73**: 64-67 (1986b).

M.H. Irwin and R. Mayne. Use of monoclonal antibodies to locate the chondroitin sulfate chain(s) in type IX collagen. *J. Biol. Chem.* **261**: 16281–16288 (1986).

C. Junien, D. Weil, J.C. Myers, N. Van Cong, M.-L. Chu, C. Foubert, M.S. Gross, D.K. Prockop, J.C. Kaplan, and F. Ramirez. Assignment of the human proα2(I) collagen structural gen (COL1A2) to chromosome 7 by molecular hybridization. *Am. J. Hum. Genet.* **34**: 381–387 (1982).

P.D. Killen, C.A. Francomano, Y. Yamada, W.S. Modi, and S. J. O'Brien. Partial structure of the human α2(IV) collagen chain and chromosomal localization of the gene (COL4A2). *Hum. Genet.* **77**: 318–324 (1987).

T. Kimura, K.S.E. Cheah, S.D.H. Chan, V.C.H. Lui, M.-G. Mattei, M. van der Rest, K. Ono, E. Solomon, Y. Ninomiya, and B.R. Olsen. The human α2(XI) collagen (COL11A2) chain. Molecular cloning of cDNA and genomic DNA reveal characteristics of a fibrillar collagen with differences in genomic organization. *J. Biol. Chem.* **264**: 13910-13916 (1989a).

T. Kimura, M.-G. Mattei, J.W. Stevens, M.B. Goldring, Y. Ninomiya, and B.R. Olsen. Molecular cloning of rat and human type IX collagen cDNA and localization of the α1(IX) gene on the human chromosome 6. *Eur. J. Biochem.* **179**: 71–78 (1989b).

I. Kiss, F. Deák, R.G. Holloway, Jr., H. Delius, K.A. Mebust, E. Frimberger, W.S. Argraves, P.A. Tsonis, N. Winterbottom, and P.F. Goetinch. Structure of the gene for cartilage matrix protein, a modular protein of the extracellular matrix. Exon/intron organization, unusual splice sites, and relation to α chains of β_2 integrins, von Willebrand factor, complement factors B and C2, and epidermal growth factor. *J. Biol. Chem.* **264**: 8126–8134 (1989).

I. Kiss, F. Deak, S. Mestric, H. Delius, J. Soos, K. Dekany, W.S. Argraves, K.J. Sparks, and P.F. Goetinck. Structure of the chicken link protein gene: exons correlate with the protein domains. *Proc. Natl. Acad. Sci. U.S.A.* **84**: 6399–6403 (1987).

K. Kohno, G.R. Martin, and Y. Yamada. Isolation and characterization of a cDNA clone for the amino-terminal portion of the pro-α1(II) chain of cartilage collagen. *J. Biol. Chem.* **259**: 13668–13673 (1984).

K. Kohno, M. Sullivan, and Y. Yamada. Structure of the promoter of the rat type II procollagen gene. *J. Biol. Chem.* **260**: 4441–4447 (1985).

H. Konomi, J.M. Seyer, Y. Ninomiya, and B.R. Olsen. Peptide specific antibodies identify the α2 chain as the proteoglycan subunit of type IX collagen. *J. Biol. Chem.* **261**: 6742–6746 (1986).

R.A. Kosher and M. Solursh. Widespread distribution of type II collagen during embryonic chick development. *Dev. Biol.* **131**: 558–566 (1989).

R.A. Kosher, S.W. Gay, J.R. Kamanitz, W.M. Kulyk, B.J. Rodgers, S. Sai, T. Tanaka, and M.L. Tanzer. Cartilage proteoglycan core protein gene expression during limb cartilage differentiation. *Dev. Biol.* **118**: 112–117 (1986a).

R.A. Kosher, W.M. Kulyk, and S.W. Gay. Collagen gene expression during limb cartilage differentiation. *J. Cell Biol.* **102**: 1151–1156 (1986b).

D. Kravis and W.B. Upholt. Quantitation of type II procollagen mRNA levels during chick limb cartilage differentiation. *Dev. Biol.* **108**: 164–172 (1985).

T. Krusius, K.R. Gehlsen, and E. Ruoslahti. A fibroblast chondroitin sulfate proteoglycan core protein contains lectin-like and growth factor-like sequences. *J. Biol. Chem.* **262**: 13120–13125 (1987).

K. Kuhn. The classical collagens: types I, II, and III. In: *Structure and Function of Collagen Types*, Eds R. Mayne and R.E. Burgeson. Academic Press, Orlando. pp. 1–42, 1987.

H. Kuivaniemi, G. Tromp, M. L. Chu, and D. J. Prockop. Structure of a full-length cDNA clone for the prepro $\alpha 2(I)$ chain of human type procollagen. Comparison with the chicken gene confirms unusual patterns of gene conservation. *Biochem. J.* **252**: 633–640 (1988).

W.M. Kulyk, B.J. Rodgers, K. Greer, and R.A. Kosher. Promotion of embryonic chick limb cartilage differentiation by transforming growth factor-β. *Dev. Biol.* **135**: 424–430 (1989).

M.L. Law, L. Tung, H.G. Morse, R. Berger, C. Jones, K.S. Cheah, and E. Solomon. The human type II collagen gene (COL2A1) assigned to 12q14.3. *Ann. Hum. Genet.* **50 (Part 2)**: 131–137 (1986).

B. Lee, H. Vissing, F. Ramirez, D. Rogers, and D. Rimoin. Identification of the molecular defect in a family with spondyloepiphyseal dysplasia. *Science* **244**: 978–980 (1989).

G. Liau, M. Mudryj, and B. de Crombrugghe. Identification of the promoter and first exon of the mouse $\alpha 1(III)$ collagen gene. *J. Biol. Chem.* **260**: 3773–3777 (1985).

T.F. Linsenmayer, E. Gibney, and T.M. Schmid. Sequential appearance of type X collagen in the developing avian notochord. *Dev. Biol.* **113**: 467–473 (1986).

R.H. Lovell-Badge, A. Bygrave, A. Bradley, E. Robertson, R. Tilly, and K.S. Cheah. Tissue-specific expression of the human type II collagen gene in mice. *Proc. Natl. Acad. Sci. U.S.A.* **84**: 2803–2807 (1987).

R.H. Lovell-Badge, A.E. Bygrave, A. Bradley, E. Robertson, M.J. Evans, and K.S. Cheah. Transformation of embryonic stem cells with the human type-II collagen gene and its expression in chimeric mice. *Cold Spring Harbor Symp. Quant. Biol.* **50**: 707–711 (1985).

G. Lozano, Y. Ninomiya, H. Thompson, and B.R. Olsen. A distinct class of vertebrate collagen genes encodes chicken type IX collagen polypeptides. *Proc. Natl. Acad. Sci. U.S.A.,* **82**: 4050–4054 (1985).

L.N. Lukens, A.M. Frischauf, P.J. Pawlowski, G.T. Brierley, and H. Lehrach. Construction and characterization of type II collagen complementary deoxyribonucleic acid clones. *Nucleic Acids Res.* **11**: 6021–6039 (1983).

P. LuValle, Y. Ninomiya, N. Rosenblum, and B.R. Olsen. The type X collagen gene. *J. Biol. Chem.* **263**: 18378–18385 (1988).

P. LuValle, M. Hayashi, and B.R. Olsen. Transcriptional regulation of type X collagen during chondrocyte maturation. *Dev. Biol.* **133**: 613–616 (1989).

P.F. Mallein-Gerin, R.A. Kosher, W.B. Upholt, and M.L. Tanzer. Temporal and spatial analysis of cartilage proteoglycan core protein gene expression during limb development by in situ hybridization. *Dev. Biol.* **126**: 337–345 (1988).

R. Mayne. Cartilage collagens. What is their function, and are they involved in articular disease. *Arthritis Rheum.* **32**: 241–246 (1989).

D. McCormick, M. van der Rest, J. Goodship, G. Lozano, Y. Ninomiya, and B.R. Olsen. Structure of the glycosaminoglycan domain in the type IX collagen-proteoglycan. *Proc. Natl. Acad. Sci. U.S.A.* **84**: 4044–4048 (1987).

S.A. McDonald and R.S. Tuan. Expression of collagen type transcripts in chick embryonic bone detected by in situ cDNA-mRNA hybridization. *Dev. Biol.* **133**: 221–234 (1989).

M. Mendler, S.G. Eich-Bender, L. Vaughan, K.H. Winterhalter, and P. Bruckner. Cartilage contains mixed fibrils of collagen types II, IX, and XI. *J. Cell Biol.* **108**: 191–197 (1989).

E.J. Miller. The structure of fibril-forming collagens. *Ann. N.Y. Acad. Sci.* **460**: 1–13 (1985).

M. Mina, W.B. Upholt, and E.J. Kollar, Stage-related chondrogenic potential of avian mandibular ectomesenchymal cells. *Differentiation* in press (1991).

J. Mollenhauer and K. von der Mark. Isolation and characterization of a collagen-binding glycoprotein from chondrocyte membranes. *EMBO J.* **2**: 45–50 (1983).

N.P. Morris and H.P. Bachinger. Type XI collagen is a heterotrimer with the composition (1α, 2α, 3α) retaining non-triple-domains. *J. Biol. Chem.* **262**: 11345–11350 (1987).

Y. Muragaki, T. Kimura, Y. Ninomiya, and B.R. Olsen. The complete primary structure of two distinct forms of human $\alpha 1(IX)$ collagen chains. *Eur. J. Biochem.* **192**: 703–708 (1990).

L.W. Murray, J. Bautista, P.L. James, and D.L. Rimoin. Type II collagen defects in the chondrodysplasias. I. Spondyloepiphyseal dysplasias. *Am. J. Hum. Genet.* **45**: 5–15 (1989).

H.-D. Nah, B.J. Rodgers, W.M. Kulyk, B.E. Kream, R.A. Kosher, and W.B. Upholt. In situ hybridization analysis of the expression of the type II collagen gene in the developing chicken limb bud. *Collagen Relat. Res.* **8**: 277–294 (1988).

P.J. Neame, J.E. Christner, and J.R. Baker. The primary structure of link protein from rat chondrosarcoma proteoglycan aggregate. *J. Biol. Chem.* **261**: 3519–3535 (1986).

P.J. Neame, J.P. Perin, F. Bonnet. J.E. Christner, P. Jolles, and J.R. Baker, An amino acid sequence common to both cartilage proteoglycan and link protein. *J. Biol. Chem.* **260**: 12402–12404 (1985).

Y. Ninomiya and B.R. Olsen. Synthesis and characterization of cDNA encoding a cartilage-specific short collagen. *Proc. Natl. Acad. Sci. U.S.A.* **81**: 3014–3018 (1984).

Y. Ninomiya, M. van der Rest, R. Mayne, G. Lozano, and B.R. Olsen. Construction and characterization of cDNA encoding the α2 chain of chicken type IX collagen. *Biochemistry* **24**: 4223–4229 (1985).

Y. Ninomiya, A.M. Showalter, M. von der Rest, N.G. Seidah, M. Cretien, and B.R. Olsen. Structure of the carboxyl propeptide of chicken type II procollagen determined by DNA and protein sequence analysis. *Biochemistry* **23**: 617–624 (1984).

Y. Ninomiya, P. Castagnola, D. Gerecke, M. Gordon, O. Jacenko, P. LuValle, M. McCarthy, Y. Muragaki, I. Nishimura, S. Oh, N. Rosenblum, N. Sato, S. Sugrue, R. Taylor, G. Vasios, N. Yamaguchi, and B.R. Olsen. The molecular biology of collagens with short triple-helical domains. In: *Extracellular Matrix Genes* Eds L.J. Sandell and C.D. Boyd. Academic Press, Orlando. pp. 80–114, 1990.

Y. Ninomiya, M. Gordon, M. van der Rest, T. Schmid, T. Linsenmayer, and B.R. Olsen. The developmentally regulated type X collagen gene contains a long open reading frame without introns. *J. Biol. Chem.* **261**: 5041–5050 (1986).

I. Nishimura, Y. Muragaki, and B.R. Olsen. Tissue-specific forms of type IX collagen-proteoglycan arise from the use of two widely separated promoters. *J. Biol. Chem.* **264**: 20033–20041 (1989).

C. Niyibizi and D.R. Eyre. Identification of the cartilage α1(XI) chain in type V collagen from bovine bone. *FEBS Lett.* **242**: 314–318 (1989).

A. Noro, K. Kimata, Y. Oika, T. Shinomura, N. Maeda, and S. Suzuki. Isolation and characterization of a third proteoglycan (PG-Lt) from chick embryo cartilage which contains disulfide-bonded collagenous polypeptide. *J. Biol. Chem.* **258**: 9323–9331 (1983).

A.M. Nunez, C. Francomano, M.F. Young, G.R. Martin, and Y. Yamada. Isolation and partial characterization of genomic clones coding for a human pro-α1 (II) collagen chain and demonstration of restriction fragment length polymorphism at the 3′ end of the gene. *Biochemistry* **24**: 6343–6348 (1985).

A.M. Nunez, K. Kohno, G.R. Martin, and Y. Yamada. Promoter region of the human pro-α1(II)-collagen gene. *Gene* **44**: 11–16 (1986).

C.M. O'Donnell, K. Kaczman-Daniel, P.F. Goetinck, and B.M. Vertel. Nanomelic chondrocytes synthesize a glycoprotein related to chondroitin sulfate proteoglycan core protein. *J. Biol. Chem.* **263**: 17749–17754 (1988).

D. Ogilvie, P. Wordsworth, E. Thompson, and B. Sykes. Evidence against the structural gene encoding type II collagen (COL2A1) as the mutant locus in achondroplasia. *J. Med. Genet.* **23**: 19–22 (1986).

A. Oldberg, P. Antonsson, and D. Heinegard. The partial amino acid sequence of bovine cartilage proteoglycan, deduced from a cDNA clone, contains numerous Ser-Gly sequences arranged in homologous repeats. *Biochem. J.* **243**: 255–259 (1987).

O. Oshima, P.S. Leboy, S.A. McDonald, R.S. Tuan, and I.M. Shapiro. Developmental expression of genes in chick growth cartilage detected by in situ hybridization. *Calcif. Tissue Int.* **45**: 182–192 (1989).

A. Palotie, P. Vaisanen, J. Ott, L. Ryhanen, K. Elima, M. Vikkula, K. Cheah, E. Vuorio, and L. Peltonen. Predisposition to familial osteoarthrosis linked to type II collagen gene. *Lancet* **1** (8644): 924–927 (1989).

L. Patthy. Detecting homology of distantly related proteins with consensus sequences. *J. Mol. Biol.* **198**: 567–577 (1987).

M. Paulsson and D. Heinegard. Matrix proteins bound to associatively prepared proteoglycans from bovine cartilage. *Biochem. J.* **183**: 539–545 (1979).

M. Paulsson and D. Heinegard. Purification and structural characterization of a cartilage matrix protein. *Biochem. J.* **197**: 367–375 (1981).

M. Paulsson and D. Heinegard. Radioimmunoassay of the 148-kilodalton cartilage protein. *Biochem. J.* **207**: 207–213 (1982).

M. Paulsson and D. Heinegard. Noncollagenous cartilage proteins—current status of an emerging research field. *Collagen Relat. Res.* **4**: 219–229 (1984).

J.-P. Perin, F. Bonnet, J. Jolles, and P. Jolles. Sequence data concerning the protein core of the cartilage proteoglycan monomers. Characterization of a sequence allowing the synthesis of an oligonucleotide probe. *FEBS Lett.* **176**: 37–42 (1984).

J.-P. Perin, F. Bonnet, V. Pizon, J. Jolles, and P. Jolles. Structural data concerning the link proteins from bovine nasal cartilage proteoglycan complex. *FEBS Lett.* **119**: 333–336 (1980).

T. Pihlajaniemi, K. Tryggvason, J.C. Myers, M. Kurkinen, R. Lebo, M.-C. Cheung, D.J. Prockop, and C.D. Boyd. cDNA clones coding for the proα1(IV) chain of human type IV procollagen reveal unusual homology of amino acid sequences in two halves of the carboxyl-terminal domain. *J. Biol. Chem.* **260**: 7681–7687 (1985).

S.J. Perkins. A.S. Nealis, J. Dudhia, and T.E. Hardingham. Immunoglobulin fold and tandem repeat structures in proteoglycan N-terminal domains and link protein. *J. Mol. Biol.* **206**: 737–753 (1989).

A.R. Poole, I. Pidoux, A. Reiner, L. Rosenberg, D. Hollister, L. Murray, and D. Rimoin. Kniest dysplasia is characterized by an apparent abnormal processing of the C-propeptide of type II cartilage collagen resulting in imperfect fibril assembly. *J. Clin. Invest.* **81**: 579–589 (1988).

F.M. Pope, K.S. Cheah, A.C. Nicholls, A.B. Price, and F.G. Grosveld. Lethal osteogenesis imperfecta congenita and a 300 base pair gene deletion for an α1(I)-like collagen. *Br. Med. J.* **288**: 431–434 (1984).

F. Ramirez. Organization and evolution of the fibrillar collagen genes. In: *Collagen Volume IV: Molecular Biology,* Eds B.R. Olsen and M.E. Nimni. CRC Press, Boca Raton, FL. pp. 21–30, 1989.

F. Ramirez, M. Bernard, M.L. Chu, L. Dickson, F. Sangiorgi, D. Weil, W. De Wet, C. Junien, and M. Sobel. Isolation and characterization of the human fibrillar collagen genes. *Ann. N.Y. Acad. Sci.* **460**: 117–129 (1985).

E. Retief, M.I. Parker, and A.E. Retief. Regional chromosome mapping of human collagen genes α2(I) and α1(I) (COL1A2 and COL1A1). *Hum. Genet.* **69**: 304–308 (1985).

C. Rhodes, K. Doege, M. Sasaki, and Y. Yamada. Alternative splicing generates two different mRNA species for rat link protein. *J. Biol. Chem.* **263**: 6063–6067 (1988).

F. Ruggiero, M. Pfaffle, K. von der Mark, and R. Garrone. Retention of carboxypropeptides in type-II collagen fibrils in chick embryo chondrocyte cultures. *Cell Tissue Res.* **252**: 619–664 (1988).

E. Ruoslahti. Structure and biology of proteoglycans. *Annu. Rev. Cell Biol.* **4**: 229–255 (1988).

M.C. Ryan and L.J. Sandell. Differential expression of a cysteine-rich domain in the amino-terminal propeptide of type II (cartilage) procollagen by alternative splicing of mRNA. *J. Biol. Chem.* **265**: 10334–10339 (1990).

M.C. Ryan, M. Sieraski, and L.J. Sandell. The human type II procollagen gene: identification of an additional protein-coding domain and location of potential regulatory sequences in the promoter and first intron. *Genomics* **8**: 41-48 (1990).

S. Sai, T. Tanaka, R.A. Kosher, and M.L. Tanzer. Cloning and sequence analysis of a partial cDNA for chicken cartilage proteoglycan core protein. *Proc. Natl. Acad. Sci. U.S.A.* **83**: 5081–5085 (1986).

M. Sandberg and E. Vuorio. Localization of types I, II, and III collagen mRNAs in developing human skeletal tissues by in situ hybridization. *J. Cell Biol.* **104**: 1077–1084 (1987).

L.J. Sandell. Characterization of a human type II procollagen gene. *Orthop. Trans.* **10**: 160 (1985).

L.J. Sandell. Molecular biology of proteoglycans and link proteins. In: *Biology of Proteoglycans.* Academic Press, Orlando. pp. 27–57, 1987.

L.J. Sandell, H.L. Prentice, D. Kravis, and W.B. Upholt. Structure and sequence of the chicken type II procollagen gene. Characterization of the region encoding the carboxyl-terminal telopeptide and propeptide. *J. Biol. Chem.* **259**: 7826–7834 (1984).

L.J. Sandell, Y. Yamada, A. Dorfman, and W.B. Upholt. Identification of genomic DNA coding for chicken type II procollagen. *J. Biol. Chem.* **258**: 11617–11621 (1983).

F.O. Sangiorgi, V. Benson-Chanda, W.J. de Wet, M.E. Sobel, P. Tsipouras, and F. Ramirez. Isolation and partial characterization of the entire human pro α1(II) collagen gene. *Nucleic Acids Res.* **13**: 2207–2225 (1985a).

F.O. Sangiorgi, V. Benson-Chanda, W.J. de Wet, M.E. Sobel, and F. Ramirez. Analysis of cDNA and genomic clones coding for the pro α1 chain of calf type II collagen. *Nucleic Acids Res.* **13**: 2815–2826 (1985b).

H. Sawada. The fine structure of the bovine Descemet's membrane with special reference to biochemical nature. *Cell Tissue Res.* **226**: 241–255 (1982).

H. Sawada, H. Konomi, and Y. Nagai. The basement membrane of bovine corneal endothelial cells in culture with β-aminopropionitrile: biosynthesis of hexagonal lattices composed of a 160 nm dumbell-shaped structure. *Eur. J. Cell Biol.* **35**: 226–234 (1984).

S.A. Saxe, L.N. Lukens, and P.J. Pawlowski. Changes in the nuclear and cytoplasmic levels of type I and type II collagen RNAs during growth of chondrocytes in 5-bromo-2′-deoxyuridine. *J. Biol. Chem.* **260**: 3812–3819 (1985).

T.M. Schmid. Ph.D. thesis, University of Illinois, Urbana (1980).

T.M. Schmid and T.F. Linsenmayer. A short chain (pro)collagen from aged endochondral chondrocytes. Biochemical characterization. *J. Biol. Chem.* **258**: 9504–9509 (1983).

T.M. Schmid and T.F. Linsenmayer. Immunohistochemical localization of short chain cartilage collagen (type X) in avian tissues. *J. Cell Biol.* **100**: 598–605 (1985a).

T.M. Schmid and T.F. Linsenmayer. Developmental acquisition of type X collagen in the embryonic chick tibiotarsus. *Dev. Biol.* **107**: 373–381 (1985b).

T.M. Schmid and T.F. Linsenmayer. Type X collagen. In: *Structure and Function of Collagen Types*, Eds R. Mayne, and R.E. Burgeson. Academic Press, Orlando. pp. 223–259, 1987.

T.M. Schmid, R. Mayne, J.J. Jeffrey, and T.F. Linsenmayer. Type X collagen contains two cleavage sites for a vertebrate collagenase. *J. Biol. Chem.* **261**: 4184–4189 (1986).

A. Schmidt, Y. Yamada, and B. de Crombrugghe. DNA sequence comparison of the regulatory signals at the 5′ end of the mouse and chick α2 type I collagen genes. *J. Biol. Chem.* **259**: 7411–7415 (1984).

V.C. Sheffield and W.B. Upholt. Copy number of the chicken type II procollagen gene. *Collagen Relat. Res.* **5**: 1–8 (1985).

T.B. Shows, L. Tikka, M.G. Byers, R.L. Eddy, L.L. Haley, W.M. Henry, D.J. Prockop, and K. Tryggvason. Assignment of the human collagen α1(XIII) chain gene (COL13A1) to the q22 region of chromosome 10. *Genomics* **5**: 128–133 (1989).

E. Solomon, V. Hall, and M. Kurkinen. The human α2(IV) collagen gene, COL4A2, is syntenic with the α1(IV) gene, COL4A1, on chromosome 13. *Ann. Hum. Genet.* **51**: 125–127 (1987).

E. Solomon, L.R. Hiorns, D. Sheer, and D. Rowe. Confirmation that the type I collagen gene on chromosome 17 is COL1A1 (α1(I)) using a human genomic probe. *Ann. Hum. Genet.* **48**: 39–42 (1984).

E. Solomon, L.R. Hiorns, N. Spurr, M. Kurkinen, D. Barlow, B.L.M. Hogan, and R. Dalgleish. Chromosomal assignments of the genes coding for human types II, III, and IV collagen: a dispersed gene family. *Proc. Natl. Acad. Sci. U.S.A.* **82**: 3330–3334 (1985).

I. Stamenkovic, M. Amiot, J.M. Pesando, and B. Seed. A lymphocyte molecule implicated in lymph node homing is a member of the cartilage link protein family. *Cell* **56**: 1057–1062 (1989).

N.S. Stirpe and P.F. Goetinck. Gene regulation during cartilage differentiation: temporal and spatial expression of link protein and cartilage matrix protein in the developing limb. *Development* **107**: 23–33 (1989).

N.S. Stirpe, W.S. Argraves, and P.F. Goetinck. Chondrocytes from the cartilage proteoglycan-deficient mutant, nanomelia, synthesize greatly reduced levels of the proteoglycan core protein transcript. *Dev. Biol.* **124**: 77–81 (1987).

N.G. Stoker, K.S. Cheah, J.R. Griffin, F.M. Pope, and E. Solomon. A highly polymorphic region 3′ to the human type II collagen gene. *Nucleic Acids Res.* **13**: 4613–4622 (1985).

C.M. Strom. Molecular studies in achondroplasia using COL2A1 probes. *Basic Life Sci.* **48**: 59–60 (1988a).

C.M. Strom. A three allele restriction fragment length polymorphism within the human COL2A1 gene. *Nucleic Acids Res.* **16**: 9077 (1988b).

C.M. Strom and W.B. Upholt. Isolation and characterization of genomic clones corresponding to the human type II procollagen gene. *Nucleic Acids Res.* **12**: 1025–1038 (1984b).

C.M. Strom, R.L. Eddy, and T.B. Shows. Localization of human type II procollagen gene (COL2A1) to chromosome 12. *Somatic Cell. Mol. Genet.* **10**: 651–655 (1984a).

M.W. Su, V. Benson-Chanda, H. Vissing, and F. Ramirez. Organization of the exons coding for pro $\alpha 1$(II) collagen N-propeptide confirms a distinct evolutionary history of this domain of the fibrillar collagen genes. *Genomics* **4**: 438–441 (1989a).

M.-W. Su, B. Lee, F. Ramirez, M. Machado, and W. Horton. Nucleotide sequence of the full length cDNA encoding for human type II procollagen. *Nucleic Acids Res.* **17**: 9473 (1989b).

S.P. Sugrue, M.K. Gordon, J. Seyer, B. Dublet, M. van der Rest, and B.R. Olsen. Immuno-identification of type XII collagen in embryonic tissues. *J. Cell Biol.* **109**: 939–945 (1989).

M.D. Sussman, R.C. Ogle, and G. Balian. Biosynthesis and processing of collagens in different cartilaginous tissues. *J. Orthop. Res.* **2**: 134–142 (1984).

K.K. Svoboda, I. Nishimura, S. P. Sugrue, Y. Ninomiya, and B.R. Olsen. Embryonic chicken cornea and cartilage synthesize type IX collagen molecules with different amino-terminal domains. *Proc. Natl. Acad. Sci. U.S.A.* **85**: 7496–7500 (1988).

B.J. Swalla, W.B. Upholt, and M. Solursh. Analysis of type II collagen RNA localization in chick wing buds by in situ hybridization. *Dev. Biol.* **125**: 51–58 (1988).

B. Sykes. A high frequency Hind III restriction site polymorphism within a collagen gene. *Dis. Markers* **1**: 141–146 (1983).

B.C. Sykes, D.J. Ogilvie, and B.P. Wordsworth. Lethal osteogenesis imperfecta and a collagen gene deletion. Length polymorphism provides an alternative explanation. *Hum. Genet.* **70**: 35–37 (1985b).

B. Sykes, R. Smith, S. Vipond, C. Paterson, K. Cheah, and E. Solomon. Exclusion of the $\alpha 1$(II) cartilage collagen gene as the mutant locus in type IA osteogenesis imperfecta. *J. Med. Genet.* **22**: 187–191 (1985a).

T. Tanaka, R. Har-el, and M.L. Tanzer. Partial structure of the gene for chicken cartilage proteoglycan core protein. *J. Biol. Chem.* **263**: 15831–15835 (1988).

P. Thorogood. The developmental specification of the vertebrate skull. *Development* **103** (Suppl.): 141–153 (1988).

P. Thorogood, J. Bee, and K. von der Mark, Transient expression of collagen type II at epitheliomesenchymal interfaces during morphogenesis of the cartilaginous neurocranium. *Dev. Biol.* **116**: 497–509 (1986).

G. Tromp, H. Kuivaniemi, A. Stacey, H. Shikata, C.T. Baldwin, R. Jaenisch, and D.K Prockop. Structure of a full-length cDNA clone for the prepro $\alpha 1$(I) chain of human type I procollagen. *Biochem. J.* **253**: 919–922 (1988).

P.A. Tsonis and P.F. Goetinck. Expression of cartilage-matrix genes and localization of their translation products in the embryonic chick eye. *Exp. Eye Res.* **46**: 753–764 (1988).

W.B. Upholt. The type II collagen gene. In: *Collagen Volume IV: Molecular Biology,* Eds B.R. Olsen and M. Nimni. CRC Press, Boca Raton, FL. pp. 31–49, 1989.

W.B. Upholt and L.J. Sandell. Exon/intron organization of the chicken type II procollagen gene: intron size distribution suggests a minimal intron size. *Proc. Natl. Acad. Sci. U.S.A.* **83**: 2325–2329 (1986).

W.B. Upholt, C.M. Strom, and L.J. Sandell. Structure of the type II collagen gene. *Ann. N.Y. Acad. Sci.* **460**:130–140 (1985).

W.B. Upholt, B.M. Vertel, and A. Dorfman. Translation and characterization of messenger RNAs in differentiating chicken cartilage. *Proc. Natl. Acad. Sci. U.S.A.* **76**: 4847–4851 (1976).

P. Vaisanen, K. Elima, A. Palotie, L. Peltonen, and E. Vuorio. Polymorphic restriction sites of type II collagen gene: their location and frequencies in the Finnish population. *Hum. Hered.* **38**: 65–71 (1988).

M. van der Rest, R. Mayne, Y. Ninomiya, N.G. Seidah, M. Chretien, and B.R. Olsen. The structure of type IX collagen. *J. Biol. Chem.* **260**: 220–225 (1985).

G. Vasios, I. Nishimura, H. Konomi, M. van der Rest, Y. Ninomiya, and B.R. Olsen. Cartilage type IX collagen proteoglycan contains a large amino-terminal globular domain encoded by multiple exons. *J. Biol. Chem.* **263**: 2324–2329 (1988).

L. Vaughan, K.H. Winterhalter, and P. Bruckner. Proteoglycan Lt from chicken embryo sternum identified as type IX collagen. *J. Biol. Chem.* **260**: 4758–4764 (1985).

L. Vaughan, M. Mendler, J. Huber, P. Bruckner, K. H. Winterhalter, M. H. Irwin, and R. Mayne. D-periodic distribution of collagen type IX along cartilage fibrils. *J. Cell Biol.* **106**: 991–997 (1988).

B.M. Vertel and Y. Hitti. Biosynthetic precursors of cartilage chondroitin sulfate proteoglycan. *Collagen Relat. Res.* **7**: 57–75 (1987).

B.M. Vertel, W.B. Upholt, and A. Dorfman. Cell-free translation of messenger RNA for chondroitin sulphate proteoglycan core protein in rat cartilage. *Biochem. J.* **217**: 259–263 (1984).

M. Vikkula and L. Peltonen. Structural analyses of the polymorphic area in type II collagen gene. *FEBS Lett.* **250**: 171–174 (1989).

H. Vissing, M. D'Alessio, B. Lee, F. Ramirez, M. Godfrey, and D.W. Hollister. Glycine to serine substitution in the triple helical domain of pro-alphal(II) collagen results in a lethal perinatal form of short-limbed dwarfism. *J. Biol. Chem.* **264**: 18265–18267 (1989).

K. von der Mark, M. van Menxel, and H. Wiedemann. Isolation and characterization of new collagens from chick cartilage. *Eur. J. Biochem.* **124**: 57–62 (1982).

K. von der Mark, R. von der Mark, R. Timpl, and R.L. Trelstad. Immunofluorescent localization of collagen types I, II, and III in the embryonic eye. *Dev. Biol.* **59**: 75–85 (1977).

E. Vuorio, L. Sandell, D. Kravis, V.C. Sheffield, T. Vuorio, A. Dorfman, and W.B. Upholt. Construction and partial characterization of two recombinant cDNA clones for procollagen from chicken cartilage. *Nucleic Acids Res.* **10**: 1175–1192 (1982).

E. Vuorio, A. Superti-Furga, K. Elima, and B. Steinmann. Variable methylation patterns of the human type II collagen gene in: Development and Diseases of Cartilage and Bone Matrix. *UCLA Symp. Mol. Cell. Biol.* **46**: 97–106 (1987).

E.J. Weaver and R.G. Knowlton. A PvuII polymorphism near the 5′ and of the type II procollagen gene [COL2A1]. *Nucleic Acids Res.* **17**: 6429 (1989).

D. Weil, M. Bernard, S. Gargano, and F. Ramirez. The proα2(V) collagen gene is evolutionarily related to the major fibrillar-forming collagens. *Nucleic Acids Res.* **15**: 181–198 (1987).

D. Weil, M.G. Mattei, E. Passage, V.C. Nguyen, D. Pribula-Conway, K. Mann, R. Deutzmann, R. Timpl, and M.-L. Chu. Cloning and chromosomal localization of human genes encoding the three chains of type IV collagen. *Am. J. Hum. Genet.* **42**: 435–445 (1988).

E.H. Weiss, K.S.E. Cheah, R.G. Grosveld, H.H.M. Dahl, E. Solomon, and R.A. Flavell. Isolation and characterization of a human collagen α1(I)-like gene from a cosmid library. *Nucleic Acids Res.* **10**: 1981–1994 (1982).

T.N. Wight and R.P. Mecham. *Biology of Proteoglycans*. Academic Press, Orlando, 1987.

L. Wood, N. Theriault, and G. Vogeli. Complete nucleotide sequence of the N-terminal domains of the murine $\alpha 1$ type III collagen chain. *Gene* **61**: 225–230 (1987).

D. Woodbury, V. Benson-Chanda, and F. Ramirez. Amino-terminal propeptide of human pro-alpha2(V) collagen conforms to the structural criteria of a fibrillar procollagen molecule. *J. Biol. Chem.* **264**: 2735–2738 (1989).

P. Wordsworth, D. Ogilvie, L. Priestly, R. Smith, R. Wynne-Davies, and B. Sykes. Structural and segregation analysis of the type II collagen gene (COL2A1) in some heritable chondrodysplasias. *J. Med. Genet.* **25**: 521–527 (1988).

D.W. Wright and R. Mayne. Vitreous humor of chicken contains two fibrillar systems: an analysis of their structure. *J. Ultrastruct. Mol. Struct. Res.* **100**: 224–234 (1988).

Y. Yamada, M. Mudryj, and B. de Crombrugghe. A uniquely conserved regulatory signal is found around the translation initiation site in three different collagen genes. *J. Biol. Chem.* **258**: 14914–14919 (1983).

N. Yamaguchi, P.D. Benya, M. van der Rest, and Y. Ninomiya. The cloning and sequencing of $\alpha 1$(VIII) collagen cDNAs demonstrate that type VIII collagen is a short chain collagen and contains triple-helical and carboxyl-terminal non-triple-helical domains similar to those of type X collagen. *J. Biol. Chem.* **264**: 16022–16029 (1989).

M.F. Young, G. Vogeli, A.M. Nunez, M.P. Fernandez, M. Sullivan, and M.E. Sobel. Isolation of cDNA and genomic DNA clones encoding type II collagen. *Nucleic Acids Res.* **11**: 4207–4227 (1984).

D.R. Zimmerman and E. Ruoslahti. Multiple domains of the large fibroblast proteoglycan, versican. *EMBO J.* **8**: 2975–2981 (1989).

2

Physicochemical Aspects of Cartilage Extracellular Matrix

WAYNE D. COMPER
Biochemistry Department
Monash University
Victoria, Australia

Glossary

A	area
A_2, A_3	second and third osmotic virial coefficients
c	concentration in moles per unit volume
C	concentration in mass per unit volume
D	translational diffusion coefficient
f_{ij}	frictional coefficient between 1 mole of i with j
J	flux in molar units
j	flux in mass units
J_v	volume flow
k	specific hydraulic conductivity
l	length
l_s	distance over which solute-solvent exchange takes place near the pore of membrane
L_p^o	osmotic permeability coefficient
M	molecular weight
P	mechanical pressure
R	universal gas constant
r	radial distance in centrifuge
S	sedimentation coefficient
T	temperature
$\overline{V}_i$	partial molar volume of i
$\overline{v}_i$	partial specific volume of i
x	distance
η	dynamic viscosity
Π	osmotic pressure
ρ	solution density
ω	rotor speed

Introduction

There are many different types of cartilagenous tissues that are characterized by their morphology and their structure and function. At the molecular level there are gross similarities in forms of their macromolecular composition, although some tissue-specific compositional differences may exist (Heinegård and Paulsson, 1987). This chapter focuses on the physicochemical properties of some well studied cartilages, particularly articular cartilage.

Many of these tissues have a biomechanical function but represent unusual biochemical systems in that they are essentially passive extracellular matrices (cell volume ~3–5% of tissue volume) with only low levels of enzyme-catalyzed reactions. The manner in which the matrix responds to its endogenous cells (chondrocytes) and to its environment appears to be by passive physicochemical processes. These processes are governed or mediated by two types of interactions between matrix components, namely (1) enthalpic interactions involving the redistribution of heat often associated with close proximity binding and (2) entropic interactions in which molecules interact by virtue of space filling considerations through steric obstruction (excluded volume) and frictional resistance.

The extracellular matrix (referred to as the matrix) is a multicomponent system that may be described as a three-dimensional porous collagen fiber network impregnated with a gel-like material (histologic term, ground substance). The metabolism of the chondrocytes seems to be concerned mainly with the biosynthesis and secretion of matrix material. For example, the anionic polysaccharides present in the form of proteoglycans, type II collagen, and glycoproteins are all cell products. The macromolecular composition to a large extent determines the properties of the tissue in terms of its degree of hydration, ionic composition, permeability, and mechanical properties.

It follows that many physiological functions of cartilage are determined largely by the nature and organization of these matrix components. This illustrates another unusual aspect of these systems. Whereas all macromolecular components have a distinct evolutionary status by virtue of the nature of their assembly through posttranslational chemical modification and aggregation in the matrix (Mathews, 1975), matrix macromolecules are only indirectly related to the gene in both chemical and topographical terms. While it is apparent that chondrocytes must, in many instances, regulate one or more properties of their matrix environment, it is not clear at the present time, although absolutely intriguing, how the matrix censors information of the environment that will influence the cell.

Previous Reviews

Surprisingly, there are only a few reviews specifically on the physicochemical properties of cartilage. These include the extensive studies of Maroudas and co-workers on the thermodynamic and dynamic aspects of water and other solutes in cartilage (Maroudas, 1975b, 1980), the electrochemical properties (Grodzinsky, 1983; Eisenberg and Grodzinsky, 1988), the biomechanical properties (Mow *et al.*, 1984), and the structural studies of the ground substance and its constituent proteoglycans (Muir, 1980; Carney and Muir, 1988). Other related reviews on connective tissues in general, but with many aspects pertinent to cartilage, are on the structure-function properties of matrix macromolcules (Comper and Laurent, 1978; Comper, 1984) and matrix permeability and transport (Preston *et al.*, 1984).

This chapter sets out to describe a different approach to the physicochemical properties of the matrix. This will involve special emphasis on the relatively new aspects of the structure-function relationships of the macromolecular and polyelectrolyte properties of cartilage in relation to individual matrix macromolecules and water. The reader is referred to the other reviews cited above for subject areas that have been omitted from this article. Omissions are not the result of any lack of importance, but are due to the comprehensive coverage the subject areas have already received.

Matrix Organization

Biphasic Matrix

The major components of the matrix are water (and dissolved simple electrolyte NaCl), type II collagen, and a dominant (by mass) large (MW 2–3.0 $\times$ 10^6) proteoglycan (Muir, 1980; Carney and Muir, 1988; Heinegård and Paulsson, 1987). Our knowledge is poor concerning the exact organization of these components. The components have been generally considered to exist in a simple biphasic arrangement (Gersh and Catchpole, 1970; Mow *et al.*, 1980) consisting of a porous network of insoluble collagen fibers filled with a soluble phase of water and proteoglycans.

In articular cartilage, the size of the collagen fibers, consisting of essentially type II collagen, varies throughout the matrix and occurs in the diameter range of 30 to 200 nm (Mayne and von der Mark, 1983). The collagen fibers appear to be randomly organized into a tight network in the middle and deeper zones of cartilage. At the surface the fibers are finer and less random, lying more or less parallel to the surface. Model analysis suggests that the distance between fibers is in the range of 60 to 200 nm (Byers *et al.*, 1983), although systematic electron microscopical studies have not been performed to confirm these estimates. The biphasic matrix can then be simply considered as consisting of collagen modules with openings or "pores" of 60 to 200 nm in diameter containing polyanionic proteoglycan and water.

Compositional Analysis

Articular cartilage is not uniform in composition but varies within a single joint from the surface to the deep zone. The proportions of its major constituents are also known to vary between different types of cartilage with aging and with disease. On a wet weight basis the water content may lie in the range of 65 to 80%, collagen in the range of 15 to 25%, and proteoglycans in the range of 3 to 10% (Maroudas, 1979).

More specifically, the relatively few studies on chemically determined cartilage composition give for normal femoral head cartilage 75% water, 18% collagen, and a total glycosaminoglycan (GAG) content of 3 to 7% (Maroudas, 1980), and for pig cartilage 68% water, 19% collagen, and 2.2% GAG (Brandt and Muir, 1969).

For routine determination of polyanionic GAG content, Maroudas and coworkers have used a method to determine total anionic charge content of the tissue by equilibrating pieces of tissue in solutions of very low NaCl concentration containing radioactive Na^{22}. The uptake of Na^{22}, as a counterion to the anionic charge of the tissue, then yields a quantity described as the fixed charge density of the tissue. The method does not discriminate different sources of anionic charge in the tissue, but agreement between charge content of the GAG chain and chemically determined polysaccharide content has been claimed (Maroudas and Thomas, 1970). Using this method on slices of cartilage, Maroudas (1979) has found that the fixed charge density may vary 0.12 to 0.2 meq/g wet weight tissue for hip cartilage, 0.06 to 0.17 meq/g tissue for knee cartilage, and 0.04 to 0.15 meq/g tissue for ankle cartilage. Using a conversion factor of 3.2 times milliequivalent = grams GAG (Urban *et al.*, 1979) and a concentration of water as 0.75 g/g wet weight of tissue gives a range of GAG concentration in these cartilages of 17 to 85 mg ml^{-1} (tissue water).

The assessment of GAG concentration on the basis of tissue water content has been based on the assumption that all the water is available to dissolve the polysaccharide. This assumption may be too simplistic as it is evident that a small fraction of this water will be intracellular and unavailable for the proteoglycan. More recently, estimates of intrafibrillar water, water that is specifically within the domain of collagen fibers, has been found to be significant (Maroudas and Bannon, 1981). It has also been assumed that this water is unavailable for GAG on the basis that the large proteoglycan molecule will be excluded from the fiber. The GAG chains may, however, penetrate the aqueous spaces of the fiber. In any case, there are questions concerning the quantitative aspects of intrafibrillar water measurements.

The notion of intrafibrillar water has come from the discrepancies revealed of a comparision of osmotic pressures[1] generated from cartilage pieces of known fixed charge density (based on total tissue water content) and of the

[1] Proteoglycan and cartilage tissue osmotic pressure have been measured by an equilibrium dialysis technique at 37°C against concentrated solutions of polyethyleneglycol (PEG) (up to 350 mg ml^{-1}) in 0.15 *M* NaCl (Urban *et al.*, 1979). The osmotic pressure of these solutions were not directly estimated but interpolated from virial coefficients obtained at low concentration (<20 mg ml^{-1}) at 4°C in H_2O (Edmond and Ogston, 1968) and are therefore likely to be subject to error. Further, due to its high volume fraction (up to 0.27), the PEG will significantly affect the NaCl partition (and therefore osmotic pressures) between the PEG solution and the proteoglycan/tissue compartment. This is likely to artifically increase the apparent osmotic pressure of the proteoglycan compartment.

corresponding proteoglycan compartment (Maroudas and Bannon, 1981). An intrafibrillar water content of 0.7 g H_2O per gram collagen was estimated from albumin partition experiments on proteoglycan-depleted cartilage. These results imply that intrafibrillar water is not subject to the significant osmotic pull by ambient proteoglycan. However, evidence indicates fiber morphology is affected by surrounding proteoglycan (Katz *et al.*, 1986). At the present time conclusions of this work show that while intrafibrillar water may be important, the quantitative estimates have been questioned (Bert and Pearce, 1984) and require analysis by independent means. Further, the distribution of glycosaminoglycan chains has to be assessed.

Compositional Heterogeneity

Apart from the gross considerations of matrix organization considered *vide supra*, it is becoming more evident that there are important tissue-specific structural/organizational heterogeneities of the matrix whose physicochemical significance is not known. These include the peculiar environment of the immediate pericellular matrix of the chondrocyte, called the chondron, which probably has distinct chemical composition and organization (Poole *et al.*, 1987) as compared to the bulk matrix. The heterogeneities of the macromolecular components include the minor collagens of cartilage (for review see Mayne and Irwin, 1986) and the ''small'' proteoglycans (for reviews see Heinegård and Paulsson, 1987; Carney and Muir, 1988) which probably have a functional interaction with collagen fibers in the matrix (Scott, 1988).

Dynamics of Matrix Components

Basic Considerations

The dynamics of matrix components are of major significance in understanding structure-function relationships in cartilage, particularly those relating to tissue hydration and compressibility. Matrix molecules move through collisions with water in the tissue. The thermal energy of the solvent is transduced to kinetic energy of the matrix molecules. This energy will account for the various vibrational modes of the molecule: the most important, in terms of tissue properties, is energy associated with translational motion relative to water. This type of motion is called translational diffusion.[2] Other motions including rotational diffusion, which do not relate to the relative

[2] There are a number of different terms used in the literature to describe translational diffusional motion; these include mutual diffusion, segment diffusion, tracer self-diffusion, and cooperative diffusion. All these types of diffusion measure displacement of the diffusing molecule relative to solvent but may vary under the conditions with which they are measured and the nature of the concentration gradient and absolute concentration. The diffusional motion of biological significance is that representing the motion of the molecule with respect to water and includes the equivalent terms mutual, segment, and cooperative diffusion coefficients.

movement of the molecule with respect to water, have no *a priori* relationship to translational motion.

The translational diffusional process in an isothermal system is related to the concentration gradient of the molecule described by Fick's Law., i.e.,

$$J_1 = D_1(\partial c_1/\partial x) \tag{1}$$

where J_1 is the flux of component 1 (which is the proteoglycan in most cases considered here), c_1 the concentration of proteoglycan (moles per unit volume), and x the distance over which diffusion takes place. The relationship between the flux of component 1 and its concentration gradient is given by the translational diffusion coefficient D_1. This equation then states that molecules diffuse spontaneously through thermal motion to relax the concentration gradient.

The important factors governing translational motion will then be the effective concentration of the molecule and the manner in which it moves through water, i.e., the frictional interaction with water. The diffusion equation also assumes that the diffusing molecules are ideal (their colligative properties (e.g., osmotic pressure) vary linearly with concentration). For matrix molecules, particularly proteoglycans, this is not the case as their solution properties are markedly nonideal. These nonidealities arise from excluded volume effects, charge effects, and macromolecular interactions, and may affect both the effective concentration of the material and its frictional interaction with water. With these considerations in mind, we can rewrite Equation 1 as

$$J_1 = (D_{eff})_1 \frac{\partial (c_{eff})_1}{\partial x} \tag{2}$$

with effective quantities for diffusion coefficient and concentration. We can relate the effective diffusion coefficient to the hydrodynamic quantity, the frictional coefficient between 1 mole of proteoglycan with water (component 2) (f_{12}) (Spiegler, 1958)

$$(D_{eff})_1 = RT/f_{12} \tag{3}$$

where R is the universal gas constant and T the temperature. The effective concentration is directly related to the osmotic pressure (Π)

$$(c_{eff})_1 \simeq \Pi/RT \tag{4}$$

so that substituting Equations 3 and 4 into Equation 2 gives

$$J_1 \simeq \frac{1}{f_{12}} \frac{\partial \Pi}{\partial x} \quad (5)$$

where

$$D_1 \simeq \frac{M_1}{f_{12}} \frac{\partial \Pi}{\partial C_1} \quad (6)$$

where C_1 is the concentration of 1 in terms of mass per volume units and M_1 the molecular weight of 1: therefore, the flux of component 1 is directly related to its effective concentration or osmotic pressure gradient and its frictional coefficient. Each of these parameters is discussed in detail in the following sections.

Volume Expansion Concept

In response to the translational movement of component 1 relative to water, there will be an equal and opposite volume flow of water, i.e., the osmotic gradient of the molecule results in a reverse volume flow of water. The volume flow of component 1 is given by the product of the molar flux by its partial molar volume $\overline{V}_1$ so that

$$J_1 \overline{V}_1 = -J_2 \overline{V}_2 \quad (7)$$

the translational diffusion can then be regarded as the volume expansion of the distribution of each component.

Osmotic Pressure

The osmotic pressure of macromolecules can be described by a polynomial expression called a virial expansion such that

$$\Pi = RT\left[\frac{C_1}{M_1} + A_2 C_1^2 + A_3 C_1^3 + \ldots\ldots\ldots\right] \quad (8)$$

where A_2 and A_3 are the second and third virial coefficients, respectively (Tanford, 1961). For chondroitin sulfate proteoglycan the virial terms are large positive quantities which make Π increase rapidly with concentration as compared to the uncharged polysaccharide dextran (Figure 1). The magnitude of Π above 4 to 5 mg ml^{-1} is determined primarily by the virial coefficients in Equation 8. These, in turn, are determined by the activities of the microcounterions of the polyanionic chains and excluded volume effects associated with polyion-polyion interaction (Comper and Preston, 1974; Urban *et al.*, 1979; Zamparo and Comper, 1990). The total osmotic pressure of the polyanionic proteoglycan compartment is lowered because of the Donnan distribution which results in the polyelectrolyte excluding the simple electrolyte NaCl. This exclusion is seen in cartilage as well (Table 1). The exclusion of NaCl is more pronounced at higher proteoglycan concentration. In any case, at concentrations of proteoglycans that exist in cartilage (see "Compositional Analysis" *vide supra*) the osmotic pressures may be in the range of several atmospheres (Urban *et al.*, 1979).[3] The influence of proteoglycan structure on osmotic pressure would appear insignificant (Comper and Preston, 1974) as both chondroitin sulfate (CS) chains, proteoglycan subunit (PGS) and proteoglycan aggregate (PGA), have similar osmotic pressures, i.e., the altered macroion distribution within structures of protein cores or hyaluronate cores has no effect (Urban *et al.*, 1979). Note, however, that this comparison is not a true reflection of the activity of these components in the extracellular matrix. Although the networks that confine proteoglycans in tissues are not clearly defined, it appears that CS chains and even proteoglycan monomers are not retained in the matrix and may be readily transported out of the tissue.

Matrix proteoglycans offer significant excluded volume not only to other molecules (see "Steric Exclusion", *vide infra*) but to themselves. Such interaction tends to increase their effective concentration, and this contributes to the nonideality of osmotic pressure as embodied in Equation 8. This component has been estimated to contribute 20 to 50% of the total osmotic pressure (Comper and Preston, 1974; Urban *et al.*, 1979; Zamparo and Comper, 1990). Entropic interactions contributing to osmotic pressure may increase in cartilage if proteoglycans undergo excluded volume interactions with other macromolecules (Laurent and Ogston, 1963; Meyer *et al.*, 1971; Comper and Preston, 1974).

Segmental Mobility of Chondroitin Sulfate Chains

The diffusion coefficients defined by Equation 1 of native PGA (MW $\sim$200 $\times$ 10^6), its constituent proteoglycan monomer (MW $\sim$2.6 $\times$ 10^6), and CS (MW $\sim$30,000) measured by ultracentrifugal techniques are markedly different in dilute solution but show molecular weight independence above concentration of 5 to 10 mg ml^{-1} (Figure 2). This type of diffusion coefficient does not strictly represent the hydrodynamic properties of these materials, as it is determined by use of the absolute concentration (Equation 1), and as such is a complex function of both osmotic pressure and hydrodynamic frictional coefficient (Equation 6). In any case, since the osmotic pressure is a molecular weight independent quantity, it is evident that the hydrodynamic frictional coefficient is also molecular weight independent. The marked increase in the diffusion coefficient with concentration is due to the predomi-

[3] Pressure conversion units are: 1 atm = 14.69 lb in^{-2} = 1.013 $\times$ 10^6 dyne cm^{-2} = 1033 cm H_2O (4°C) = 1.013 $\times$ 10^5 Nm^{-2} (or Pascal) = 0.0394 osmolar at 37°C.

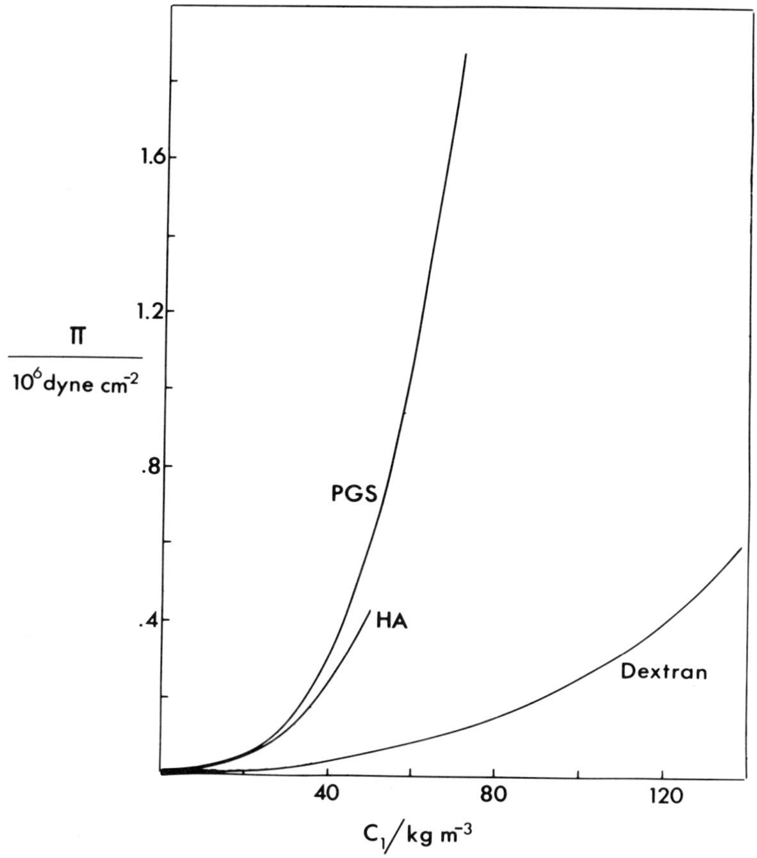

Figure 1. The osmotic pressure as a function of concentration for Swarm rat chondrosarcoma PGS[4] in PBS (Williams and Comper, 1990), hyaluronate (Wik, 1979) in 0.2 *M* NaCl, and dextran in H_2O. (From Zamparo, O. and Comper, W.D. Hydraulic conductivity of chondroitin sulfate proteoglycan solutions. *Arch. Biochem. Biophys.* **274**: 259–269 (1989). With permission.)

nance of the osmotic term (and of its microion contribution, see "Osmotic Pressure," *vide supra*) over the hydrodynamic frictional term (Equation 6).

Independent estimates of the frictional interaction of proteoglycans with water can be made directly from sedimentation velocity measurements. In this technique water is made to move relative to the proteoglycan by sedi-

[4] Frequent reference is made in this article to physicochemical studies performed on proteoglycans isolated from Swarm rat chondrosarcoma (Comper and Williams, 1987). The assumption made that these proteoglycans behave in a similar way to the large cartilage proteoglycan is probably a reasonable one in the light of their similar chemical structures (Hascall and Kimura, 1981) (although the major difference is that chondrosarcoma proteoglycan lacks keratan sulfate) and aggregation functions.

Table 1
Distribution of NaCl Between Femoral Head Cartilage and 0.15 *M* NaCl Solution[7]

Fixed charge density (meq/g tissue water)	NaCl tissue concentration (*M*)
0.105	0.101
0.120	0.103
0.145	0.092
0.154	0.091
0.162	0.086
0.170	0.081
0.178	0.081

Data from Maroudas (1979)

mentation of the proteoglycan. The sedimentation coefficient (S_1) so obtained can be directly related to the specific hydraulic conductivity (k)[5] of the solution. The following equations apply:

$$k \simeq \eta_2 \, S_1/C_1 \, (1 - \bar{v}_1 \, \rho) \tag{9}$$

where η_2 is the dynamic viscosity of solvent, $\bar{v}_1$ the partial specific volume, ρ the solution density, and where the sedimentation coefficient

$$S_1 \simeq \frac{M_1}{f_{12}} \tag{10}$$

represents the frictional coefficient between unit mass of proteoglycan and water. Hydraulic conductivities have been measured up to 80 mg ml^{-1} CS, which is comparable to that found in cartilagenous tissues (see "Compositional Analysis", *vide supra*). Measurements are made at rotor speeds corresponding to mechanical pressures greater than 10 atm that far outweigh the osmotic pressure of the materials.

Specific hydraulic conductivities measured in this way indicate that they are molecular weight independent (Figure 3). These studies demonstrate that the core structures and the macroscopic viscosities of the monomer and aggregate do not influence the relative movement of water to proteoglycan. Rather, this movement is governed by the interaction of water with a segment of the CS chain. Therefore, the translational movement of such enormous molecules as PGA and PGS is governed by, and behaves like, CS chain segments.

[5] The volume flow of fluids through porous beds is described by Darcy's law (1856). For water flow we have $J_v = k\Delta P \, A/l\eta_2$ where k is the specific hydraulic conductivity, A the area, l the length, η_2 the dynamic viscosity, and ΔP the pressure difference. Other definitions of conductivity are hydraulic conductance, $k'' = k/l\eta_2$; hydraulic conductivity, $k' = k/\eta_2$.

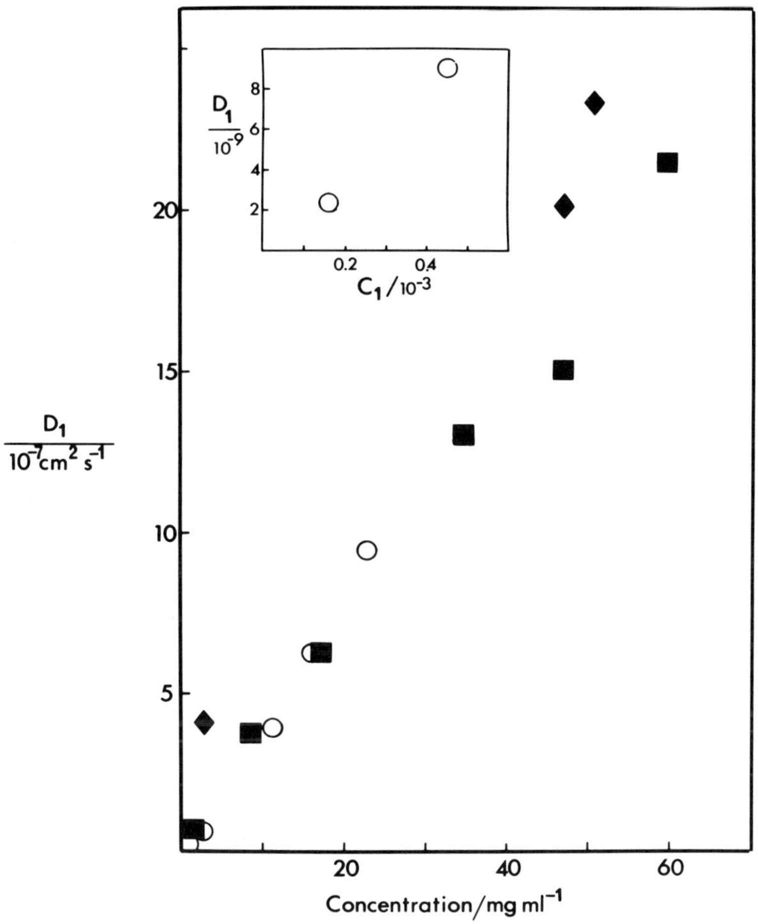

Figure 2. The mutual diffusion coefficients, D_1 of Swarm rat chondrosarcoma native PGA (○), PGS (■) and CS chains (♦) in PBS. Inset: data for PGA at low concentrations. (From Comper, W.D. and Williams, R.P.W., Hydrodynamics of concentrated proteoglycan solutions. *J. Biol. Chem.*, **262**: 13464–13471 (1987). With permission.)

It is of interest to note that in another related study, Torchia *et al.*, (1977) established from ^{13}C nmr studies that CS motion in cartilage is "relatively free," i.e., independent of aggregation and placement of the proteoglycans within the collagen network of cartilage. While there is no direct relationship between the segment motion measured by ^{13}C nmr to translational motion (Torchia *et al.*, 1977, interpreted the measurements on the basis of rotational diffusion), it is clear that both types of motion show molecular weight independence.

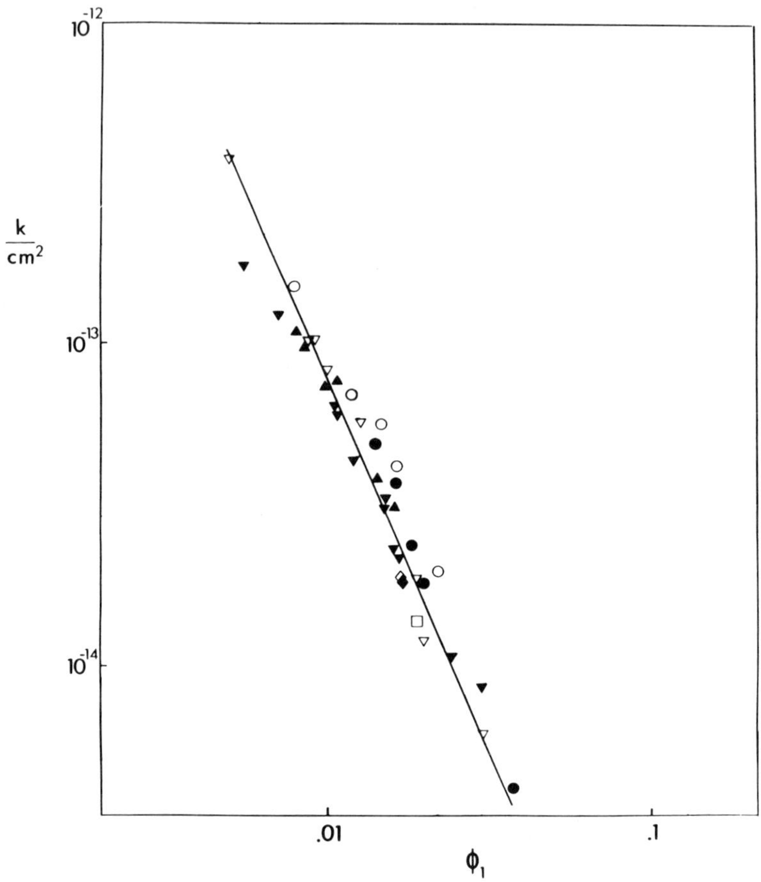

Figure 3. Specific hydraulic conductivity, k, of Swarm rat chondrosarcoma PGA (□), PGS (▽) in PBS, and PGS in 1 *M* NaCl (○); whale cartilage C4-S (▲), and shark cartilage C6-S (▼) in PBS; and C6-S in varying NaCl concentrations (*M*) 0.035 (◇), 0.07 (◆), and 1.0 (●). (From Zamparo, O. and Comper, W.D. Hydraulic conductivity of chondroitin sulfate proteoglycan solutions. *Arch. Biochem. Biophys.* **274**: 259–269 (1989). With permission.)

Compressibility (?) of Proteoglycans

The sedimentation of proteoglycans to the bottom of the centrifuge tube at sufficient rotor speeds to yield a sediment "volume" has been discussed in terms of compression of proteoglycan (Hascall, 1977). It has been inferred from these experiments and widely cited in the literature that pressures generated by the ultracentrifugal field directly made the proteoglycan compress, like compressing a spring, and that this compression is reversible when the load is released. These conclusions were drawn from varying sediment volume with rotor speed. Similar experiments were earlier performed on hyaluronate

(Balazs, 1965). The resistance to compression offered by the proteoglycan in this manner has been frequently used in discussions of proteoglycan function in cartilage matrices. It appears that proteoglycans do function to confer resistance to compression in cartilage but not by the means that has been inferred from these experiments.

The thermodynamic analysis of sedimentation-equilibrium in these types of experiments is quite explicit. Fujita (1962) has rigorously derived equations associated with this equilibrium. It represents the balance between the tendency of the proteoglycan to sediment and the tendency of the proteoglycan to move in the opposite direction by diffusion associated with the concentration gradient of the sediment volume. The proteoglycan flux j_1 (in mass units) is given as

$$j_1 \simeq \frac{M_1}{f_{12}} \left[C_1 \, \omega^2 \, (1 - \rho \, \bar{v}_1) \frac{\partial r^2/2}{\partial r} - \frac{\partial \Pi}{\partial r} \right] \tag{11}$$

where ω is the rotor speed and r is the radial distance from the center of the rotor.

The first term on the right side of Equation 11 describes the sedimentation motion, and the second term represents the diffusional motion of the proteoglycan in the opposite direction to sedimentation. When equilibrium is reached, i.e., $j_1 = O$, an apparent sediment volume is obtained where there is a balance between sedimentation and diffusion (osmotic force). In terms of the properties of the proteoglycan, the force that will ultimately govern the sediment volume will be the osmotic pressure of the proteoglycan. As the speed is decreased, the concentration is decreased in the sediment volume due to the relative influence of diffusion over sedimentation, and therefore the osmotic pressure required to balance the ultracentrifugal pressure is less. The actual pressures involved in these experiments (Hascall, 1977) are quite small (<0.1 atm) due to the low concentration of proteoglycans used (<3.0 mg ml^{-1}) and low rotor speeds.

The conclusions from these types of experiments should be carefully drawn, as there is no direct relationship between proteoglycan structure and its conformation and the pressure applied. It has been discussed previously in "Osmotic Pressure" that the osmotic pressure at concentrations greater than 5 mg ml^{-1} is molecular weight independent and therefore independent of molecular conformation. Also implied in these analyses is that since osmotic pressure is directly related to the free energy of mixing (Tanford, 1961), concentration dependent conformational charges contributing to configurational entropy, and therefore changes in excluded volume interactions, do not significantly affect the osmotic pressure. This is further demonstrated from the similarity of osmotic pressures of CS and proteoglycan (Urban and Maroudas, 1979) and from our sedimentation and diffusional experiments at high concentrations, approaching those in cartilage, where factors contributing to the sediment volume will be the same for CS and PGA.

The overall conformation and structure of the proteoglycans is not important in this regard. Changes in conformations with increased polysaccharide concentration must occur as the hydrodynamic volume of the molecule in dilute solution (dilute hydrodynamic volume) must be reduced by virtue of space-filling problems (see ''Proteoglycan Conformation,'' *vide infra*). However, these entropic changes appear to have little influence on osmotic pressure and therefore sediment volume. It is important to note that previous experiments investigating proteoglycan ''compression'' (Hascall, 1977; Harper and Preston, 1987) have been performed at high ionic strength where counterion colligative properties are significantly reduced and the relative influence of configurational entropy through excluded volume interactions is enhanced.

At present, it is difficult to conclude anything from the centrifugation experiments concerning the role of compressive rigidity of the proteoglycan, as distinct from the properties of its constituent CS chains, in cartilage biomechanics.

Specific Hydraulic Conductivity

This section addresses the relationship of cartilage proteoglycan to tissue hydraulic permeability in more detail, as it is clearly an important parameter governing the biomechanical properties of the tissue.

Maroudas (1975b) has established an inverse correlation between articular cartilage proteoglycan content and cartilage hydraulic conductivity[5] (Figure 4). Direct evidence for the specific role of proteoglycan on tissue conductivity comes from hydraulic conductivity studies from sedimentation analysis and classical ultrafiltration studies (Comper and Williams, 1987; Zamparo and Comper, 1989).

Comparison of the hydraulic conductivity data on proteoglycan in solution (Figure 3; see ''Segmental Mobility of Chondroitin Sulfate Chains,'' *vide supra*) and intact cartilage (Figure 4a) show they are the same order of magnitude (Figure 4b). The differences may correspond to inaccuracies in the estimate of cartilage proteoglycan concentration based on total tissue water content rather than extracellular and extrafibrillar water (see ''Compositional Analysis'', *vide supra*), the role of the collagen network (Levick, 1987), and the distribution of proteoglycan in cartilage. A further question concerning this comparison is that the average fixed charge density (proteoglycan concentration) in cartilage was obtained on unstressed tissue, whereas hydraulic flow measurements may correspond to higher proteoglycan concentrations due to stress effects on the distribution of proteoglycan in the tissue. This may be an important factor considering the marked concentration dependence of k; for a tenfold change in k from 10^{-13} cm^2 to 10^{-14} cm^2 corresponds to a 2.7-fold change in proteoglycan concentration from 18.8 mg ml^{-1} to 50 mg ml^{-1} (Figure 3).

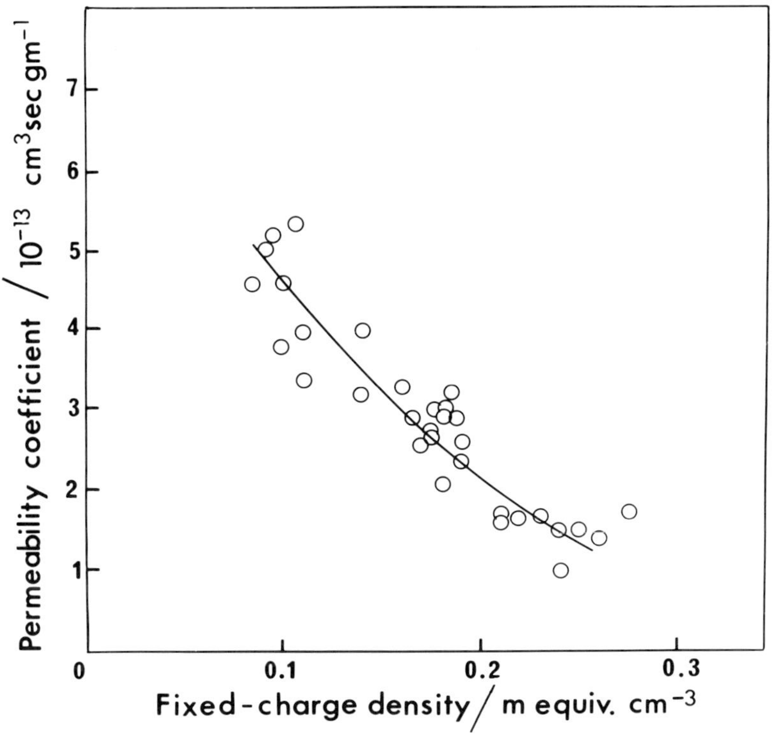

Figure 4A. Variation of "permeability coefficient conductivity" ($k\eta_2$) (units of $cm^3s\ g^{-1}$)[5] as a function of fixed charge density in milliequivalent ml^{-1} (tissue water). (From Maroudas, A., Biophysical chemistry of cartilagenous tissue with special reference to solute and fluid transport. *Biorheology* **12**: 233–248 (1975b). With permission.)

The factors affecting the very high resistance to flow afforded by the proteoglycan are currently being investigated. The influence of high NaCl concentrations on hydraulic conductivity for PGS is small with an increase in k of 1.2 to 1.6-fold (Figure 3) and even smaller for CS. Analysis of k for CS at ~36 mg ml^{-1} reveals also that there are only small changes over the NaCl concentration range of 0.035 to 1 *M*. These results are consistent with the observations by Maroudas (1975a) that the hydraulic conductivity of cartilage is not influenced significantly by increasing the ambient NaCl concentration from 0.15 to 1.0 *M*. Studies where ionic conditions have affected the mechanical properties of cartilage (Sokoloff, 1963; Grodzinsky *et al.*, 1981) could be interpreted on the basis of change in the osmotic properties of the tissue rather than a major effect on k.

The mechanism for the high resistance to flow afforded by the proteoglycan in cartilage appears to be due to the dynamic interaction of water with the backbone segments of the CS chains. Our studies have demonstrated that the presence or position of the sulfate groups on CS has little influence on the

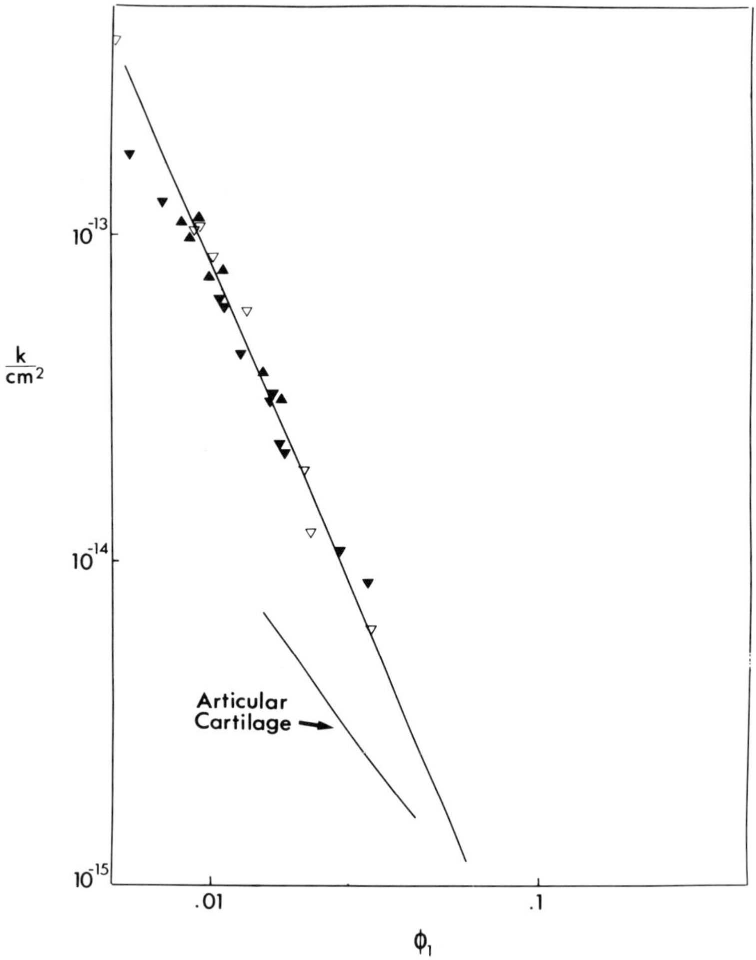

Figure 4B. The specific hydraulic conductivity, k, as a function of proteoglycan volume fraction ϕ_1 in PBS as compared to data from Maroudas (1975b) (Figure 4A) on cartilage slices with varying proteoglycan content (the conversion factor of fixed charge density to proteoglycan concentration was from Urban et al. (1979). (From Zamparo, O. and Comper, W.D. Hydraulic conductivity of chondroitin sulfate proteoglycan solutions. *Arch. Biochem. Biophys.* **274**: 259–269 (1989). With permission.)

magnitude of k as C4-S, C6-S, chondroitin (desulfated CS) and hyaluronate have very similar hydraulic conductivities (Figures 3 and 5). The nature of the glycosidic linkage may be of importance. Hyaluronate and chondroitin sulfate, with alternating β-1,4 and β-1,3 linkages, have considerably lower values as compared to neutral dextran with α-1,6 linkages, and highly charged heparin with β-1,4 and α-1,4 linkages (Figure 5) (recent results indicate that

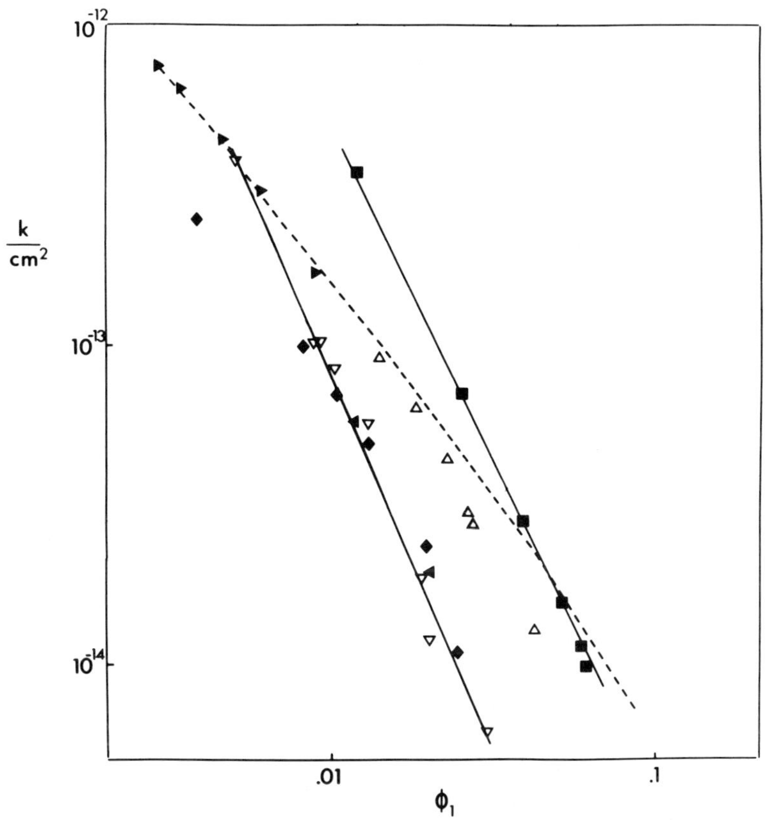

Figure 5. The specific hydraulic conductivity, k, as a function of volume fraction for various materials in PBS. PGS (▽), hyaluronate (◀), desulfated chondroitin sulfate (◆), heparin (△), and dextran (■). The dotted line represents the predicted value of k from the Happel-Brenner theory (1965) for fibers of cylindrical radius of 0.55 nm. Data from Zamparo and Comper (1989) and Comper and Zamparo (1989). The data for hyaluronate (▶) in thermodynamic equilibrium with buffered 0.2 *M* NaCl is from Preston *et al.* (1965).

desulfated heparin and heparin in 1 *M* NaCl behave very similarly to heparin in phosphate buffered saline (PBS), Zamparo and Comper, 1989).

A number of studies (Ethier, 1986; Levick, 1987) have used the early data of the sedimentation of hyaluronate in dilute solution (Preston *et al.*, 1965) as indicative of the influence of sulfated glycosaminoglycans on hydraulic conductivity on various extracellular matrices including cartilage. Data for hyaluronate from Preston *et al.*, (1965) and our studies up to a concentration of 30 mg ml^{-1} demonstrate that they are not significantly different from CS (Figure 5). Levick (1987) pointed out the apparent agreement of the Happel-Brenner model (1965), derived for the flow of water through a random dis-

Table 2
Predicted Flow of Water in 1 s through a 200 nm Cube Proteoglycan Compartment of Cartilage Under a Load of 147 lb in^{-2} (10 atm)[3]

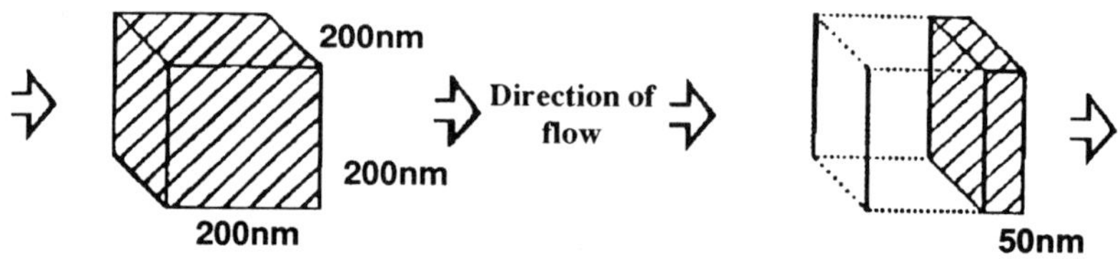

	Uniform proteoglycan distribution (shaded area)	Stress-induced proteoglycan distribution
Proteoglycan concentration	60 mg ml^{-1}	240 mg ml^{-1}
No. of PG subunit molecules	111	111
Specific hydraulic conductivity	6×10^{-15} cm^2	2.5×10^{-16} cm^2*
Osmotic pressure	~1 atm	~6 atm+
Volume flow through compartment in 1 s	1.1×10^{-10} ml	8×10^{-12} ml
Volume flow (1 s)/compartment volume	13,500	1,000

* Extrapolated from Figure 3.
\+ From Urban *et al.* (1979).

tribution of neutral cylindrical fibers, with fibers of radius of 0.55 nm in dilute solution. In more concentrated solutions the model offers a poor prediction for not only hyaluronate but uncharged dextran (Figure 5).

The Happel-Brenner model does point to the quantitatively important consideration of surface-volume ratio of the fiber as a function of volume fraction; the narrower the fiber the lower the k value. This, in turn, focuses on the differences in the polysaccharides seen in Figure 5 as being a combination of glycosidic linkage and chain conformation giving rise to a characteristic surface-volume ratio that may govern the respective k value. The low k value of CS chains, therefore, appears to indicate a distinct structure-function relationship for this polysaccharide in cartilage.

It is informative to consider a calculation of volume flow through a 200 nm cube proteoglycan compartment as shown in Table 2. For a uniform distribution of proteoglycan in the compartment, a large flow of water is obtained as compared to the volume of the compartment; the ratio is 13,500 for a flow over 1 s. If there is a stress induced redistribution of proteoglycan (concentration polarization) in the compartment as estimated in Table 2 we may increase the proteoglycan concentration 4-fold and thereby increase its osmotic pressure 6-fold, decrease the conductivity 24-fold and decrease the rate of flow by 13-fold. The marked concentration dependence of k is offset by the small distance over which flow the proteoglycan compartment operates. In fact, the ratio of flow through a compartment to volume of the compartment will be $k\Delta P-\Delta\Pi)/\eta_2 1^2$ on the basis of this model. This demonstrates the critical role of the distance over which flow takes place. With pressures of the order of 10 atm, k $\sim 10^{-15}$ cm^2, flow will be $\sim 10^{-6}/l^2$. When distances are greater than 100 μ the flow will be negligible. The relative influence of osmotic pressure and hydraulic conductivity of cartilage in response to mechanical load will depend on proteoglycan concentration, concentration polarization, and effective distances over which water flow takes place.

Osmotic Flow

Osmotic flow is an important process in cartilage hydrodynamics. Apart from its resistance to mechanical pressures that drive water out of the tissue, it provides the only mechanism by which water can be reimbibed into the compressed tissue once mechanical load is released. The mechanism of osmotic flow has been discussed over the past 70 years and was the subject of a symposium about 10 years ago (Hammel, 1979). We have recently addressed the mechanism of this process in detail (Williams and Comper, 1987, 1990; Comper and Williams, 1987, 1990).

The conclusion from our studies is that for osmotic flow to occur across a membrane that separates water from a solution of osmotically active solute impermeable to the membrane, a series flow is required consisting of (1) water flow across the membrane and (2) the diffusional exchange of osmotically active solute with solvent on the solution side of the membrane at the mouth of the pore (Figure 6). The diffusional exchange is a translational motion (see "Basic Considerations", *vide supra*) associated with the concentration gradient of solute near the membrane or the reverse chemical potential gradient of water. The usual relationship between osmotic volume flow, J_v, of water and the osmotic pressure difference ($\Delta\Pi$) across the membrane is

$$J_v = -L_p^o \Delta\Pi \tag{12}$$

where L_p^o is the osmotic permeability coefficient which has normally been regarded as a constant describing the interaction of water with the membrane. Prior to our studies, there had been no account made of the influence of the dynamic properties of the osmotically active solute on the magnitude of L_p^o. Previous studies by Van Damme *et al.*, (1982) had established that polymer transport across a free liquid boundary, where high effective polymer concentration gradients (or osmotic pressure gradients) exist, proceeds by a diffusion process. This translational diffusion could equally well be described by water diffusion (volume flow) in the opposite direction. This water flow is driven by identical osmotic pressure gradients that drive osmotic flow across membranes. Another more familiar system associated with the identification of osmotic flow is the swelling of gels. The gels imbibe water through the differences in osmotic pressure between the gel phase and surrounding solvent. The rate of imbibition is governed by the diffusion (or volume expansion;

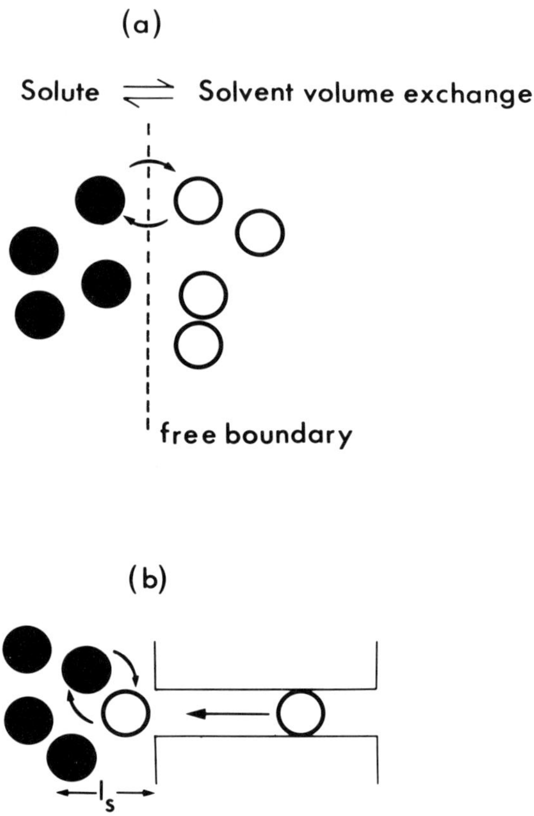

Figure 6. Schematic diagram of solute (●)-solvent volume exchange at (a) a free liquid boundary and (b) over a length l_s near the surface of a semipermeable membrane.

see ''Volume Expansion Concept,'' *vide supra*) of the gel-polymer network, and this is identical with polymer diffusion in semidilute polymer solutions (Buckley and Berger, 1962; Tanaka and Filmore, 1979; Candau *et al.*, 1982).

Our studies (Williams and Comper, 1987, 1990) on osmotic flow across polyester tortuous membranes and polycarbonate membranes of controlled straight cylindrical pores of known pore size and density have demonstrated that (1) the osmotic flow can be described at high solute concentration as a rate limiting process associated with solvent-solute exchange over a distance l_s[6] in the solution adjacent to the membrane pore of total area (A) in the

[6] The distance l_s is difficult to quantitate which makes measurement of volume flow by either ultrafiltration or osmotic flow of semiquantitative value only. One study (Comper and Williams, 1990) using controlled cylindrical pore membranes of varying porosity indicated l_s distances of approximately 40 nm for proteoglycan solutions.

membrane, (2) osmotic flow is proportional to r_p^2 (where r_p is the pore radius) and not to r_p^4, which suggested dependence on a diffusional process, and (3) the osmotic permeability coefficient can be represented by the following equation:

$$L_p^o \simeq \frac{A}{l_s}\frac{M_1}{f_{12}} \tag{13}$$

The marked influence of M_1/f_{12}, which is a highly concentration-dependent quantity for the proteoglycan (Figure 3 as $k\alpha M_1/f_{12}$) on $L_p{}^o$ is demonstrated in Figure 7 for osmotic flow studies or proteoglycan. Here it is seen that L_p^o decreases rapidly with concentration in the same manner as M_1/f_{12} (Williams and Comper, 1990). The separate influence of the colligative properties of the proteoglycan solution as governed by its constituent microcounterions (and their high mobility) is clearly distinguished from the frictional coefficient of the whole electroneutral proteoglycan kinetic unit that determines the rate of flow. Similar observations have been made for dextran sulfate (Figure 7). Another aspect of our osmotic flow studies, not predicted previously, is that osmotic flow should be faster for low molecular weight material as compared to high molecular weight material in dilute solution, and this has been demonstrated (Williams and Comper, 1987). However, in semidilute solution where colligative and dynamic properties governing volume expansion are molecular weight independent, PGS, PGA, and CS will behave identically due to similarities in Π and M_1/f_{12} providing the membrane is impermeable to these solutes.

Balance of Osmotic and Mechanical Pressures

In relation to the mechanical properties of cartilage, its most important component is water. The load-bearing properties of the tissue result largely from its ability to retain water (an essentially incompressible fluid) under the application of load. At the molecular level this has to be interpreted in relation to the balance of osmotic and mechanical pressures on the proteoglycan-containing system. This can be described by an equation analogous to that describing sedimentation equilibrium (Equation 11) in the form of Darcy's law[5] such that for semidilute solutions:

$$J_v \simeq \bar{v}_1 \frac{A}{l_s}\frac{M_1}{f_{12}} (\Delta P - \Delta\Pi) \tag{14}$$

where the rates of osmotic and hydraulic flow are governed by the same factor, M_1/f_{12}. The significance of the various terms in Equation 14 can be appreciated in the following model of cartilage. A simplified cartilage structure could be viewed as an array of porous spaces, each space surrounded by a

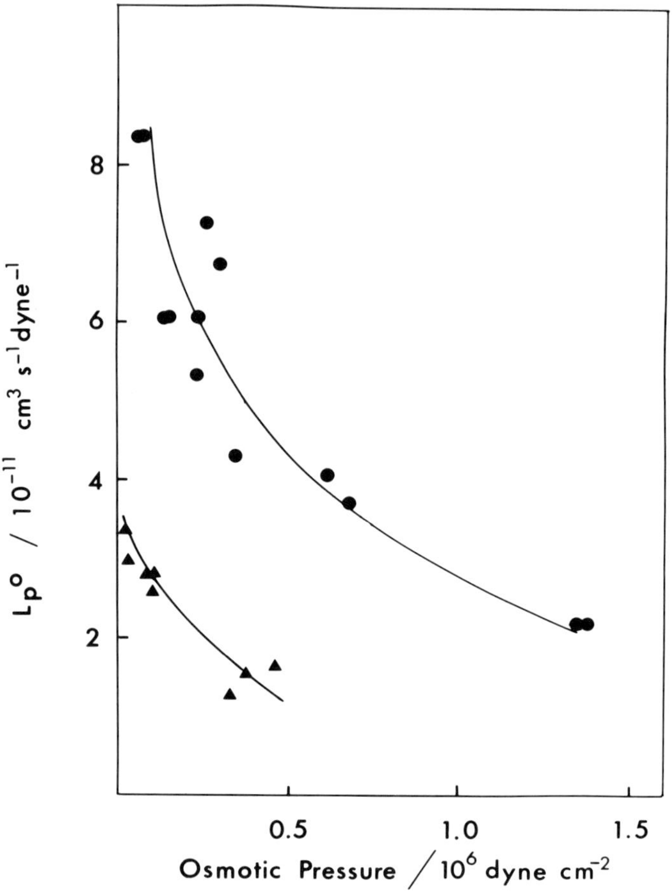

Figure 7. Variation of L_p^o as a function of osmotic pressure for proteoglycan subunit (▲) and dextran sulfate (●) in PBS. (From Williams, R.P.W. and Comper, W.D. Osmotic flow caused by polyelectrolytes. *Biophys. Chem.* **36**: 223–234 (1990). With permission.)

network of collagen fibers which retains but does not immobilize the proteoglycan within the space. When these units are subject to mechanical compression (e.g., $\Delta P \gg \Delta \Pi$) water will be ultrafiltered out of the space in a particular direction (of least resistance) which will simultaneously give rise to a buildup of proteoglycan concentration on the face of the porous wall (Table 2). The permeability of the proteoglycan compartment will further decrease in accord with the variation of k. The magnitude of the volume flux is critically dependent on the length of the compartment l_s over which filtration takes place (see "Specific Hydraulic Conductivity", *vide supra*). The application of mechanical load will give a pattern of concentration gradients in the tissue corresponding to the array of porous cells. When the load is released (i.e.,

$\Delta P = O$), the proteoglycan concentration gradients relax by diffusion (therefore the requirement for proteoglycan diffusive mobility), and the volume expanding property results in osmotic imbibition of water into the porous network and tissue as a whole.

Collagen-Proteoglycan Matrix

The mutal coexistence and interpenetration of collagen-polysaccharide matrices leading to the ability to resist pressure was early recognized in the elegant ultracentrifugal studies of Fessler (1960) and Disalvo and Schubert (1960). These studies alluded to the intimate relationship between the dynamics of proteoglycan and that of the collagen matrix: as an osmotic-swelling system they both have to be considered together. At present, not much is known of the dynamic properties of the collagen network. It is likely that osmotic volume expansion of the proteoglycan occurs as the primary "event" and expansion of the collagen network ensues. This is a cross-diffusion mechanism. Significant mobility of the proteoglycans in cartilage appears possible as both intact monomers and aggregates can be leached from tissue studies in culture when stabilized by cycloheximide[7] (Bolis *et al.*, 1989). Questions could then be raised as to the nature of the molecular properties of collagen that allows its movement to take place. What type of fibers participate in swelling? There may be fiber heterogeneity as some may be subject to volume expansion and others not—vectorial factors may also be involved. Another mechanism associated with the volume expansion of the proteoglycan-collagen network could be that dynamically they are a hybrid single component subject to thermal motion. The interaction of the large proteoglycan with collagen fibrils has been shown to be weak (for review, see Scott, 1988) possibly involving electrostatic or hydrogen bonds (Bolis *et al.*, 1989). If the proteoglycans bind to immobile fibers (immobility arising from fiber thickening and interfiber crosslinking), then conceivably their osmotic activity would be affected. The osmotic flow model requires an effective concentration of solute (osmotic

[7] *In vitro* studies of cartilage often assume that the tissue is stable. In spite of the low cell population of many cartilages, this assumption may not be valid. There is now considerable evidence to suggest that cartilage proteoglycans, at least, undergo degradation. Sandy *et al.*, (1978) have demonstrated that during organ culture of adult rabbit articular cartilage, proteoglycans undergo limited proteolysis and are readily leached from the tissue. After 24 h incubation, greater than 20% of the total proteoglycan content (as measured by uronic acid content) is lost from tissues. Significant inhibition of proteoglycan loss, and therefore enhanced stability of the tissue *in vitro* can be obtained with the inclusion of various metabolic inhibiting agents in the culture medium such as cycloheximide or puromycin or colchicine or enzyme inhibitors or by lowering the temperature of the culture system to 4°C (Comper *et al.*, 1981). The rate of proteoglycan loss may be tissue specific. Bovine articular cartilage proteoglycan is more slowly leached and may have half lives in the range of 6 to 11 d depending on the culture conditions. This half life is increased to 24 and >300 d for cyclohexamide or 4°C treatment, respectively (Campbell *et al.*, 1984).

pressure which is essentially provided by the sodium counterions on the CS chains) and a characteristic kinetic unit of the solute (CS chain segment) to move in relation to water. If proteoglycan immobility is derived from immobilization of the hyaluronate core or protein core of the aggregate or subunit, respectively, then this would severely reduce the distance over which the osmotic behavior of the CS segments could have an influence. Any immobilization of the CS chain segments themselves would have a drastic influence on the osmotic activity of the proteoglycan. In addition, even weak binding to mobile fibers may affect the effectiveness of the proteoglycan compartment to resist hydraulic flow; the resistance will decrease as it will be unable to increase in concentration or participate in concentration polarization effectively (Table 2) but this could be offset by the effective distance over which the relative flow of water to proteoglycan takes place. These considerations do focus on the need to understand the microorganization of proteoglycan in collagen networks to explain hydrodynamic behavior.

Proteoglycan Conformation

An important aspect of proteoglycan hydrodynamics, especially in relation to proteoglycan retention and organization in matrix, is the effective size and conformation of the molecule. In dilute solution the proteoglycans are noninteracting individual molecules. They can be regarded as microscopic gels with characteristic kinetic segments of CS chains and elastic restriction of the proteoglycan structure. The volume expansion properties of the CS segments is high, and as such the individual proteoglycan molecules are highly hydrated in dilute solution. The proteoglycan subunit has an effective hydrodynamic radius in the range of 50 to 80 nm and that of the aggregate 1000 to 1600 nm (Comper and Williams, 1987).[4] It is emphasized that the large hydrodynamic volume is a result of the diffusional volume expansion (see ''Volume Expansion Concept'', *vide supra*) associated with the thermal motion of CS segments. The factors governing the diffusional volume expansion have been discussed in Equations 1 to 7. The driving force for expansion is the osmotic pressure of the CS segment. Contrary to popular belief this expansion is not the result of repulsive charge interactions between CS chains. This is only a minor factor in the osmotic pressure term. In addition, the water properties are not altered in this domain (for review, see Comper and Laurent, 1978; Comper and Williams, 1987) nor in proteoglycan compartments in cartilage (Maroudas and Venn, 1977; Maroudas, 1980). Additionally, there is no evi-

[4] Frequent reference is made in this article to physicochemical studies performed on proteoglycans isolated from Swarm rat chondrosarcoma (Comper and Williams, 1987). The assumption made that these proteoglycans behave in a similar way to the large cartilage proteoglycan is probably a reasonable one in the light of their similar chemical structures (Hascall and Kimura, 1981) (although the major difference is that chondrosarcoma proteoglycan lacks keratan sulfate) and aggregation functions.

Table 3
Molecular Volumes Corresponding to the Dilute Hydrodynamic Volume and Fully Extended Core Structures

	Volume dilute hydrodynamioc (10^{-16} ml)	Fully extended cylindrical structure (10^{-16} ml)
PGS	21.8	2.46
PGA	1.75×10^5	2.51×10^3

From Comper and Williams (1987).

dence that the resilience of cartilage results from the ''water structuring property'' of proteoglycan as has been stated in recent literature (Caplan, 1984).

In concentrating proteoglycans there will be a concentration where their hydrodynamic domains will consistently make contact. This is called the critical concentration, which corresponds to ~5 mg ml^{-1} for PGS and 0.06 mg ml^{-1} for PGA (Comper and Williams, 1987). For concentrations above the critical concentration it is necessary, purely through space filling requirements, that the proteoglycan must occupy a reduced volume as compared to its hydrodynamic volume at infinite dilution. The nature of the required conformational changes in the proteoglycan are not clear, although two factors will be important: (1) the flexiblity of the protein core for PGS and hyaluronate core for PGA and (2) the balance between the internal and external osmotic pressures in relation to the domains of each molecule.

We know from the structure of chondrosarcoma PGS (Hascall and Kimura, 1981) that the mean concentration of CS chains in the dilute hydrodynamic volume is ~1.2 mg ml^{-1}. A fully extended molecule (along its protein core structure) has a cylindrical structure, with a reduced volume (Table 3) and an average cylindrical domain CS concentration of ~11 mg ml^{-1}. From osmotic considerations it is likely that with increasing concentration the proteoglycan will initially prefer to take on a more restrictive extended conformation without changing the average cylindrical domain concentration of CS. This is consistent with the dominant role that the colligative properties of CS microcounterions have on total osmotic pressure (see ''Osmotic Pressure'' *vide supra*) as compared to entropic contributions from the configurational entropy of the core structure. The osmotic asymmetry within the molecule leading to decreased flexibility of the protein core may also explain the restricted flexibility of the protein core in PGS units in cartilage as measured by ^{13}C nmr (Torchia *et al.*, 1977). However, such conformational changes may not have been experienced under conditions where macromolecular compression has been interpreted such as in sedimentation-equilibration experiments in 0.5 *M* GuHCl (Hascall, 1977) or in light scattering/chromatography in polymer solvents containing 4 *M* GuHCl. The salt concentration used in these experiments would minimize the ionic contribution to osmotic pressure. Conformational changes resulting in increasing the cylindrical domain CS concentration through chain folding or the intermeshing of CS chains from different molecules are more likely to occur at higher concentrations.

Partitioning of Macromolecules at Cartilage/Fluid Interfaces

The discussion of macromolecular partitioning in cartilage in this section is made with special reference to articular cartilage. Histological staining techniques have demonstrated that the major route by which extracartilagenous macromolecules would enter articular cartilage is through its surface exposed to synovial fluid (Stockwell, 1979). The conditions of entry of a macromolecule will essentially depend on factors governing its partition between cartilage and joint space, the hyaluronate coat on the surface of cartilage and on the nature of the collagen-proteoglycan matrix in the tissue.

Steric Exclusion

Steric exclusion is a simple concept that has wide ramifications in the properties of extracellular matrices. When two or more particles cannot occupy the same space at the same time they therefore exclude a certain space for each other. This exclusion occurs at a molecular level and involves entropic interactions without heat effects between interacting components. Steric exclusion phenomena offered by matrix macromolecules has been reviewed extensively (Comper and Laurent, 1978; Comper, 1984). Prior to a discussion of their effects in cartilage two salient points must be considered. Firstly, the magnitude of exclusion is intimately related to the size and shape of the interacting molecules. For example, consider the excluded volumes of molecules of identical molecular weight and molecular volume when one is globular with a radius of 5 nm and the other is a rod of length 1000 nm and cylindrical radius 0.4 nm. For sphere-sphere interaction (interactions between globular particles) the molar covolume (i.e., the volume excluded by one molecule to another) is 0.26×10^6 cm^3, for sphere-rod interaction the molar covolume is 5.5×10^6 cm^3, and for rod-rod interaction the covolume is 76×10^6 cm^3. In this case, the magnitude and effectiveness of exclusion is greatly enhanced by the presence of a rod-like structure in the interacting system. The second point is that steric exclusion of a compound from a part of a solution by the presence of another molecule (again, it may be an identical or different molecule) leads to an increase in its concentration in the remaining solution. In thermodynamic terms, this results in an increase in activity or effective concentration of the compound. This may affect many properties of the system such as chemical equilibrium, osmotic pressures, solubilities, driving force in diffusion, polymer conformation, receptor binding, and partitioning between different compartments (for review, see Comper, 1984).

The exclusion properties of the connective tissue polysaccharides, hyaluronate and CS proteoglycans, to proteins are similar; at concentrations of 10 mg ml^{-1} these will exclude ~50% of the solution space for albumin. The exclusion is a nonlinear function of polysaccharide concentration (Comper, 1984). This was also observed by Snowden and Maroudas (1976) and Maroudas (1979) who studied albumin exclusion by cartilage slices of varying fixed charge density (Figure 8). Over a range of fixed charge of ~0.025 to 0.08 meq g^{-1} (wet tissue) the partition coefficient varied enormously, from 0.3 to 0.01. These experiments demonstrate that while albumin may be excluded from high proteoglycan concentration compartments in the deeper zones of cartilage, a substantial amount of albumin is able to penetrate the more superficial layers associated with lower fixed charged density. Recent evidence suggests that the exclusion properties of connective tissue polysaccharides, often regarded as rod-like in an interacting system, depends on the nature of the glycosidic linkage in the polysaccharide chain (Zamparo and Comper, 1990).

The excluded volume offered by collagen to albumin (approximately 4 cm^3 g^{-1} collagen [Pearce and Laurent, 1977]) which will vary with fiber radius (Bert and Pearce, 1984) is significant because of its high concentration in cartilage. We can predict that in pig cartilage whose macromolecular composition is known (*vide supra*) that collagen at 0.19 g/g tissue would exclude albumin in the range of 0.4 to 0.8 cm^3/g tissue as compared to the polysaccharide component, which at a concentration of 0.022 g/g tissue would exclude albumin in the range of 0.5 to 0.75 cm^3/g tissue.

Exclusion of low molecular weight components will also be a function of collagen concentration, cartilage depth, and proteoglycan content. There is considerable evidence that the partition of molecules decreases markedly with increasing molecular weight (Maroudas, 1979). The effective "pore" size as calculated from hydraulic flow measurements also indicates a severe limitation to solutes. Estimates of pores in the proteoglycan network in the range of 1.5 to 2.0 nm diameter have been made (Bernich *et al.*, 1972; Maroudas, 1975a) as compared to the 60 to 200 nm pores in the collagen network (Byers, 1983).

Cationic Proteins

Recent evidence suggests that the cartilage proteoglycans participate in charge-charge interactions with other macromolecules, and that these interactions may have important physiological and pathological consequences. The basis for this interest arises from the qualitative findings that cationic macromolecules or neutral molecules made cationic by chemical means are able to diffuse into and bind to matrix material after having been injected intra-articularly. This occurs specifically due to the cationic nature of the molecule. Cationized forms of bovine serum albumin and other proteins with pI >8.5 show a strong affinity for cartilage and penetrate deeply into the matrix (Van den Berg *et al.*, 1985, 1986) as shown by histochemical and autoradiographic techniques. Using the same techniques it was found that negatively charged native albumin showed little or no affinity for cartilage. Some caution should accompany these studies as quantitative partition studies

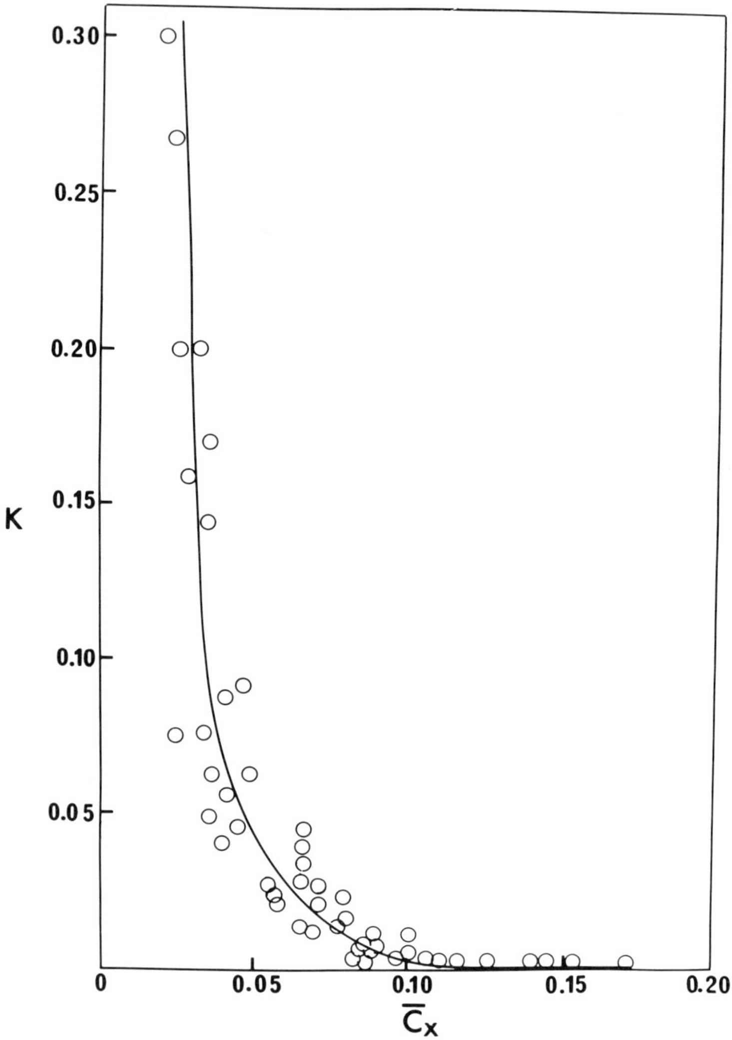

Figure 8. Variation in the partition coefficient of serum albumin (K) with fixed charge content (meq g^{-1} tissue). (From Maroudas, A., in *Adult Articular Cartilage,* Ed M.A.R. Freeman. Pitman Medical, Tunbridge Wells, England. p. 215–290, 1979. With permission.)

by Maroudas (1979) (Figure 8) have demonstrated that only dense hyaline articular cartilage is effectively impermeable for macromolecular solutes such as albumin and immunoglobulins, and that cartilage with lower proteoglycan content is permeable to albumin. Experiments designed to quantitate a direct comparison of the partition of cationic and native albumin are required to establish the preference of cationic material.

Notwithstanding these doubts, further studies of intra-articular injected material have demonstrated that other cationized proteins up to 150,000 (IgG) in molecular weight readily penetrated cartilage, cationic catalase (MW ~ 240,000) showed superificial penetration, but penetration of cationic ferritin (MW ~ 440,000) was not demonstrated, suggesting that the molecular weight range of 240,000 to 440,000 represents the upper molecular weight range of penetration of cationic glubolar proteins (Schalkwijk *et al.*, 1985). The actual mechanism for possible facilitated transfer of these materials is not known.

Several naturally occurring proteins are able to penetrate cartilage and some are known to affect matrix structure and chondrocyte function. For example, elastase, a major leukocyte neutral protease with a high pI, has been shown to penetrate articular cartilage and to localize in the matrix (Janoff *et al.*, 1976) and in the pericellular region of the chondrocyte (Bartholomew *et al.*, 1987). Elastase has been implicated in cartilage breakdown in arthritis. Another enzyme, lysozyme, a bacteriolytic enzyme found in many types of mammalian tissues and body secretions and also characterized by a high pI, has been found in cartilage (Sorgente *et al.*, 1972). It was demonstrated that chondroitinase, which effectively removes CS from cartilage slices, does not remove the lysozyme. This may indicate that while the CS may not eventually immobilize lysozyme in the matrix, it may have facilitated its initial entry and localization. It is known that lysozyme found in cartilage is from extracartilagenous sources (Thonar *et al.*, 1988).

Further considerations for cationic protein penetration may be with inflammation where cationic proteins such as elastase and cathepsin b may be released locally in the joint cavity and subsequently penetrate and perhaps mediate destructive processes in the cartilage. Cationized antibodies to albumin, antigens, and immune complexes are known to bind to articular cartilage; this binding may mediate an inflammatory response (Rios and Mannick, 1987). Charge-charge interactions may predispose cartilage to antigen-induced arthritis through the binding of antigen-antibody complexes found in cartilage of patients with rheumatoid arthritis. There is also the possibility of other unidentified cationic proteins present in cartilage that may accumulate over the course of time and periods of inflammation and injury.

The conditions in the proteoglycan-containing matrix would appear to favor charge-charge interactions of cationic proteins with proteoglycans—as the concentration of NaCl is significantly lowered in proteoglycan compartments (Table 1). The mechanism of polyanion-polycation binding in these systems is not known, although in other GAG-containing systems it is evident that the sulfate groups play an important role (Comper, 1981). These conclusions have come from studies on model binary interacting systems of polycation-anionic polysaccharide-water/NaCl[8] and in the analysis of basophilic cell

[8] Many studies of the interaction of various GAGs with proteins under so called physiological conditions, namely 0.15 *M* NaCl pH 7.4, may not be valid when interactions with GAGs at high concentration in tissues are concerned. The lower NaCl concentration in the GAG compartment, as a result of Donnan distribution (Table 1), may indeed favor interaction that may not be prevalent in 0.15 *M* NaCl.

granules where enzymes are bound to sulfated polysaccharides in an inactive form. The stereochemistry of the sulfate isomers on CS may also offer selectivity of binding as molecular modeling demonstrates that the C6-S groups protrude markedly from the polysaccharide chain, whereas the C4-S groups tend to be buried within the hexosamine sugar ring. The charge distribution along the CS chain is also not uniform as it seems unlikely that any one CS chain is sulfated exclusively at either the C4 or C6 position but rather the chains are copolymers of the two isomers (for review, see Carney and Muir, 1988). Sulfation, too, may not be uniform along the length of the chain. It is speculated that the sulfate groups may play an informational role in terms of binding proteins (see also "Summary of Structure-Function for Cartilage-Proteoglycans", *vide infra*).

Molecular Transport in Extracellular Matrices

Restricted Transport

Many studies on the diffusion and flow of various solutes have shown restricted movement in cartilage (Maroudas, 1980) and in polysaccharide solutions (Laurent *et al.*, 1963, 1964; Comper and Laurent, 1978; Preston *et al.*, 1984). These studies have demonstrated that polysaccharides can clearly regulate molecular transport in the matrix. Some of the factors that have been used to explain the reduction in transport include the size and shape of the particule, local viscosity, and volume excluded by the matrix. The interrelationships of these parameters have yet to be fully understood (Ogston *et al.*, 1973). In many instances the reduction in transport corresponds with the change in partition coefficient (see "Partitioning of Macromolecules at Cartilage/Fluid Interfaces" *vide supra*). Few studies, however, have been done on concentrated proteoglycan solutions. Ogston *et al.*, (1973) have reported albumin transport to be reduced to ~50% in 20 mg ml^{-1} CS proteoglycan solutions.

Multicomponent Systems

A multicomponent system is defined as one which contains solvent (water) together with at least two dissolved solutes. Diffusion of components associated with simple concentration gradients in these multicomponent systems can spontaneously generate gravity-driven convection processes even when the system may appear to be gravitationally stable (with the heavier, more dense solution initially below the lighter, less dense solution). These types of instabilities have been known for many years (Anderson, 1955; Brakke, 1955; Cussler, 1976) and treated as an experimental nuisance rather than being regarded as a transport phenomenon with possible biological significance. To illustrate how the effect occurs, consider the ternary system shown in Figure 9. In this experiment (Comper and Preston, 1981), the solute b, say CS proteoglycan is initially at uniform concentration with no initial con-

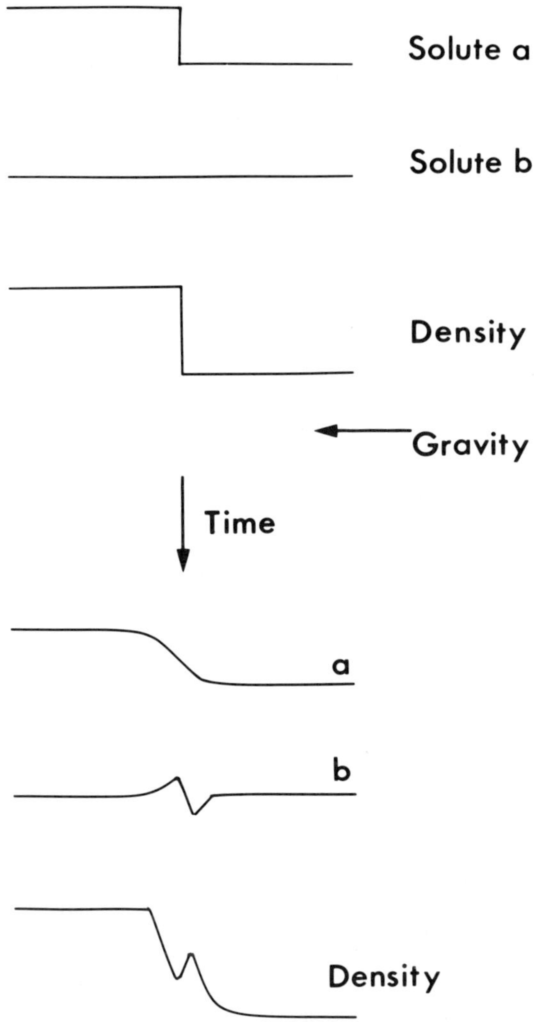

Figure 9. Schematic diagram of the concentration and density gradients occurring in a ternary system with time.

centration gradient and solute a, say CS, exists with an initial concentration gradient which also yields a density gradient. As the experiment proceeds the movement of CS creates a movement of proteoglycan (recognize that it too has a concentration gradient as its CS chains see other free CS chains) which produces a concentration profile in the proteoglycan distribution. This ternary effect produces a density profile with a maximum and minimum due to different partial specific volumes of proteoglycan and CS (due to protein content of proteoglycan). The density distribution represents a laminar of solution of lower density lying below a laminar of fluid of higher density. This situation may spontaneously generate fingers or columns of fluid with light fingers of fluid moving upwards and heavier fingers moving downwards in the vertical plane. The series of ordered, countercurrent, interdigitated, vertical fluid flows create aesthetic patterns and organizations (Preston *et al.*, 1980) and may result in extremely rapid molecular transport (up to 100,000 times their normal diffusional rates) (Laurent *et al.,* 1983; Comper *et al.*, 1987a, b).

The phemonenon occurs with a wide variety of solute distributions in both polymer and biopolymer systems. Only small concentration gradients are required and nonspecific types of molecular interactions between components (namely excluded volume and frictional effects) are sufficient to potentiate the phenomenon (Comper and Preston, 1981; Harper *et al.,* 1984; Comper *et al.,* 1987a, b).

These studies have shown that virtually *all* multicomponent systems have the potential to form flows including enzyme gradients that produce gradients of substrate and product that may occur in turnover of matrix components (Comper *et al.,* 1987a) and orientation of collagen fibers (Ghosh and Comper, 1988). While the molecular distributions and concentration gradients in matrices are not known, it is thought that favorable conditions for flow are likely in the pericellular environment around the cell (chondron) and in remodeling dynamic matrices that occur during embryonic development and wound healing. The flows are decreased significantly with increasing viscosity but can be driven at essentially any angle, in tubes down to 200 μ in diameter (Comper *et al.,* 1987a). The flows are particularly suited to the rapid transport of large macromolecules (Laurent *et al.,* 1983) such as PGA and cells (Comper *et al.,* 1987a).

Another passive physicochemical process that may be of biological importance, and which creates unusual organizations and structures in aqueous solution, namely fractals (dendritic type structures), can be generated by injecting a viscous polysaccharide solution through a fine syringe needle into water (Nittman *et al.,* 1985). It is conceivable that such processes may operate in cartilage matrices when subject to mechanical load.

Summary of Structure-Function for Cartilage Proteoglycans

There has been a poor understanding of the structure-function of cartilage proteoglycans and their role in the mechanical properties of the matrix. Common descriptions of the structure-function of the ''large'' proteoglycan often focus on its ''water binding'' capacity, its high osmotic activity, its excluded volume, and ion exchange properties, and its compressibility function in that it can act ''like a spring'' or a ''balloon'' to confer to the tissue resistance to compression. These concepts and others have been discussed in detail in this chapter. The conclusion is that some of these interpretations have no physical basis at all whereas as others, such as osmotic pressure, microion interaction, and excluded volume, represent only part of the story in relation to the functional properties of these molecules. There is now emerging intriguing new evidence that puts some of these physical properties in a more rigorous and consistent perspective.

Comper and Laurent (1978) posed the question as to "how these molecules (i.e., proteoglycans) act in specific ways in relation to their physicochemical properties." The cartilage "large" proteoglycans are clearly multifunctional molecules, but as molecules informationally distant from the gene they represent a novel and interesting situation in terms of structure-function concepts. Even though the molecules exist as complicated structures with hierarchical branching organizations, their physical properties are now known to be primarily determined by their constituent CS chains. The branched structure of the proteoglycan subunit and aggregate, apart from creating a larger macromolecule which is more likely to be retained in the tissue, clearly enables a more efficient solution packing mechanism to maintain relatively high levels of CS chains and their constituent anionic charges for a lower viscosity. Intrinsic viscosity of chondrosarcoma PGS and PGA in PBS is 265 ml g^{-1} and 1015 ml g^{-1}, respectively (unpublished data) as compared to say linear hyaluronate of the same molecular weight as PGS (intrinsic viscosity 3981 ml g^{-1} interpolated from data of Cleland and Wang, 1970).

The mechanism for the high resistance to flow afforded by the proteoglycan in cartilage appears to be due to the dynamic interaction of water with the CS segments of the proteoglycan (see "Specific Hydraulic Conductivity," *supra*). The shape and structure of the polysaccharide backbone of CS appears to confer specificity on the proteoglycan with regard to hydraulic flow resistance and ultimately the weight-bearing properties of the tissue under large mechanical load. Aggregation of the proteoglycan may be important in conferring viscoelastic properties to the tissue (Mow *et al.*, 1984). The anionic charges of the CS chain, particularly the carboxyl group of the uronic acid residue, appears important for determining the relatively high osmotic pressure of the proteoglycan—its osmotic pressure is not significantly different from hyaluronate (Figure 1). The osmotic pressure of the proteoglycan is required for tissue recovery after mechanical compression. Therefore, dynamic water interaction with the CS, which is important for hydraulic flow and osmotic flow, depends on the glycosidic linkage and the uronic acid content of the polysaccharide. On the basis of these conclusions, it is likely that the sulfate groups on the polysaccharide chain appear to specifically participate in the enthalpic electrostatic binding interactions with other macromolecules. The sulfate groups acting in this way appear to be crucial in a metabolic sense, as their substitution as a posttranslational modification reaction represents an intense energy requiring process; the equivalent of two molecules for ATP for every molecular addition of sulfate is required. The information contained in the distribution of anionic charge groups along the GAG chain may lend further significance to the original suggestion by Meyer (1960) that they "act as templates or organizers in the development of the architecture of the organized fibrous elements of connective tissues."

Acknowledgments

This work was supported by grants from the Australian Research Committee, the Arthritis Foundation of Australia, and the Monash Special Research fund. I acknowledge the contributions to the studies reported here by my postgraduate students R. P. W. Williams and O. Zamparo.

References

N.G. Anderson. Studies on isolated cell components. *Exp. Cell Res.* **9**: 446–459 (1955).

E.A. Balazs. Sediment volume and viscoelastic behaviour of hyaluronic acid solutions. *Fed. Proc.* **25**: 1817–1822 (1965).

J.S. Bartholomew and D.A. Lowther. Receptor-mediated binding of leukocyte elastase by chondrocytes. *Arthritis Rheum.* **30**: 431–438 (1987).

E.M. Bernich, P. Lotke, and R. Rubenstein. Membrane characteristics of human articular cartilage. *Biochim. Biophys. Acta* **266**: 732–736 (1972).

J.L. Bert and R.H. Pearce. The interstitium and microvascular exchange. In: *Handbook of Physiology*. The Cardiovascular System, Vol. IV, Microcirculation, Part 1, Eds. E.M. Renkin and C.C. Michel, American Physiological Society, Bethesda. pp. 521–548, 1984.

S. Bolis, C.J. Handley, and W.D. Comper. Passive loss of proteoglycan from articular cartilage explants. *Biochim. Biophys. Acta* **993** 157–167 (1989).

M.K. Brakke. Zone electrophoresis of dyes, proteins and viruses in density-gradient columns of sucrose solutions. *Arch. Biochem. Biophys.* **55**; 175–190 (1955).

K.D. Brandt and H. Muir. Characterization of protein-polysaccharides of articular cartilage from immature pigs. *Biochem. J.* **114**: 871–876 (1969).

D.J. Buckley and M. Berger. The swelling of polymer systems in solvents. II. Mathematics of diffusion. *J. Polym Sci.,* **56**: 175–185 (1962).

R. Byers, M.T. Bayliss, A. Maroudas, J. Urban, and B. Weightman. Hypothesizing about joints. In: *Studies in Joint Disease*. Eds A. Maroudas and J. Holborrow. Pitman Medical, London. pp. 241–276, 1983.

M.A. Campbell, C.J. Handley, V.C. Hascall, R.A. Campbell, and D.A. Lowther. Turnover of proteoglycans in cultures of bovine articular cartilage. *Arch. Biochem. Biophys.* **234**: 275–289 (1984).

S. Candau, J. Bastide, and M. Delsanti. Structural, elastic and dynamic properties of swollen polymer networks. *Adv. Polym. Sci.* **44**: 27–71 (1982).

A.I. Caplan. Cartilage. *Sci. Am.* **251**: 82–90 (1984).

S.L. Carney and H. Muir. The structure and function of cartilage proteoglycans. *Physiol. Rev.* **68**: 858–910 (1988).

R.L. Cleland and J.L. Wang. Ionic polysaccharides. III. Dilute solution properties of hyaluronic acid fractions. *Biopolymers* **9**: 799–810 (1970).

W.D. Comper. *Heparin (and Related Polysaccharides)*. Polymer Monographs, Vol. 7. Gordon and Breach, New York, 1981.

W.D. Comper. Interstitium. In: *Edema*. Eds N.C. Staub and A.E. Taylor. Raven Press, New York. pp. 229–262, 1984.

W.D. Comper, M. DeWitt, and D.A. Lowther. Effects of antiinflammatory drugs on proteoglycan degradation as studied in rabbit articular cartilage organ culture. *Biochem. Pharmacol.* **30**: 459–468 (1981).

W.D. Comper and T.C. Laurent. Physiological function of connective tissue polysaccharides. *Physiol. Rev.* **58**: 255–315 (1978).

W.D. Comper, L. Pratt, C.J. Handley, and G. Harper. Cell transport in model extracellular matrices. *Arch. Biochem. Biophys.* **252**: 60–70 (1987a).

W.D. Comper and B.N. Preston. Model connective tissue systems: a study of polyion-mobile ion and of excluded volume interactions of proteoglycans. *Biochem. J.* **143**: 1–9 (1974).

W.D. Comper and B.N. Preston. On the spontaneous self-organization in various biopolymer systems to form structured flows. *Biochem. Int.* **3**: 558–564 (1981).

W.D. Comper and R.P.W. Williams. Hydrodynamics of concentrated proteoglycan solutions. *J. Biol. Chem.* **262**: 13464–13471 (1987).

W.D. Comper and R.P.W. Williams. Osmotic flow caused by chondroitin sulfate proteoglycan across well defined Nuclepore membranes. *Biophys. Chem.* **36**: 215–222 (1990).

W.D. Comper, R.P.W. Williams, G.J. Checkley, and B.N. Preston. Stability analysis in diffusion-convection systems with and without cross diffusion. *J. Phys. Chem.* **91**: 993–1000 (1987b).

W.D. Comper and O. Zamparo. Hydraulic conductivity of polymer solutions. *Biophys. Chem.* **34**: 127–135 (1989).

E.L. Cussler. *Multicomponent Diffusion*. Elsevier, Amsterdam. 1976.

H. Darcy. Les fontaines publiques de la ville de Dijon. (1956).

J. Disalvo and M. Schubert. Interaction during fibril formation of soluble collagen with cartilage proteinpolysaccharides. *Biopolymers* **4**: 247–258 (1960).

E. Edmund and A.G. Ogston. An approach to the study of phase separation in ternary aqueous systems. *Biochem. J.* **109**: 569–576 (1968).

S.R. Eisenberg and A.J. Grodzinsky. Electrokinetic micromodel of extracellular matrix and other polyelectrolyte networks. *PCH Physicochem. Hydrodynamics* **10**: 517–539 (1988).

C.R. Ethier. The hydrodynamic resistance of hyaluronic acid: estimates from sedimentation studies. *Biorheology* **23**: 99–113 (1986).

J.H. Fessler. A structural function of mucopolysaccharide in connective tissue. *Biochem. J.* **76**: 124–132 (1960).

H. Fujita. *Mathematical Theory of Sedimentation Analysis*. Academic Press, London. 1962.

I. Gersh and H.R. Catchpole. The nature of the ground substance of connective tissue. *Perspect. Biol. Med.* **3**: 282–219 (1960).

S. Ghosh and W.D. Comper. Oriented fibrillogenesis of collagen *in vitro* by ordered convection. *Connect. Tissue Res.* **17**: 33–41 (1988).

A.J. Grodzinsky. Electromechanical and physicochemical properties of connective tissues. *CRC Crit. Rev. Biomed. Eng.* **9**: 133–199 (1983).

A.J. Grodzinsky, V. Roth, E.R. Myers, W. Grossman, and V.C. Mow. The significance of electromechanical and osmotic forces on the nonequilibrium swelling behaviour of articular cartilage in tension. *J. Biomech. Eng.* **103**: 221–231 (1981).

H.T. Hammel. Forum on osmosis. *Am. J. Physiol.* **237:** R95–R125 (1979).

J. Happel and H. Brenner. *Low Reynolds Number Hydrodynamics*. Prentice-Hall, Englewood Cliffs, NJ, 1983.

G.S. Harper, W.D. Comper, and B.N. Preston. Dissipative structures in proteoglycan solutions. *J. Biol. Chem.* **259**: 10582–10589 (1984).

G.S. Harper and B.N. Preston. Molecular shrinkage of proteoglycans. *J. Biol. Chem.* **262**: 8088–8095 (1987).

V.C. Hascall. Interaction of cartilage proteoglycans with hyaluronic acid. *J. Supramol. Struct.* **7**: 101–120 (1977).

V.C. Hascall and G.K. Hascall. Proteoglycans. In: *Cell Biology of Extracellular Matrix*. Ed E.D. Hay. Plenum Press, New York. pp. 39–65, 1981.

V.C. Hascall and J.H. Kimura. Biosynthesis, excretion and aggregation of proteoglycans by rat chondrosarcoma chondrocytes. *Ala. J. Med. Sci.* **18**: 29–34 (1981).

D. Heinegård and M. Paulsson. Cartilage. *Methods Enzymol.* **145**: 336–363 (1987).

D. Heinegård and Y. Sommarin. Proteoglycans: an Overview. *Methods Enzymol.* **144**: 305–315 (1987).

A. Janoff, G. Feinstein, C.J. Malemud, and J.M. Elias. Degradation of cartilage proteoglycan by human leukocyte granule neutral proteases: a model of joint injury. I. Penetration of enzyme into rabbit articular cartilage and release of $^{35}SO_4$-labelled material from the tissue. *J. Clin. Invest.* **57**: 615–624 (1976).

E.P. Katz, E.J. Wachtel, and A. Maroudas. Extracellular proteoglycans osmotically regulate the molecular packing of collagen in cartilage. *Biochim. Biophys. Acta* **882**: 136–139 (1986).

T.C. Laurent, I. Björk, A. Pietruszkiewicz, and H. Persson. On the interaction between polysaccharides and other macromolecules. II. The transport of globular particles through hyaluronic acid solutions. *Biochim. Biophys. Acta* **78**: 351–359 (1963).

T.C. Laurent and H. Persson. The interaction between polysaccharides and other macromolecules. VII. The effect of various polymers on the sedimentation rates of serum albumin and α-crystallin. *Biochim. Biophys. Acta* **83**: 141–147 (1964).

T.C. Laurent and A.G. Ogston. The interaction between polysaccharides and other macromolecules. IV. The osmotic pressure of mixtures of serum albumin and hyaluronic acid. *Biochem. J.* **89**: 249–253 (1963).

T.C. Laurent, B.N. Preston, W.D. Comper, G. J. Checkley, K. Edsman, and L.O. Sundelof. Kinetics of multicomponent transport by structured flow in polymer solutions. I. Studies on a polyvinylpyrrolidone-dextran system. *J. Phys. Chem.* **87**: 648–654 (1983).

J.R. Levick. Flow through interstitium and other fibrous matrices. *Q. J. Exp. Physiol.* **72**: 409–438 (1987).

A. Maroudas. Fluid transport in cartilage. *Ann. Rheum. Dis.* **34** (Suppl.): 77–81 (1975a).

A. Maroudas. Biophysical chemistry of cartilagenous tissue with special reference to solute and fluid transport. *Biorheology* **12**: 233–248 (1975b).

A. Maroudas. In: *Adult Articular Cartilage*. Ed M.A.R. Freeman. Pitman Medical, Tunbridge Wells, Kent, England. pp. 215–290, 1979.

A. Maroudas. Physical chemistry of articular cartilage and the intervertebral disc. In: *The Joints and Synovial Fluid,* Vol. II. Ed L. Sokoloff. Academic Press, London. pp. 239–291, 1980.

A. Maroudas and C. Bannon. Measurement of swelling pressure in cartilage and comparison with the osmotic pressure of constituent proteoglycans. *Biorheology* **18**: 691–632 (1981).

A. Maroudas and H. Thomas. A simple physicochemical method for determining fixed anionic groups in connective tissue. *Biochim. Biophys. Acta* **215**: 241–218 (1970).

A. Maroudas and M. Venn. Chemical composition and swelling of normal and osteoarthritis femoral head cartilage. *Ann. Rheum. Dis.* **36**: 399–403 (1977).

M.B. Mathews. *Connective Tissue. Macromolecular Structure and Evolution*. Springer Verlag, Berlin. 1975.

R. Mayne and M.H. Irwin. Collagen types in cartilage. In: *Articular Cartilage Biochemistry*. Eds K.E. Kuettner, R., Scheyerback, and V.C. Hascall. Raven Press, New York, pp. 23–35, 1986.

R. Mayne and K. von der Mark. Collagens of cartilage. In: *Cartilage, Structure, Function and Biochemistry*. Vol. 1. Academic Press, New York. pp. 181–214, 1983.

F.A. Meyer, W.D. Comper, and B.N. Preston. Model connective tissue systems. A physical study of gelatin gels containing proteoglycans. *Biopolymers* **10**: 1351–1364 (1971).

K. Meyer. Nature and function of mucopolysaccharides of connective tissue. In: *Molecular Biology, Elementary Processes of Nerve Conduction and Muscle Contraction*. Ed D. Nachmansohn. Academic Press, London. pp. 69–76, 1960.

V.C. Mow, M.H. Holmes, and W.M. Lai. Fluid transport and mechanical properties of articular cartilage: a review. *J. Biomech.* **17**: 377–394 (1984).

V.C. Mow, S.C. Kuei, W.M. Lai, and C.G. Armstrong. Biphasic creep and stress relaxation of articular cartilage: theory and experiments. *J. Biomech. Eng.* **102**: 73–84 (1980).

V.C. Mow, M.F. Mak, W.M. Lai, L.C. Rosenberg, and L.-H. Tang. Viscoelastic properties of proteoglycan subunits and aggregates in varying solution concentrations. *J. Biomech.* **17**: 325–338 (1984).

H. Muir. The chemistry of the ground substance of joint cartilage. In: *The Joints and Synovial Fluid.* Vol. II. Ed L. Sokoloff. Academic Press, London. pp. 27–94, 1980.

J. Nittman, G. Daccord, and H.E. Stanley. Fractal growth of viscous fingers: quantitative characterization of a fluid instability phenomenon. *Nature (London)* **314**: 141–144 (1985).

A.G. Ogston, B.N. Preston, and J.D. Wells. On the transport of compact particles through solutions of chain polymers. *Proc. R. Soc. London Ser. A* **333**: 297–316 (1973).

C.A. Poole, M.H. Flint, and B.W. Beaumont. Chondrons in cartilage: ultrastructural analysis of the pericellular environment in adult human articular cartilages. *J. Orthop. Res.* **5**: 509–522 (1987).

R.H. Pearce and T.C. Laurent. Exclusion of dextrans by meshworks of collagen fibers. *Biochem. J.* **163**: 617–625 (1977).

B.N. Preston, M. Davies, and A.G. Ogston. The composition and physicochemical properties of hyaluronic acids prepared from ox synovial fluid and from a case of mesothelomia. *Biochem. J.* **96**: 449–474 (1965).

B.N. Preston, T.C. Laurent, and W.D. Comper. Transport in polysaccharide solutions. In: *Molecular Biophysics of the Extracellular Matrix.* Eds S. Arnott, D.A. Rees, and E.R. Morris. Humana Press, Clifton, NJ, pp. 119–184, 1984.

B.N. Preston, T.C. Laurent, W.D. Comper, and G.J. Checkley. Rapid polymer transport in concentrated solutions through the formation of ordered structures. *Nature (London)* **287**: 499–503 (1980).

E.Z. Rios and M. Mannik. Charge-charge interactions between articular cartilage and cationic antibodies, antigens, and immune complexes. *Arthritis Rheum.* **30**: 1265–1273 (1987).

J.D. Sandy, H.L.G. Brown, and D.A. Lowther. Degradation of proteoglycan in articular cartilage. *Biochim. Biophys. Acta* **543:** 536–544 (1978).

J. Schalkwijk, W.B. van den Berg, L.B.A. van de Putte, L.A.B. Joosten, and L. van den Bersselaar. Cationization of catalase, peroxidase, and superoxide dismutase. Effect of improved intra-articular retention on experimental arthritis in mice. *J. Clin. Invest.* **76**: 198–205 (1985).

J.E. Scott. Proteoglycan-fibrillar collagen interactions. *Biochem. J.* **252**: 313–323 (1988).

J. Snowden, and A. Maroudas. Distribution of serum albumin in human normal and degenerate cartilage. *Biochim. Biophys. Acta* **428**: 726–740 (1976).

L. Sokoloff. Elasticity of articular cartilage: effect of ions and viscous solutions. *Science* **141**: 1055-1056 (1963).

N. Sorgente, V.C. Hascall, and K.E. Kuettner. Extractability of lysozyme from bovine nasal cartilage. *Biochim. Biophys. Acta* **284:** 441–450 (1972).

K.S. Spiegler. Transport processes in ionic membranes. *Trans. Faraday Soc.* **54**: 1408–1426 (1958).

R.A. Stockwell. *Biology of Cartilage Cells.* Cambridge University Press, London. 1979.

T. Tanaka and D. J. Fillmore. Kinetics of swelling of gels. *J. Chem. Phys.* **70**: 1214–1218 (1979).

C. Tanford. *Physical Chemistry of Macromolecules.* John Wiley & Sons, New York. 1961.

E.J.-M.A. Thonar, S.B. Feist, K. Fassbender, M.E. Lenz, B.L. Matijevitch, and K.E. Kuettner. Quantification of hen egg white lysozyme in cartilage by an enzyme-linked immunosorbent assay. *Connect. Tissue Res.* **17**: 181–198 (1988).

D.A. Torchia, M.A. Hasson, and V.C. Hascall. Investigation of molecular motion of proteoglycans in cartilage by ^{13}C magnetic resonance. *J. Biol. Chem.* **252**: 3617–3625 (1977).

J.P.G. Urban, A. Maroudas, M.T. Bayliss, and J. Dillon. Swelling pressures of proteoglycans at the concentrations found in cartilagenous tissues. *Biorheology* **16**: 447–464 (1979).

M.-P. van Damme, W.D. Comper, and B.N. Preston. Experimental measurements of polymer unidirectional fluxes in polymer and solvent systems with non-zero chemical-potential gradients. *J. Chem. Soc. Faraday Trans. 1* **78**: 3357–3367 (1982).

W.B. van den Berg and L.B.A. van de Putte. Electrical charge of the antigen determines its localization in the mouse knee joint. *Am. J. Pathol.* **121**: 224–234 (1985).

W.B. van den Berg, P.L.E.M. van Lent, L.B.A. van de Putte, and W.A. Zwarts. Electrical charge of hyaline articular cartilage: its role in the retention of anionic and cationic proteins. *Clin. Immunol. Immunopathol.* **39**: 187–197 (1986).

K.-O. Wik. *Physicochemical Studies on Hyaluronate*. Thesis. Uppsala University, Uppsala, Sweden (1979).

K.-O. Wik and W.D. Comper. Hyaluronate diffusion in semidilute solutions. *Biopolymers* **21**: 583–599 (1982).

R.P.W. Williams and W.D. Comper. Osmotic flow caused by nonideal marcomolecular solutes. *J. Phys. Chem.* **91**: 3443–3448 (1987).

R.P.W. Williams and W.D. Comper. Osmotic flow caused by polyelectrolytes. *Biophys. Chem.* **36**: 223–234 (1990).

O. Zamparo and W.D. Comper. Hydraulic conductivity of chondroitin sulfate proteoglycan solutions. *Arch. Biochem. Biophys.* **274**: 259–269 (1989).

O. Zamparo and W.D. Comper. Model anionic polysaccharide matrices exhibit lower charge selectivity than is normally associated with kidney ultrafiltration. *Biophys. Chem.* **38**: 167–178 (1990).

3

Energy Metabolism in Cartilage

IRVING M. SHAPIRO, TOSHIKAZU TOKUOKA,* AND SUSAN F. SILVERTON
Departments of Biochemistry and Oral Medicine
School of Dental Medicine
University of Pennsylvania
Philadelphia, Pennsylvania

Introduction: Mechanisms for Energy Transduction in Tissues

The objective of this chapter is to examine mechanisms of chondrocyte energy generation and utilization and, where possible, to relate tissue energy status to cartilage function. As other chapters of the book attest, cartilage serves a number of eclectic functions in the organism. For example, in embryonic long bone, cartilage maintains the spatial integrity of the mineralizing tissue, while in the epiphysis, the cartilaginous plate provides a mechanism for rapid bone growth. Thus, the function of the tissue varies according to its anatomical location and the developmental status of the organism.

In this review, we will examine energy metabolism using tissue paradigms that include mature articular cartilage and the mineralizing growth plate. In both of these systems, considerable attention will be directed at delineating the relationship between energy metabolism and the vascular supply. Knowledge of the vascular system is important, not just as an indicator of the tissue oxygen tension but as a major determinant of mitochondrial function. To provide a background to these relationships, an overview of mechanisms of metabolic energy production is presented and concepts that are used in subsequent portions of the chapter are defined.

It is possible to define the energy status of any cell or system using rigorous thermodynamic criteria ($-\Delta G$). However, it is frequently more convenient to describe the metabolic status in terms of the synthesis or degradation of high energy compounds. In cartilage, the most important compounds are adenosine triphosphate (ATP), the reduced form of nicotinamide adenine dinucleotide (NADH), and the dephosphorylated derivatives of ATP (adenosine diphosphate, ADP, and adenosine monophosphate, AMP). From a metabolic viewpoint, irrespective of their anatomical location, chondrocytes maintain their phenotypic state in part, by generating adequate levels of these compounds.

* Dr. Tokuoka is currently on leave from Showa University, Japan.

The major metabolic pathways for energy production by chondrocytes are similar to those of other tissues. Most cells can oxidize simple sugars, amino acids, and long chain fatty acids to generate metabolic energy. Cells release and conserve much of the chemical energy of these nutrients by a complex series of exquisitely controlled reactions in the cytosolic and mitochondrial compartments of the cell. Although ATP can be generated in both compartments, in many tissues, much of the ATP is produced by mitochondrial oxidation of metabolic intermediates. The mechanism of energy production is tightly regulated and closely linked to energy expenditure.

In the cytosol, large multienzyme complexes of the glycolytic pathway generate some ATP directly from ADP. When oxygen is limiting, then hexoses are degraded to lactate (in this case, the metabolic pathway is called anerobic glycolysis); if the oxygen tension is adequate, then pyruvate is formed. This end product of glycolysis is utilized by mitochondrial enzymes of the tricarboxylic acid (TCA) cycle to generate NADH, the reduced flavoprotein of succinate dehydrogenase ($FADH_2$), and other reduced coenzymes. In most cells, cytosolic NADH cannot traverse the mitochondrial membrane, and special shuttle mechanisms have evolved to transfer the reducing power associated with NAD to the mitochondrion.

Mitochondria are complex organelles that oxidize a wide range of metabolic intermediates. In a series of elegant studies, Johnson and colleagues (1981) showed that in living cells, these organelles had a characteristic appearance that was dependent on the tissue type and the oxidative state of the cell. Multienzyme complexes located in both the inner mitochondrial membrane and the mitochondrial matrix oxidize TCA cycle intermediates derived from primary nutrients. As these compounds undergo progressive dehydrogenation by mitochondrial enzymes (malate-, oxaloacetate-, pyruvate-, isocitrate-, α-ketoglutarate, and succinate dehydrogenases), the coenzymes NAD, FAD, and possibly NADP are reduced to form energy-rich molecules (NADH, $FADH_2$, and NADPH). It is important to note that NAD and NADH also serve as important components in polyadenylation reactions in the cell; these will be discussed in some detail later. Transfer of electrons from these molecules and formation of the corresponding oxidized substrate provides the proton-motive force for the subsequent synthesis of ATP.

Although some details of ATP generation are still obscure, it is now believed that ATP is generated by the activity of an electrogenic proton pump that spans the inner mitochondrial membrane (Mitchell, 1979). The energy for synthesis of ATP is derived from two components: a pH gradient and the electric potential across the mitochondrial membrane. The pump becomes activated during oxidative metabolism by electrons flowing through the membrane and the concomitant transport of protons. As the inner membrane is only selectively permeable to protons, an asymmetrical gradient is developed across the mitochondrial membrane. This energy gradient can then be utilized to drive reactions that result in the phosphorylation of ADP (as well as other energy requiring processes). The transfer of energy through a series of linked electron and proton acceptor molecules (the respiratory chain) is amazingly efficient. Greater than 70% of the energy available as NADH is conserved as ATP. Protons that are utilized by the system are eventually consumed by combining with oxygen.

The respiratory chain consists of three protein complexes: NADH-coenzyme Q reductase (also called NADH dehydrogenase), cytochrome reductase, and cytochrome oxidase. These proteins are linked together by ubiquinone and a protein (cytochrome c). Dehydrogenation of NADH-linked intermediates, such as isocitrate or malate, provide protons that reduce all of the components of the respiratory chain and generate three molecules of ATP. Electrons gained by the oxidation of $FADH_2$-linked substrates (by such enzymes as succinate-, glycerol phosphate-, and acetylCoA-dehydrogenases) enter the respiratory chain at the level of ubiquinone and thereby bypass the first of the multienzyme complexes. In the latter case only two molecules of ATP are generated.

Generation of ATP in the mitochondrion is linked to cell energy demands. Events that occur at the level of the electron transport system are controlled by the adenine nucleotide concentration (ATP, ADP, and AMP) as well as by the availability of molecular oxygen. Limitation in the molecular oxygen tension, dissolution of the proton motive force for non-ATP synthesizing processes, and the presence of ionophores that can dissipate the ion gradient, all serve to reduce the rate of ATP synthesis.

The energy state of the cell can be defined in terms of its adenine nucleotide content: the ATP/ADP ratio; the phosphorylation potential (ATP/ADP:Pi); the energy charge ratio (ATP + 0.5ADP/ATP + ADP + AMP). While the adenine nucleotide ratio is higher in the cytosol than in the mitochondrion, the values for both compartments are maintained by electrogenic exchange of ATP for ADP—a process which is linked to the mitochondrial membrane potential. Another important guide to the metabolic state is the ratio of NAD/NADH (also called the reduction/oxidation ratio or redox status). A low ratio is indicative of a cell in a reduced state. Finally, the ratio of NADP/NADPH provides a measure of nonmitochondrial oxidative activity in the cell. Generation of NADPH provides a second measure of the "reductive reserve" of a tissue as well as indicating the activity of the pentose phosphate shunt pathway.

Nutrient Requirements and Energy Generation by Cartilage

Surprisingly, detailed information about the metabolic fuels required by cartilage is scanty. This paucity of information is in stark contrast to current knowledge of metabolic needs of most of the tissues of the body. Information collected during the past 30 years has documented the nutrient requirements of the major tissues, such as heart, brain, liver, and kidney. Moreover, for these and many other tissues, detailed information exists on the regulation of energy metabolism in a variety of physiological and disease states.

Based on what is known in other systems, and the results of studies that have been performed using intact tissues and isolated cells, we may assume that cartilage cells can utilize a number of different substrates: the most important of these are carbohydrates and amino acids. Glucose is the major metabolic fuel of most types of connective tissue cells (Borle *et al.*, 1960a, b; Picard and Cartier, 1960; Cohn and Forscher, 1962; Flanagan and Nichols, 1964; Neuman *et al.*, 1978). The results of a considerable number of other experiments indicate that while the TCA cycle is active in cartilage, as much as 80% of glucose is metabolized to lactate by anerobic glycolysis. Moreover, hexose metabolism is regulated by the phosphate, vitamin D, and cortisone status of the organism (Kunin and Krane, 1965; Balogh and Kunin, 1968; Meyer and Kunin, 1969, 1973). Studies of cells in culture also indicate that isolated chondrocytes are dependent on anerobic glycolysis. In this respect, chondrocytes are similar to other cell types. Indeed, results of many investigations clearly document that cultured cells generate much of their ATP through glycolysis and produce lactate as an end-product.

Evidence that cartilage cells can utilize oxidative metabolism to produce chemical energy is supported by the observation that chondrocytes contain mitochondrial dehydrogenases; that isolated mitochondria can utilize TCA cycle substrates (these are discussed in greater detail later); and that the mitochondria contain enzymes required for oxidative phosphorylation and electron transport (Balogh and Kunin, 1968, Meyer and Kunin, 1969, 1973; Lee and Shapiro, 1974, and Yamamoto and Gay, 1988). A study of mitochondrial enzymes of avian cartilage suggests that there is high NADP-linked dehydrogenase activity. This observation is of particular interest, as Hermann-Erlee (1966) has previously shown that the synthesis of NADP is regulated by parathyroid hormone. If this is the case, an interesting link may exist between the effects of parathyroid hormone on mineralized connective tissue and oxidative metabolism. This observation is worthy of further investigation.

There is substantial evidence to indicate that chondrocytes can degrade glucose by the NADP-dependent pentose phosphate shunt pathway. Cohn and Forscher (1962) reported that glucose labeled in the 1-C position can be metabolized by this route, while Dixit (1972) noted that the activity of glucose-6-phosphate dehydrogenase was dependent on the vitamin D status of the organism. A considerable number of workers have shown the presence in cartilage of enzymes required for pentose phosphate shunt function (see Stockwell, 1983; Sampson and Cannon, 1986; Silverton *et al.*, 1989b). The relationship between shunt activity, glycolysis, and insulin status was measured by Hough *et al.* (1983). These workers demonstrated that in the rat, about 15% of glucose catabolism is channeled through the pentose phosphate pathway. Interestingly, while the imposition of a diabetic state impairs glycolysis, the pentose phosphate shunt activity is increased. Moreover, insulin treatment fails to restore normal shunt activity, although it does normalize glycolysis. The authors of this study consider that overutilization of glucose by the shunt is similar to the response seen in other tissues where insulin is not required for glucose uptake (Gonzalez *et al.*, 1978; Socher *et al.*, 1979). It is unfortunate that the mechanism of glucose uptake by chondrocytes has not been explored.

Studies with isolated cells and slices indicate that cartilage requires amino acids for growth and as an energy source. This observation is in line with studies of other tissues where it has been shown that glutamine, leucine, and isoleucine are rapidly taken up by a wide variety of cell types; other amino acids that are metabolized include valine, methionine, cysteine, threonine, tyrosine, and phenylalanine (Eagle *et al.*, 1956; Eagle, 1959). In a study of the amino acid metabolism of cultured epiphyseal chondrocytes, Ishikawa *et al.*, (1985) demonstrated that these cells, in both short- and long-term culture, consumed glutamine and to a lesser extent threonine, leucine, isoleucine, serine, and tryptophan. In long-term culture, there was some change in the pattern of amino acid utilization and a greater dependency on cysteine, tryptophan, and glycine. Chondrocytes appear to have a low requirement for basic amino acids. Indeed, arginine is the only basic amino acid that is continuously consumed by cells in long-term culture.

The amino acid glutamine is a major source of energy for connective tissue cells, and it cannot be replaced by carbohydrates (Reitzer *et al.*, 1979). That glutamine is required in large amounts by cells is not really surprising in view of the very high serum concentration of this amino acid. In studies of glutamine metabolism by mammalian fibroblasts as well as by transformed cells, it was shown that this amino acid can provide from 30 to 80% of the cellular energy needs (Donnelly and Scheffler, 1976; Zielke *et al.*, 1976; Reitzer *et al.*, 1979). Studies by Zielke *et al.*, (1984), using human fibroblasts, showed that glutamine was oxidized 50 times faster than other substrates.

The major pathways for glutamine metabolism are shown in Figure 1. Glutamine is required by cartilage cells as an amino donor for the synthesis of glycosoaminoglycans. Thus, glutamine stimulates glycan synthesis by costal cartilage (Handley *et al.*, 1980). In its absence, the levels of both N-acetylglucosamine 6-phosphate and UDP-N-acetylhexosamines are decreased. Oppenheimer *et al.*, (1969) has shown that glutamine is required for the synthesis of hexosamine nucleotides for glycoproteins. Unlike liver and kidney, cartilage appears to lack, or have low levels of glutamine synthetase, and hence the amino acid glutamate cannot serve as a substrate for glutamine synthesis. Handley *et al.* (1980) reported that they could not measure glutamine synthetase activity in mature avian or mammalian articular cartilage, while Biggers *et al.* (1961) were unable to show enzyme activity in embryonic cartilage. Thus, for all intents and purposes, glutamine is an essential amino acid for cartilage cells.

Glutamine can be deamidated to glutamate and then to α-ketoglutarate

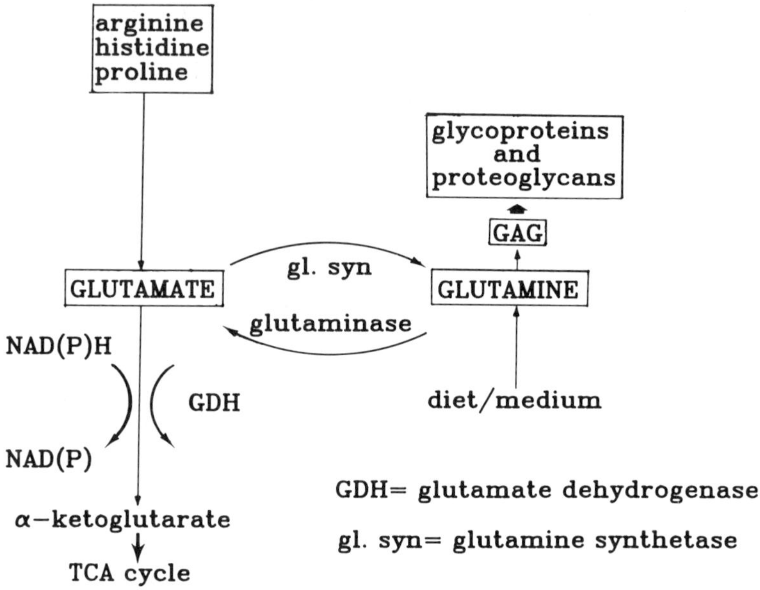

Figure 1. Metabolism of glutamine and glutamate in cartilage.

(Figure 1). Further oxidative decarboxylation by the enzymes of the TCA cycle would sequentially degrade α-ketoglutarate to CO_2. In terms of energetics, two molecules of ATP are produced for each molecule of glutarate oxidized (by experiment, about 11% of cell lactate is probably derived from this pathway). The interaction between glucose and glutamine has received recent study. Although glucose serves as primary fuel for glycolysis while glutamine serves as a major substrate for oxidative metabolism, glucose has a glutamine sparing effect. Zeilke *et al.* (1984) showed that the rate of glutamine degradation is dependent on the glucose concentration of the external medium. Thus, if the glucose level is high, the rate of utilization of glutamine is low. These workers noted that the apparent absolute need for glucose or other carbohydrates is more related to the role of these compounds in anabolic reactions than as sources of metabolic energy.

Very few studies have been performed to determine the value of fatty acids for energy generation by cartilage slices or cultured cells. One investigation of the mode of action of diphosphonates offers some limited indirect evidence that articular chondrocytes require fatty acids as a metabolic fuel (Guenther *et al.*, 1979). Also, cytochemical studies by Dunham *et al.* (1986) indicate that articular chondrocytes contain low, but significant, levels of hydroxyacyl dehydrogenase. Thus, it is likely that chondrocytes can metabolize fatty acids as an energy source; however, the importance of these compounds in the overall energy economy of the tissue has not been determined.

Oxygen Tension and Energy Metabolism

In all tissues, oxygen serves as the final electron acceptor for mitochondrial cytochrome oxidase. In addition, oxygen serves as a substrate for a very large number of other enzymes (oxygenases and oxidases). Tissue oxygen tensions vary and the actual value is dependent on a number of factors. These include the number of cells per unit volume and the rate of cellular metabolism, oxygen diffusibility through the extracellular matrix, and the vascular supply.

In most soft tissues, the vascular supply is sufficient to provide the cells with oxygen in excess of that required for normal cellular energy metabolism. Nevertheless, within a single complex tissue, oxygen gradients have been shown to exist. In liver, for example, the intercellular oxygen gradient along a sinusoid can vary by two orders of magnitude. As far as subcellular organelles are concerned, Jones et al. (1985) report that there are differences in the oxygen utilizing systems of the individual subcellular compartments. Studies with isolated mitochondria indicate that the K_M for oxygen is very low, about 1 mmHg or 1 μM.[1] Using liver cells DeGroot and Noll (1987) have shown that oxygen gradients of up to 7 mm Hg exist toward the mitochondrial compartment and, as might be expected, become maximal when mitochondria polarize toward one side of the cell.

Before considering whether changes in oxygen tension can modulate selective aspects of the chondrogenic response or chondrocyte metabolism, it is necessary to consider how oxygen deprivation can affect energy metabolism. In general, cells become very sensitive to the oxygen tension when it falls below a specific threshold value. When deprivation occurs, there is decreased oxygen uptake and an associated inhibition of energy metabolism. For example, DeGroot and Noll (1987) showed that for hepatocytes, the rate of oxygen consumption is constant between 5 and 100 mmHg. Below 5 mm Hg, the rate decreases rapidly, the energy charge is lowered, there is an increase in the NADH/NAD ratio and a change in the reductive state of cytochromes c and aa_3. The results of these experiments suggest that a cellular mechanism exists which regulates oxidative metabolism in response to the local oxygen tension.

Most mammalian tissues can tolerate periods of hypoxia despite major changes in the concentration of adenine and pyridine nucleotides. During these episodes of ''neahypoxia'', cells maintain mitochondrial pH and membrane potential. Metabolically, there is stimulation of glycolysis, increased lactate formation, and a decrease in the activity of Na^+/K^+ ATPase. Clearly, while the functional state of the tissue is limited, activities essential for cell survival are maintained.

[1] Oxygen vapor pressure is expressed as millimeters of mercury, while oxygen concentration is given as micromolars. At low oxygen tensions, i.e., ≤ 20 mmHg, the numerical values are nearly equal, e.g., 20 mmHg is approximately 20 μM.

Hochachka (1986) has proposed that cells adapt to oxidative stress by undergoing "metabolic arrest". When arrest occurs, a decrease in both oxidative and nonoxidative metabolism can serve to preserve cellular function during the hypoxic interlude. If maintained for any excessive period, then "metabolic death" would ensue, probably due to failure of the mitochondrion to synthesize ATP at a rate consistent with the minimum needs of the cell (Chance *et al.,* 1985). In this case, following collapse of the mitochondrial membrane potential, there would be an increase in cytosolic calcium level. The cytosol would receive calcium from the mitochondrion (by sodium-calcium exchange), the endoplasmic reticulum, and the plasma membrane. The mechanism of calcium release from the endoplasmic reticulum is the subject of intense scrutiny, and there is evidence to indicate that this is achieved by the activity of inositol phosphates as well as other compounds derived from membrane phospholipids (Berridge and Irvine, 1989).

Oxygen Supply to the Growth Cartilage

The epiphysis is a very complex structure containing nonmineralized cartilage, calcified cartilage, and endochondral bone. The blood supply to this structure varies from species to species. In general, tissue in the premineralized regions of the growth plate is almost avascular. In contrast, tissue in the mineralizing region contains a large number of blood vessels.

The vasculature of the mammalian growth plate has been extensively reviewed by Trueta and colleagues (1953, 1960). These workers drew attention to the role of metaphyseal vessels for normal mineral deposition in bone and cartilage. The published studies questioned whether the metaphyseal vessels ended as regular capillaries or irregular sinusoids, and whether the vessels remained patent during endochondral mineralization. In a recent study, Arsenault (1987) examined the microvasculature of the epiphyseal-metaphyseal junction of the rat. He showed that the venous return occurs below the epiphysis and that the vascular tree terminated as capillary sprouts that are bordered by mineralized cartilage.

Despite the concentration of blood capillaries in the metaphyseal region, Arsenault considers that this region is hypoxic. He comments, "The red blood cells seen clustered at the tip of the sprout may represent blood stasis and indicate that these vessels are not yet circulatory."

Howlett and colleagues (1979, 1980, 1984) described the blood supply to the chick epiphysis in some detail. These workers reported that in premineralized cartilage, epiphyseal vessels, although few in number, provide oxygen and nutrients to cells in the resting and proliferating regions. A second group of vessels originating in the metaphysis, supply blood to calcified cartilage and bone. These vessels also provide oxygen to cells in the hypertrophic region and some may anastomose with the epiphyseal vessels. The metaphyseal structures (also called vascular canals and channels) are complex and have been shown to contain a number of arterioles and venules. At the periphery of the canal, perivascular cells that are morphologically and functionally distinct from neighboring hypertrophic chondrocytes have been described—these are discussed in more detail in the last section of the chapter. Similar types of canals have been described in other species (Haines, 1934; Levene, 1964; Wilsman and Van Sickle, 1972). In humans, cartilage canals contain a muscular arteriole, a venule, and capillaries. Interestingly, the blood capillaries are extremely thin and fenestrated—features that would facilitate nutrient delivery (Rodriquez, 1985). It should also be mentioned that aside from the more formalized mechanism for delivering oxygen to epiphyseal chondrocytes through vascular channels, there is some evidence of a microcirculation in the epiphysis. Boyde and Shapiro (1987) reported spaces or "fenestrae" in the matrix surrounding chondrocytes in all regions of the growth cartilage. The fenestrae would promote transport of oxygen from the blood vessels to each of the chondrocytes.

The oxygen tension in the growth plate of the rabbit and in a healing bone defect was examined by Brighton and Heppenstall (1971) using a glass electrode system. It was reported that the oxygen tension in these tissues is related to cell maturation, status of repair, and degree of vascularization. In the developing epiphysis, the highest oxygen tension is in the proliferative region where values as high as 50 mmHg have been recorded. In contrast, cells in the resting and hypertrophic region could be considered to be hypoxic, as the oxygen tension was reduced to 25 mmHg. Similarly, in fracture callus during the chondrogenic stage of repair, the oxygen tensions can decrease to 20 to 30 mmHg. When bone was present in the callus, the recorded oxygen tension increased to 40 to 50 mmHg (Heppenstall *et al.*, 1975).

Oxygen Tension in Articular Cartilage

In contrast to the epiphyseal plate, cells in articular cartilage have a very limited blood supply. Hence, it is likely that articular chondrocytes are exposed to low oxygen tensions. Although a few sparse vessels actually penetrate the cartilage, the major source of nutrients is from the synovium on one side and nutrient bone vessels on the other (Greenwald and Hayes, 1979). Thus, many chondrocytes in articular cartilage may receive oxygen and nutrients only by diffusion from the tissue fluid or from the synovium.

Articular cartilage has relatively few cells per unit volume of tissue and a low vessel density (Stockwell, 1967). It has been estimated that cells that lie deep within human articular cartilage may be 1 to 3 mm from a nutrient artery. As far as the actual oxygen tension is concerned, it was estimated by Falchuk *et al.* (1970), Lund-Oleson (1970), and Treuhaft and McCarty (1971) that at the surface of articular cartilage the oxygen tension is between 20 to 80 mmHg. Moreover, because of the asymmetry of the oxygen supply, a gradient must exist across the tissue. As a result of the gradient, chondrocytes in the layers of tissue adjacent to the synovium are probably exposed to a normoxic tension. In contrast, cells in the deepest regions of the cartilage may be exposed to a low oxygen tension. Commenting on the metabolic state of articular chondrocytes, Marcus (1973) suggests that the low rate of aerobic respiration of cells in the cartilage may be an adaptive response to the low oxygen tension.

There is some evidence to indicate that applied pressure forces tissue fluid to permeate the hydrated extracellular matrix. Poole *et al.* (1987) examined the microstructure of human articular cartilage and described a system of macro- and micropores in the "felt-like" organization of the capsular weave. These workers considered that expression of water from the pericellular matrix could generate microcirculatory currents that would facilitate transport of oxygen and nutrients to the articular chondrocytes. If cells in the tissue are dependent on this type of microcirculatory system, then it is possible that many of the cells in articular cartilage may well exist in an hypoxic environment. However, it should be remembered that this tissue contains very few cells per unit volume and that in comparison with most other tissues of the body, articular chondrocytes are nominally active. Thus, their oxygen needs are probably quite low.

There have been few measurements of oxygen consumption by chondrocytes isolated from articular cartilage. Studies performed with epiphyseal chondrocytes indicate that the cells have a relatively low rate of oxygen consumption (1.5 to 2 nmol/min/mg protein) and that the consumption rate is dependent on the medium oxygen tension (Silverton *et al.,* 1989a). In the latter study, it was found that when the oxygen tension was below 20 μM, a decrease in the rate of oxygen uptake was seen. In articular cartilage, it is likely that the rate of oxidative metabolism would be adapted to the limited oxygen availability.

There appears to be little doubt that one effect of aging is that there is occlusion of vascular canals that extend from bone to articular cartilage. This occurrence is likely to accentuate the presumed hypoxia of cartilage cells, especially those in the deepest aspects of the tissue. Occlusion probably changes the chemical composition of the cartilage matrix. Experimentally, it has been found that under conditions of oxygen lack there is increased synthesis of keratan sulfate and decreased synthesis of chondroitin sulfates in cartilage (Stockwell, 1970; Scott and Haigh, 1988). Aw and Jones (1984) demonstrated a reduction of chondroitin sulfate formation due to a decrease in the rate of formation of UDP glucuronate from UDP glucose and NAD. This effect is probably due to hypoxia-related lowering in the level of NAD and an increase in the concentration of NADH. Also there is some evidence that NADH inhibits UDP-glucose dehydrogenase.

Oxygen Tension and Growth of Isolated Chondrocytes

A large number of experiments have been conducted to examine the effects of oxygen tension on cultured chondrocytes. In many of these studies, chondrogenesis and mineralization has been linked to oxygen diffusion gradients and the development of localized microenvironments. In micromass cell and tissue culture, the development of cartilage nodules (<1 mm diameter) can be considered as ecological niches where oxygen, lactate, proton, and nutrient gradients would be present. These gradients may serve to produce local microenvironments which favor mineralization and thus may be used *in vivo*. More recent studies using a variety of cells grown on microspheres have shown that very steep metabolite gradients exist. These gradients have been linked with synthesis of specific proteins and glycosoaminoglycans, changes in cell redox status, and increased dependence on nonmitochondrial oxidative metabolism. The importance of each of these three interrelated systems is discussed elsewhere in the chapter.

Oxygen Supply to Cartilage in the Developing Bone

The very close relationship between bone formation and tissue vascularization has been described in considerable detail by Fell (1925) and more recently by Pechak *et al.* (1986a, b). In the latter study of the tibial diaphysis, cells in the central cartilage core of the developing limb bud were examined. It was shown that these cartilage cells remain at a distance from the vascular supply until marrow development takes place. The result lends credence to the idea that chondrocyte hypertrophy requires an hypoxic environment. This observation begs the following question in long bone development: is chondrocyte hypertrophy a manifestation of lowered oxygen tension? A similar question can be asked of hypertrophic chondrocytes in the growth cartilage (see discussion below).

Pyridine Nucleotides and Chondrogenesis

In view of the importance of pyridine nucleotides in almost every aspect of cell function, it is not surprising that these compounds and their derivatives have been associated with both chondrogenesis and calcification. A brief overview of the metabolism of ribose nucleotide derivatives of nicotinamide and nicotinate is provided in Figure 2. In this section, the relationship between the vascular system and the local tissue concentrations of NAD will be discussed. Subsequently, NAD and nicotinamide requirements for both cell proliferation and chondrogenesis will be reviewed. A later section will accent the importance of the redox ratio in controlling chondrocyte activity. It is important to note that NAD can be utilized for both redox- and nonredox-related systems. (Pentose phosphate shunt utilization of NADP has already been mentioned.) The relationship between NAD/NADH, gluconeogenesis and phosphate transport is outside the scope of this review (see Hammerman *et al.*, 1983).

Caplan viewed the differential development of the vascular system as a means of providing cells of the developing embryo with chemical cues. In this way, the vascular system could promote accumulation of metabolites such

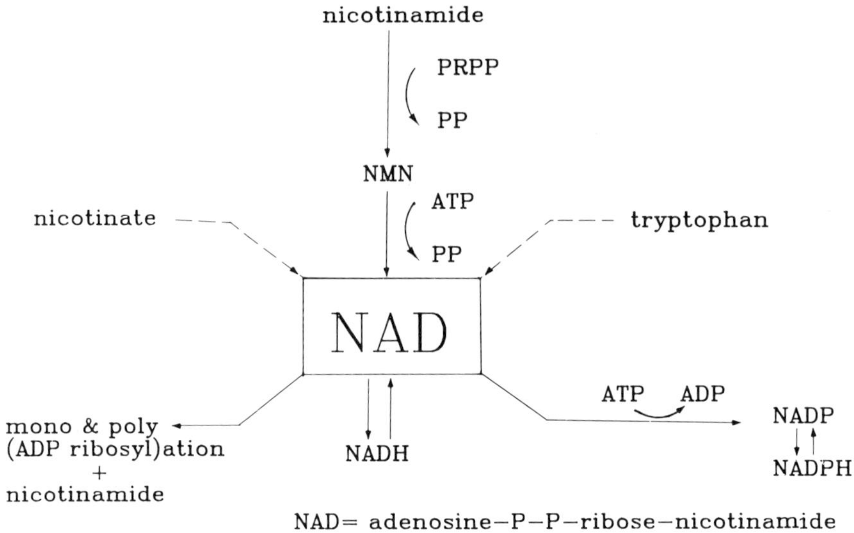

Figure 2. Major pathways for the biosynthesis of NAD and related compounds.

as NAD at selected sites and thereby direct myo-, chondro-, and osteogenesis. It was hypothesized that the local oxygen tension would influence cell NAD concentrations, leading to low NAD and high NADH levels in highly vascularized zones. Indeed, Caplan stated that control of cellular changes during differentiation is mediated by the intracellular concentration of NAD (the "NAD hypothesis"). Support for this hypothesis was found in studies of limb bud cultures where it was found that when the cellular NAD levels were high, there was inhibition of cartilage formation. In those regions where NAD concentration was low, there was stimulation of chondrogenesis (Caplan and Rosenberg, 1975). Caplan and Koutroupas (1973) were able to relate the distribution of NAD to regional vascularity. They showed that the NAD content was low in the relatively avascular prechondrogenic central core region of the developing limb bud. In contrast, in the more superficial vascularized regions of the bud, the NAD levels were high. Moreover, NAD antagonists caused inhibition of normal chondrogenesis, with abnormal limb development and mineralization defects (Caplan, 1972; Misawa *et al.*, 1981). Thus, an attractive theory for cartilage development was propounded linking vascular supply and, by inference, nutrient flow and oxygen tension to the energetic state of the tissue and the expression of the chondrocyte phenotype.

Follow-up experiments using cells grown in a chemically defined medium could not confirm that there existed a direct link between NAD levels and chondrogenesis (Midura *et al.*, 1985). However, these workers noted a transient increase in ADP ribosylation activity during maximal cell proliferation.

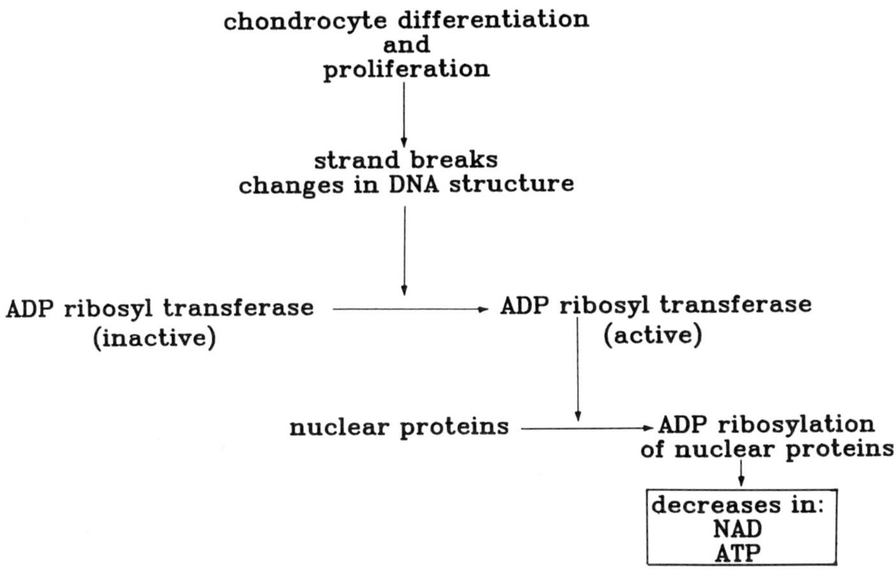

Figure 3. The effect of ADP ribosylation on adenine nucleotides in cartilage.

Also, the cultures became very sensitive to compounds such as nicotinamide or 3 amino-benzamide. As a result, the NAD hypothesis was modified to read "cellular nicotinamide levels directly control the activity of poly(ADP ribose) synthetase activity which itself must be associated with chromatin transitions involved in the differentiation phenomena" (Caplan, 1985).

Measurement of poly(ADP ribose) synthetase and ADP ribosylated protein levels in chondrogenic cultures and during limb bud development support the modified hypothesis. Cherney *et al.* (1985) reported a sharp increase in synthetase activity at the onset of the initial expression of the cartilage or muscle phenotype. Based on the earlier observations of Juarea-Salinas *et al.* (1979), Durkacz *et al.* (1980), and Shall (1982), Cherney *et al.* (1985) theorized that this activity is due to an increase in excision repair processes during cellular differentiation. In their words, the enzyme may act in "a fail-safe system as a sentinel to monitor DNA integrity during the dynamic structural transitions associated with terminal cell differentiation." They commented that if there was extensive DNA damage, the activated synthetase would deplete the NAD pool, thereby promoting a "suicide" process. In this way, cells that are chromosomally defective can be eliminated from the developing tissue.

The investigation also reexamined the NAD content of myogenic and chondrogenic regions of the limb bud to ascertain whether the reported differences in NAD content are due to site-specific changes in synthetase levels. However, only modest differences in NAD levels were observed between sites. The modified NAD hypothesis was further tested using synthetase inhibitors. Unfortunately, the results of these experiments are somewhat confusing. For example, Nishio *et al.* (1983) reported that synthetase inhibitors enhanced chondrocyte differentiation. In contrast, Midura *et al.* (1985) noted that the inhibitors reduced chondrogenesis. Reasons for these apparent conflicts may relate to the type of inhibitor used, the concentration of inhibitor, and differences in the effects of the inhibitor on cell proliferation and cytodifferentiation.

How the ribosylation process modifies cellular function has been the subject of an extraordinary amount of work. It is now clear that the NAD-dependent ADP ribosylation reaction provides a major mechanism for posttranslational control of protein synthesis (for a review, see Ueda and Hayaishi, 1985). Moreover, during differentiation, proliferation, or transformation, nuclear poly(ADP-ribosyl)ation activity is increased. Following DNA damage, many cells exhibit an increase in ADP-ribosylation and a concomitant drop in NAD levels (Jump *et al.*, 1979; Smulson *et al.*, 1975). The activated transferase cleaves NAD at the N-glycosidic bond and covalently attaches single or multiple ADP-ribose groups to the acceptor protein.

ADP-ribose transfer can affect a number of cell functions. The first involves repair of DNA strand breaks. There is evidence that ribosylation changes the affinity of the macromolecular structure of chromatin to enzymes which repair DNA (Poirer *et al.*, 1982). Another function is the modulation of the activity of DNA-binding enzymes such as ligases, topoisomerases, and endonucleases (see Ueda and Hayaishi, 1985). One final activity of the transferase concerns regulation of protein synthesis. Iglewski *et al.* (1984) has shown that the cytosolic transferase inhibits the activity of protein elongation initiation factor 2 (EIF-2). The relationship between this factor, protein synthesis, and redox status is discussed towards the end of the chapter.

Although the studies described above suggest that regional variations in NAD-dependent ribosylation activity modulate cytodifferentiation, the importance of the local blood flow and oxygen tension in chondrogenesis remains undefined. Very recently, Hwang *et al.* (1988) investigated the linkage between oxygen tension, chondrogenesis, and tissue NAD levels. Using a micromass culture technique, these workers showed that high oxygen tensions had little effect on the NAD content of the cells. They concluded that factors other than oxygen tension drove chondrogenesis. Therefore, we find that little experimental support remains for the original argument that temporal or spatial changes in vascularity provide local cues for chondrocyte differentiation.

The relationship between ADP ribosylation and calcium transport by chondrocytes has not been extensively studied. However, investigations conducted with other cell types indicate that ribosylation can be regulated by the extracellular calcium concentration (Richter *et al.*, 1983). Gomperts *et al.* (1983) provided evidence that the ribosylation of a G-binding protein was associated with calcium entry into the cell. In a study reported by Duncan *et al.* (1988), using human fibroblasts, it was shown that one effect of lowering the medium

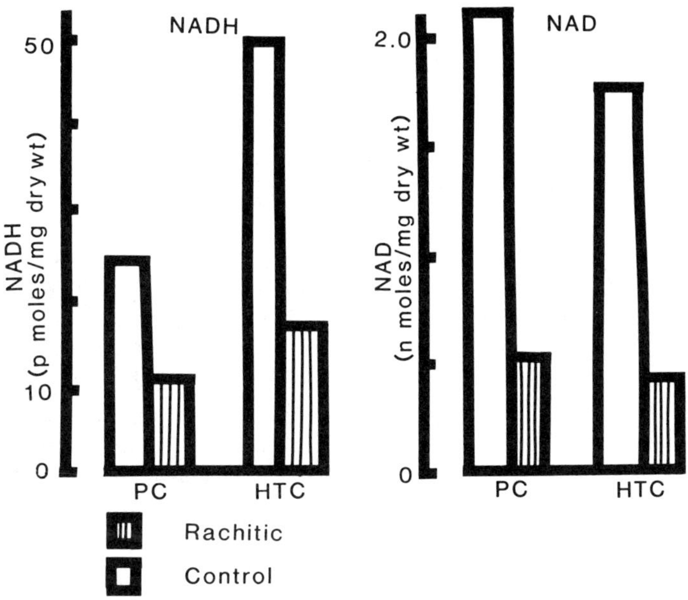

Figure 4. Mean values of NAD and NADH in normal and rachitic epiphyseal cartilage. Values shown are means for the proliferating (PC) and hypertrophic (HTC) zones.

calcium concentration was the initiation of a rapid efflux of cellular NAD. Inhibition of NAD loss by 3, 3-aminobenzamide indicated that mono- rather than poly(ADP ribosyl)ation was affected by calcium.

In a study of the rachitic growth plate, Kakuta *et al.* (1986) observed a profound decrease in the chondrocyte pyridine nucleotide concentration. As can be seen in Figure 4, in the growth cartilage, the concentration of both NAD and NADH decreases dramatically when an animal is maintained on a low calcium low vitamin D diet. The documented change in redox status may be secondary to calcium-mediated NAD depletion. Additionally, when calcium intake is restricted, the associated alterations in calcium pools, due to mobilization of calcium from mitochondria and the plasma membrane, are also dependent on ribosylation events (Duncan *et al.*, 1988). If mitochondrial calcium stores are utilized for mineral formation, then the transport process may be controlled by an NAD-dependent ADP-ribosylation.

Finally, nicotinamide may play a role in the maturation and possibly the mineralization of epiphyseal cartilage by influencing the vascularization process. Recent work by Kull *et al.* (1987) has shown that nicotinamide and a nicotinamide complex are angiogenic molecules. If these compounds are liberated from cartilage cells by the NAD-dependent ribosylation reactions that were discussed earlier, these molecules (see Figure 3) could overcome the effects of antiangiogenesis factors of the cartilage matrix. In this way, nicotinamide could promote vascularization of the maturing epiphyseal cartilage. Once vascular invasion is initiated, it is likely that mineralization would follow.

Cartilage Energy Metabolism and Calcium Transport

The energy status of epiphyseal cartilage will be described in this section and changes in energy metabolism will be related to the development of mineralization. Cellular mechanisms for the accumulation and transport of calcium will be reviewed and relevant processes that may lead to calcium ion release will be considered. Although biochemical data on this topic are limited, information obtained from parallel investigations performed on a variety of noncartilaginous tissues will also be examined. In addition, a number of basic processes that mediate ion translocation will be presented, and this information will be used to critique relevant studies performed on cartilage.

Although it may be argued that profound differences in the way soft and hard tissues utilize and transport calcium are possible, work carried out on soft tissues indicate that common mechanisms exist. In all tissues, calcium ions at micromolar levels serve a major role in modulating physiological function, probably by acting as a second messenger (Berridge and Galione, 1988). Nevertheless, total calcium flux in most soft tissues is limited. In contrast, in epiphyseal cartilage, while calcium is of major importance in regulating cellular function, it is also utilized to form the mineral component of the extracellular matrix. From this point of view, the calcium flux is likely to be orders of magnitude greater than that of soft tissues. In addition, since a major function of the hypertrophic chondrocyte is to facilitate mineral formation, systems that may only function in a rudimentary way in nonmineralizing tissues may assume considerable importance in cartilage.

Calcium Entry into Cells

Studies of a variety of tissues have demonstrated a large calcium gradient across the plasma membrane. In most cases, the calcium ion concentration external to the cell is measured in millimolar concentrations. In contrast, the cytoplasmic calcium ion values are recorded at less than micromolar levels. Thus, there is greater than a 1000-fold difference in calcium concentrations across the plasma membrane (Carafoli, 1987). In chondrocytes isolated from the bovine epiphysis, Star *et al.* (1987) showed that the cytosolic ionic calcium concentration (106 to 137 n*M*) was similar to that reported for a number of different cell types. Thus, for cartilage and many other tissues, the plasma membrane can be considered as a "calcium dam", holding back high external concentrations of calcium. Clearly, any modification in the structure of the plasma membrane can result in massive changes in the intracellular calcium concentration. Thus, permeability of the membrane to calcium is very tightly controlled.

There are three pathways that permit calcium ions to enter the cell. In many excitable tissues, gated channels that are specific for calcium, as well as certain other divalent cations, exist. Most commonly, these gates are regulated by changes in membrane potential, and their activity can be blocked by a number of drugs that interact with lipids on the inner aspect of the plasma membrane (Kokobun and Reuter, 1984). Channels of this sort have been described in bone by Chesnoy-Marchais and Fritsch (1988). Probably of greater importance than gated channels in nonexcitable tissues, such as liver and bone, is the activity of a sodium-calcium exchange system. This is a low affinity high capacity porter which can act to transport calcium ions, in exchange for sodium, from the cell. The activity of this plasma membrane system is increased by ATP which raises the affinity of the system for calcium tenfold and sodium twofold (Blaustein, 1977).

The third type of transporter is a plasma membrane calcium ATP-asc. This transport system is regulated hormonally. In some tissues there is evidence of a cAMP-dependent protein kinase which transports calcium from the cytosol to the external milieu (Campbell, 1983). The K_M for calcium is below 1 μM and the transport system is of low capacity. Thus, it is unlikely that this system is capable of influencing states where there is high calcium flux. Nevertheless, two aspects of the transporter are noteworthy. First, ATP is required for activation of one of the protein subunits. Second, it is stimulated by calmodulin which increases its calcium affinity probably by interacting directly with the protein through a specific calcium binding domain (Carafoli, 1987).

In response to drugs, hormones, and growth factors, calcium levels in the cytosol are transiently increased. In most cells, calmodulin acts as a high affinity calcium binding protein that can interact with a large number of multifunctional protein kinases. The protein has been demonstrated in cartilage by Lewinson and Boskey (1984). A recent study by Leboy *et al.* (1988) showed that there were no zone-related differences in calmodulin mRNA levels in growth cartilage. This observation is somewhat surprising in view of the marked changes in calcium levels in specific regions of the plate. Other high affinity calcium binding proteins include troponin, parvalbumin, gelsolin, calcimedin, filamin, synexin, and the vitamin D dependent calbindins. The latter group of proteins is present in cartilage cells and responsive to endocrinal signals (Minghetti and Norman, 1988).

The concentration of cytosolic calcium ions is regulated by the endoplasmic reticulum through the activation of a calcium ATP-ase (K_M for calcium 0.1 to 1 μM). Recent work suggests that an endoplasmic reticulum protein, reticuloplasmin, may be specialized for calcium buffering. A glycoprotein present in the reticuloplasm, called endoplasmin, has been shown to bind significant quantities of calcium (Koch, 1987).

Although it is known that calcium can be released from the endoplasmic reticulum by the action of phosphorylated inositol compounds (reviewed by Berridge and Galione, 1988), little is known of the importance of this latter mechanism in controlling calcium flux in chondrocytes. In contrast, there is considerable information on the role of the mitochondrion as a calcium buffer. Many workers in the field believe that this organelle may help to initiate the calcification of cartilage.

Mitochondrial Granules and Cartilage Mineralization

It has been known for more than 20 years that respiring soft tissue mitochondria can accumulate large quantities of calcium. Moreover, in the presence of phosphate ions, calcium phosphate granules are seen in the mitochondrial matrix (Brierley and Slautterback, 1964). As the calcium/phosphate ratio of the granules is 1.6 to 1.7, Rossi and Lehninger (1963a, b) suggested that the mineral was apatite. Weinbach and von Brand (1965, 1967) and Thomas and Greenawalt (1968) showed that the mineral is a subcrystalline precursor of a calcium deficient apatite. Thus, mitochondria removed from the constraints of other subcellular compartments can form a mineral that is characteristic of the inorganic phase of mineralized cartilage. These observations formed the framework for the subsequent view that mitochondria are organelles required for the initiation of mineral formation in skeletal tissues.

Following these seminal studies, Matthews and colleagues examined cartilage for the presence of intracellular mineral deposits. Using a pyroantimonate stain and microincineration techniques, they found that calcium phosphate granules are present in chondrocytes *in situ* (Matthews *et al.*, 1968). Later, an important study by Brighton and Hunt (1978a, b) indicated that mitochondrial granules exist in the hypertrophic chondrocyte. At the calcification front, loss of mitochondrial staining appeared to coincide with the appearance of mineral in matrix vesicles. Landis and Glimcher (1982) showed that mitochondrial granules are prominent in proliferating and hypertrophic chondrocytes. The granules form clusters of about of 800 to 1000 Å diameter. These authors were careful to point out that the mitochondrial granules differ significantly from the platelet, needle and rod shaped particles that are present in the extracellular matrix. Together, these observations formed a powerful framework for building a theory that linked mitochondrial energy metabolism with mineralization. The gist of this theory is that mitochondria of maturing chondrocytes accumulate calcium and phosphate ions by an energy requiring process. At the site of mineralization, a change in cellular energy metabolism results in discharge of these calcium stores, or the calcium phosphate granules, from the cell. The presence of high local concentrations of calcium and phosphate ions or calcium phosphate granules in the extracellular matrix provides the chemical potential for mineral formation (Shapiro and Greenspan, 1969; Lehninger, 1970; Brighton and Hunt, 1978a).

It should be acknowledged that granules are seen in a wide variety of mineralizing tissues. Aside from chondrocytes, they have been seen in odontoblasts, osteoblasts, osteoclasts, and deer antler cells. Granules have also been detected in cells in which there is a high calcium flux (shell gland of the chick, hepatopancreas of the molting crab, intestinal mucosa cells after calcium loading). Experiments have been conducted to determine the physical structure of the granules and to assess whether they are apatitic or preapatitic. Studies performed on tissue slices using anhydrous techniques indicate that in the nonmineralized regions of the growth plate the calcium and phosphorus molar ratio is very low (about 1.0). In the mineralizing regions of cartilage the ratio is increased (1.2 to 1.5). Landis and Glimcher (1982) reported that the mitochondrial mineral was brushite ($CaHPO_4$, 2 H_2O, Ca/P = 1.0); hence the low ratio and poor diffraction pattern. The failure to generate a coherent electron diffraction pattern may have been due to the presence of inhibitors of solid phase maturation, such as magnesium and ATP.

Wuthier *et al.* (1985) carried out studies to model mitochondrial mineral formation. It was shown that the type of mineral formed is dependent on subtle variations in solution chemistry such as the manner by which calcium ions are presented to a phosphate buffered system. The study also highlighted the observation that ATP regulates the solid phase transitions of the mineral. Thus, in the presence of this nucleotide, a relatively stable form of calcium phosphate is formed (with a low calcium:phosphate ratio). In the absence of ATP, this intermediate converts spontaneously into brushite. The findings from this study confirmed reports that adenine nucleotides, magnesium, and bicarbonate ions regulate the type of mineral formed in calcifying systems (Blumenthal *et al.*, 1977).

The idea will be subsequently developed that chondrocyte mitochondria store calcium ions in a liquid phase and not as solid phase deposits. Before this evidence is reviewed, it may be worthwhile to summarize the major concepts advanced so far. Examination of mitochondria from a wide variety of cell types that are concerned with the induction of mineralization indicate the presence of elements associated with mineral formation. From a stoichiometric viewpont, the ratio of calcium to phosphorus in these mitochondria suggests that the mineral associated with these deposits is a precursor of the extracellular mineral, and that intramitochondrial sources of ATP may be required to stabilize this precursor phase. From this point of view, mitochondria of cells in which there is high calcium flux appear to be substantially different from mitochondria of cells in which calcium serves primarily as an intracellular metabolic signal.

The studies described above do not conclusively prove that chondrocyte mitochondria contain elevated levels of calcium and other ions. As Landis and Glimcher (1982) point out, pyroantimonate staining for calcium is nonspecific; many monovalent and divalent cations react with this stain. Moreover, when a pyroantimonate calcium deposit is formed, detection by X-ray microanalysis is difficult due to overlap of calcium and antimony peaks. Any type of aqueous fixation may result in ion relocation, precipitation, and mineral phase change (Boyde and Shapiro, 1987). Freeze-drying techniques induce mineral deposition from physiological solution. Likewise, nonaqueous fixation can induce mineral precipitation. Seveus and co-workers (Seveus *et al.* 1978; Barnard, 1981) point out that particles of calcium phosphate are seen in liver and adipose tissue that are prepared using ultramicrotomy techniques. The particles are probably due to diffusion of ions during low temperature freeze-drying.

In historical perspective, application of increasingly faster techniques for freeze-trapping soft tissues have progressively lowered the amount of calcium detected in mitochondria (Somylo *et al.,* 1985). Presently, most authorities believe that in muscle, mitochondria play a minor role in controlling calcium flux (this role has now been assumed by components of the sarcoplasmic reticulum). Whether similar results will be seen in hard tissues when newer methods are employed for freeze-slamming cartilage and bone remains to be seen.

Calcium Transport by Chondrocyte Mitochondria

Aside from the morphological observations described above, experiments have been conducted to characterize chondrocyte mitochondria in terms of their ability to perform oxidative phosphorylation and to transport calcium. Early progress in this field was slow because of the problems encountered in preparing respiring mitochondria from cartilage. These difficulties were due to the presence of components in the extracellular matrix that aggregated mitochondria and other subcellular organelles. Thus, it is very difficult to prepare mitochondrial fractions of reasonable purity. Moreover, during the preparative steps, mitochondria became loaded with calcium that was released from the cartilage mineral. As a result, the organelles became uncoupled and failed to respond to agents that modulate the energetic state of the mitochondria.

A review of the early literature reveals few studies of mitochondrial function in calcified tissues. Whitehead and Weidmann (1959) and Meyer and Kunin (1969, 1973) examined dehydrogenase activity in cartilage and showed that it is distinct from that of other tissues. Arsenis (1972) noted that calcium is actively accumulated by mitochondria isolated from costochondral cartilage *in vitro*. Meyer and Kunin (1969, 1973) also described the use of isolated mitochondrial preparations. However, in common with Arsenis, they failed to provide details of the metabolic state of the preparations that were used. In 1974, Lee and Shapiro reported a method for isolating respiring mitochondria from the premineralized region of the chick growth cartilage. The respiratory characteristics of the organelles, in terms of oxygen uptake values and P/O ratios, indicated that these organelles respire in a similar fashion to mitochondria isolated from other sources. ADP stimulated oxygen metabolism while the clear ''cut off'', when ADP was limiting, showed that the organelles were not damaged by the isolation procedure. Modifications of this procedure, by including agents in the isolation medium that prevented phosphate-dependent calcium uptake into the mitochondria, permitted respiring organelles to be prepared from hypertrophic and calcifying cartilage (Lee and Shapiro, 1978).

A comparison of the respiratory activities of mitochondria from hard and soft tissues reveals profound functional differences. In cartilage, when calcium is being accumulated, there is minimum loss of mitochondrial function. And, at high calcium concentrations, cartilage mitochondria respire and exhibit coupled oxygen uptake and phosphorylation (Shapiro and Lee, 1975). In contrast, in soft tissues, if the concentration of intramitochondrial calcium becomes elevated (over 100 nmol/mg protein), there is loss of acceptor control (Rossi and Lehninger, 1963a). Even in tissues where intramitochondrial calcium accumulates pathologically, the presence of the cation causes rapid decrease in acceptor control of respiration and activation of phospholipases. Activation of these enzymes results in membrane damage and frequently cell death.

Both calcium and ADP stimulate oxidative phosphorylation. In the chick, ADP stimulation takes precedence over calcium. This effect may have some physiological relevance, as an elevated level of intramitochondrial calcium could serve to damage the phosphorylating system and thereby inhibit many essential metabolic processes. The lack of stimulation might also indicate that these mitochondria can store calcium without damage to mitochondrial function. In bovine chondrocytes, mitochondria respond to calcium. Moreover, calcium stimulation takes precedence over ADP (Iannotti *et al.*, 1985).

Using the technique described above, Lee and Shapiro (1978) provided new information on calcium levels in chondrocyte mitochondria. They showed that the calcium content of these mitochondria are 10 to 30 times greater than that of organelles prepared from noncalcifying tissues. Sucrose gradient centrifugation techniques permitted chondrocyte mitochondria to be separated on the basis of their buoyant density. In premineralized cartilage, mitochondria form a low density-low calcium (70 to 460 nmol/mg protein) band and a high density-high calcium (280 to 700 nmol/mg protein) band. The presence of heterogeneous populations of mitochondria of differing density lent support to a concept put forward by Arsenis (1972) that mitochondria with differing functions may exist in cartilage cells (Shapiro *et al.*, 1976).

At least two classes of mitochondria have been separated on the basis of their calcium loading characteristics and enzymatic activities (Arsenis, 1972; Meyer and Kunin, 1973; Shapiro *et al.*, 1976). In addition, in a recent histochemical study of the growth plate, Yamamoto and Gay (1988) demonstrated mitochondrial heterogeneity in terms of cytochrome oxidase activity. These workers showed that more cytochrome oxidase-positive mitochondria are present in the hypertrophic zone than in the proliferating region. If cytochrome oxidase activity is used as a measure of energy-supported mitochondrial calcium accumulating activity, then these findings would support the view that the cells in the hypertrophic region are more capable of accumulating calcium than the less active proliferative cells.

Iannotti *et al.* (1985) described an exciting new approach to studying chondrocyte mitochondrial function using cultured cells. Rather than homogenizing the chondrocytes, these workers permeabilized the plasma membrane, calculated the mitochondrial content by measurement of the cytochrome aa_3 absorption spectra, and monitored calcium transport using a dual-wavelength spectrometer. In this way, problems due to aggregation of organelles, ion redistribution, and elemental loading are minimized. When these methods were applied to bovine growth plate chondrocytes, it was shown that the cells contain more than 900 nmol calcium per milligram mitochondrial protein (hepatocyte mitochondrial calcium concentration is about 20 nmol/mg protein). In a parallel experiment, they demonstrated that chondrocytes of hyaline cartilage have elevated levels of cell and mitochondrial calcium. While the chondrocyte endogenous mitochondrial calcium pool is elevated, the study also indicated that the capacity of these cells for cation uptake is twofold greater than that of cells of nonmineralizing tissues. Thus, results of experiments conducted on isolated mitochondria and whole permeabilized cells provided support for the view that chondrocyte mitochondria are specialized for calcium uptake.

Aside from examining the mechanism of calcium uptake by chondrocytes, experiments have been conducted to characterize the kinetics of calcium transport by chondrocyte mitochondria. It was noted that in the presence of succinate, mitochondria accumulate 300 nmol calcium per milligram protein within 10 min (Lee and Shapiro, 1978). This rate of uptake is similar to values reported for rat liver mitochondria during state 4 respiration. In the same study, calcium extrusion from mitochondria was determined and related to the initial calcium load. It was reported that at low endogenous calcium loads (200 to 480 nmol/mg protein), the presence of respiratory inhibitors and uncouplers causes extrusion of more than 80% of the total mitochondrial calcium. The results suggest that calcium is not simply binding to molecules of high calcium affinity, but is being maintained within the mitochondria by an energy dependent process. iannotti and colleagues (Iannotti *et al.*, 1985) also studied the calcium extrusion phenomenon. They showed that bovine growth plate mitochondria can release over 50% of the accumulated calcium and thereby confirmed that much of the intramitochondrial calcium is stored in a labile state. The mitochondrial extrudate has been subjected to analysis. Its calcium/phosphate ratio was found to be almost identical to the ratio found in the extracellular fluid of the longitudinal septa of the growth plate and very close to that of hydroxyapatite (Howell *et al.*, 1968).

Although the number of experiments performed with mitochondria of cartilage cells is very few, the data suggest that during chondrocyte maturation there is an elevation in the total quantity of cell calcium. Furthermore, as the total calcium load rises, there is a dramatic increase in the concentration of calcium in the extrudable pool. At present, it is not possible to say whether calcium in this latter pool is used to initiate the mineralization process, nor is it possible to ascertain which pool of mitochondrial calcium (extrudable vs. nonextrudable) gives rise to the "granules" described by Matthews and colleagues. Nevertheless, if the calcium ions contained within the extrudable pool are to be used for mineral formation, then mechanisms must exist that promote release of these ions from mitochondria. Moreover, the release mechanism must be under tight metabolic control. It is worth noting that phosphoenolpyruvate, a product of glycolysis, causes calcium release from isolated chondrocyte mitochondria (Shapiro and Lee, 1978). The importance of this metabolite is intriguing, since it is known that as the cells in the growth cartilage mature there is a switch from respiration to glycolysis and an accumulation of products of glycolysis. Thus, a mechanism can be discerned that could regulate mitochondrial calcium discharge.

The importance of the glycolytic regulatory system is difficult to assess. Along with a number of other agents such as phosphate ions, pyridine nucleotides, prostaglandins, palmityl CoA, and oxaloacetate, phosphoenolpyruvate acts by causing a collapse of the mitochondrial transmembrane potential. If this occurs *in vivo,* it could cause irreversible damage to mitochondria. Therefore, other mechanisms that maintain mitochondrial viability must be functional. Indeed, studies of heart muscle mitochondria have shown that sodium ions promote an electroneutral calcium exchange. As this takes place in mitochondria treated with the calcium uptake blocker, ruthenium red, it is probable that this efflux pathway is separate from the electrophoretic calcium uptake route. There is now good evidence to indicate that a sodium-calcium exchange system also exists. In addition, a calcium-hydrogen antiporter is present in a number of tissues. It remains to be determined whether these systems function in cartilage and to ascertain whether they are activated at a specific stage of chondrocyte maturation.

Energy Status of Epiphyseal Cartilage and the Initiation of Mineralization

Redox Measurements

The relationship between mitochondrial redox status and calcium transport has received intense investigation. Experiments performed by Lehninger and co-workers indicated that pyridine nucleotide oxidation promoted calcium efflux from isolated mitochondria. More recent studies showed that the transport process is more complex than originally described and that the oxidation of pyridine nucleotides alone is not sufficient to release stored calcium ions (Palmer and Pfeiffer, 1981). Nevertheless, these earlier studies underlined the importance of redox related events as general modulators of calcium transport *in vivo*.

In 1982, we examined the relationship between cell redox status, chondrocyte maturation, and mineral deposition in the growth cartilage. The metabolic state of the tissue was preserved by freeze-trapping, and fluorescent signals due to reduced mitochondrial pyridine nucleotides were recorded. To determine chondrocyte redox directly, we measured the NAD and NADH concentration of growth cartilage sections using a chemical technique (for details of the methodology, see Shapiro *et al.*, 1984; Kakuta *et al.*, 1985).

Surface fluorescence scans show an area of high fluorescence in hypertrophic cartilage due to a zone-specific elevation in the concentration of NADH. Just below this zone, in the calcifying cartilage-bone forming region, the fluorescence is low due to a decrease in the level of NADH. There is also a well-defined band of intermediate fluorescence in proliferative cartilage, which suggests that proliferative chondrocytes are in a more oxidized state than the underlying hypertrophic cells. When the growth cartilage was analyzed chemically, similar changes in the NADH (and NAD) concentrations were observed. Thus, in the precalcified proliferative zone, cells are in a more oxidized state with a lower NADH content and a higher NAD-NADH ratio than the underlying hypertrophic cells. At the site of mineralization, chondrocytes are in a reduced energetic state. The abrupt change in the redox profile of cells of the hypertrophic zone is of considerable interest because it occurs in concert with other calcification-related events. These redox changes are also seen in the epiphyses of other species.

In a more recent study, we reevaluated the relationship of hypertrophic cell redox status to the vascular canals that permeate this region of the cartilage (Shapiro *et al.*, 1988). In the chick growth cartilage, a remarkably well-developed vascular system is present, and the first evidence of mineral formation is observed in the hypertrophic zone in the immediate vicinity of these vessels (Boyde and Shapiro, 1987). A microfluorimetric scan of cartilage in the early hypertrophic region reveals many areas of low fluorescence (blood vessels) scattered throughout the field. When the area of light absorption due to the blood vessels is subtracted from the surrounding low fluorescence, a perivascular zone about 40 to 60 μm wide is seen. As the fluorescence signal is dependent on the level of NADH, this result suggests that cells in this perivascular region have a higher level of oxidative metabolism than the more distant chondrocytes. The results of this study point to a specific role for perivascular cells in cartilage mineralization.

It should be emphasized that in the chick epiphyses, the cells that initiate mineralization are those that are adjacent to blood vessels. These cells are exposed to a higher oxygen tension than more distant chondrocytes. From this perspective, it is unlikely that in the chick growth cartilage the onset of mineralization is merely due to a fall in oxygen tension and the subsequent development of localized hypoxia. Indeed, the oxidized status of the perivascular cells may reflect the transfer of calcium and phosphate ions by these cells from the blood to the cartilage matrix.

Adenine Nucleotide Measurements

To complete measurement of the energy status of cartilage, we have determined nucleotide levels in selected regions of the chick epiphysis (Matsumoto *et al.,* 1988). The major finding of the investigation was that there is a dramatic change in the content of adenine and guanine nucleotides during the mineralization process. Moreover, the change in the nucleotide profiles confirm the previously stated view that there is a major switch in energy metabolism in this tissue. In the premineralized cartilage, the ATP level is high, the ATP/ADP ratio is low (about 2), and the energy charge ratio is about 0.7. The nucleotide content of the resting-proliferating zones is similar to that seen in a number of other tissues where oxygen is limiting. The analytical data suggest that cells in this region obtain much of their energy through respiratory and glycolytic activities. As the chondrocytes hypertrophy, the ATP concentration decreases, the energy charge is less than 0.7, and the ATP/ADP ratio is below 1. It is inferred from these data that hypertrophic cells are in a reduced state and the cells are obtaining chemical energy from nonmitochondrial processes.

Before completing this discussion of the energetics of growth cartilage, it is necessary to reconcile some of the findings concerning the redox status of chondrocytes and adenine nucleotide levels in cartilage sections with observations made by other workers on the metabolic state of hypertrophic cells. First, it should be pointed out that within the past 5 years our thinking on this topic has undergone a complete change in direction. At one time, it was thought that hypertrophic cells were either dead or dying chondrocytes. That view has changed, and many workers believe that these cells are alive, functional, and expressing a different phenotype. In a very elegant study, using optimum fixation methods to preserve the cytology of the hypertrophic cell, Hunziker *et al.* (1987) showed that these chondrocytes are not degenerate and inferred from morphometric analyses that hypertrophic cells are very active metabolically. Indeed, the data suggested that there was an increase in mitochondrial volume during hypertrophy. However, in comparison with the dramatic overall increase in cell size, the mitochondrial volume as a fraction of the overall cell volume actually decreased. Thus, while these morphological observations support the view that the hypertrophic cells are metabolically active, the data indicate that they are less active than proliferative chondrocytes. From this point of view, these morphometric findings support the biochemical studies of the growth cartilage.

Adenine Nucleotide Balance and the Initiation of Protein Synthesis

No studies have been performed to examine the effects of a change in redox status or energy charge on chondrocyte function. However, it is known from investigations of a variety of cell types that a decrease in the energy charge ratio has a very profound effect on the translational regulation of protein synthesis (Live and Kaminskas, 1975; Pain *et al.*, 1980; Panniers *et al.*, 1985). Indeed, minor changes in the adenine nucleotide balance result in large changes in the synthesis rate. Moreover, adenine nucleotide ratios are tightly coupled to those of the guanine nucleotides through the activities of nucleoside diphosphate kinase (Hucul *et al.*, 1985).

A change in redox status modulates protein initiation and possibly elongation reactions. Walton and Gill (1976) suggest that the GTP/GDP ratio determines the extent of formation of the ternary initiation complex through competition for the nucleotide binding site on elongation initiation factor 2 (EIF-2). Details of the mechanism at the polyribosomal level are unclear, but it is likely that phosphorylation of EIF-2 results in inhibition of synthesis (Ryazanov *et al.*, 1988). Relevant to this mechanism is the observation that in chondrocytes, insulin treatment results in dephosphorylation of chondrocyte EIF-2 with a resulting increase in protein synthesis (Towle *et al.*, 1984). With respect to the growth plate, a low energy charge would be expected to decrease the synthesis of selected proteins of the extracellular matrix. Possibly this effect might account for the change in the rate of synthesis of type II collagen and other proteins in the hypertrophic zone of the developing growth cartilage.

Summary

In this review, we have examined mechanisms of energy generation in cartilaginous tissues and their relationship to the local oxygen tension. While the amount of work in this area is very limited, the available evidence suggests that regions of low oxygen tension may exist in cartilage. Nevertheless, even in the subsurface regions of articular cartilage, chondrocytes maintain their phenotypic state and generate adequate levels of ATP. Experiments performed with isolated chondrocytes indicate that 80% of total glucose is consumed by anaerobic glycolysis. Glutamine also appears to play an important role in the energy economy of chondrocytes. In addition, since chondrocytes lack glutamine synthetase, glutamine must be considered an essential amino acid.

The respiratory characteristics of chondrocyte mitochondria are unique in that they perform oxidative phosphorylation in the presence of high endogenous calcium loads. In mineralizing cartilage, a labile calcium pool is present in mitochondria and calcium phosphate granules are seen in chondrocyte organelles. The presence of these granules is linked with the initiation of cartilage mineralization. While mitochondrial calcium may be associated with the calcification process, we cannot exclude the possibility that the high calcium loading is an artifact produced by the specimen preparation procedures.

NAD is required for ribosylation reactions associated with chondrocyte proliferation and cytodifferentiation. The relationship between pyridine nucleotide metabolism and cartilage mineralization and chondrogenesis was critiqued. During chondrogenesis, it is doubtful that the NAD level is dependent on the vascularity or local oxygen diffusion gradients. Recent studies of the mineralizing growth cartilage show that chondrocyte maturation is accompanied by a unique series of changes in energy metabolism. In the hypertrophic zone, there is an overall decrease in the energetic status of chondrocytes. A change in the energy charge ratio or a tissue oxygen deficit could regulate protein and glycosaminoglycan synthesis in hypertrophic cartilage. Finally, in avian cartilage, mineral deposition occurs preferentially in the matrix that surrounds blood vessels. The raised oxidative status of the perivascular cells suggests that these cells may regulate the energy-dependent transport of calcium and phosphate ions from the blood to the cartilage matrix. The latter finding implies that earlier ideas linking chondrocyte hypoxia with apatite deposition may not be valid.

Acknowledgments

This review was supported by NIH grants AR-34411, DE-06533, DE-08239, and DE-09684. Susan F. Silverton is supported by NIH grant IR29AR40428. The chapter was written while Irving M. Shapiro was a visiting professor in the Department of Orthopedics at Jefferson Medical College in Philadelphia. He would like to express sincere thanks to Drs. Rocky Tuan and Richard Rothman for the privileges extended to him during that period.

References

A.L. Arsenault. Microvascular organization of the epiphyseal-metaphyseal junction of growing rats. *J. Bone Miner. Res.* **2**: 143–149 (1987).

C. Arsenis. Role of mitochondria in calcification. Mitochondrial activity distribution in the epiphyseal plate and accumulation of calcium and phosphate ions by chondrocyte mitochondria. *Biochem. Biophys. Res. Commun.* **46**: 1928–1974 (1972).

T.Y. Aw and D.P. Jones. Control of glucuronidation during hypoxia. Limitation by UDP-glucose pyrophosphorylase. *Am. J. Physiol.* **219**: 707–712 (1984).

P.F. Baker. GTP and calcium release. *Nature (London)* **320**: 395 (1986).

K. Balogh and A.S. Kunin. The effects of vitamin D and dietary phosphorus on oxidative enzymes in the epiphyseal cartilage of rachitic rats. *Lab. Invest.* **18**: 782–788 (1968).

T. Barnard. Mitochondrial matrix granules, dense particles and the sequestration of calcium by mitochondria. *Scanning Electron Microsc.* **Pt. 2**: 419–433 (1981).

M.J. Berridge and A. Galione. Cytosolic calcium oscillators. *FASEB J.* **2**: 3074–3082 (1988).

M.T. Berridge and R.F. Irvine. Inositol phosphates and cell signalling. *Nature (London)* **341**: 197–205 (1989).

J.D. Biggers, R.B.L. Gwatkin, and S. Heyner. Growth of embryonic avian and mammalian tibae on a relatively simple chemically defined medium. *Exp. Cell Res.* **25**: 41–58 (1961).

M.P. Blaustein. Effects of internal and external cations and of ATP on sodium-calcium and calcium-calcium exchange in squid axons. *Biophys. J.* **20**: 79–111 (1977).

N.C. Blumenthal, F.C. Betts, and A.S. Posner. Stabilization of amorphous calcium phosphate by Mg and ATP. *Calcif. Tissue Res.* **23**: 245–250 (1977).

A.B. Borle, N. Nichols, and G. Nichols. Metabolic studies of bone in vitro. I. Normal bone. *J. Biol. Chem.* **235**: 1206–1210 (1960a).

A.B. Borle, N. Nichols, and G. Nichols. Metabolic studies of bone in vitro. II. The metabolic pattern of accretion and resorption. *J. Biol. Chem.* **235**: 1211–1214 (1960b).

A. Boyde and I.M. Shapiro. Morphological observations concerning the pattern of mineralization of the normal and the rachitic chick growth cartilage. *Anat. Embryol.* **175**: 457–466 (1987).

G.P. Brierley and D.B. Slautterback. Studies on ion transport. IV. An electron microscope study of the accumulation of Ca^{2+} and inorganic phosphate by heat mitochondria. *Biochim. Biophys. Acta* **82**: 183–86 (1964).

C.T. Brighton and R.B. Heppenstall. Oxygen tension in zones of the epiphyseal plate, the metaphysis and diaphysis. An in vitro and in vivo study in rats and rabbits. *J. Bone Jt. Surg.* **53A**: 719–728 (1971).

C.T. Brighton and R.M. Hunt. The role of mitochondria in growth plate calcification as demonstrated in a rachitic model. *J. Bone Jt. Surg.* **60-A**: 630–639 (1978a).

C.T. Brighton and R.M. Hunt. Electron microscopic pyroantimonate studies of matrix vesicles and mitochondria in the rachitic growth plate. *Metab. Bone Dis. Relat. Res.* **1**: 199–204 (1978b).

A.K. Campbell. *Intracellular Calcium, its Universal Role as Regulator*. John Wiley & Sons, New York. pp. 155–160, 1983.

A.I. Caplan. The effects of the nicotinamide-sensitive teratogen 3-acetylpyridine on chick limb mesodermal cells in culture: biochemical parameters. *J. Exp. Zool.* **180**: 351–362 (1972).

A.I. Caplan and S. Koutroupas. The control of cartilage and muscle development in the chick limb: the role of differential vascularization. *J. Embryol. Exp. Morphol.* **29**: 571–583 (1973).

A.I. Caplan and M.J. Rosenberg. Interrelationship between poly(ADP-ribose)synthesis, intracellular NAD levels, and muscle or cartilage differentiation from mesodermal cells of embryonic chick limb. *Proc. Natl. Acad. Sci. U.S.A.* **72**: 1852–1857 (1975).

A.I. Caplan. Poly(ADP-ribose) synthetase and cell differentiation. In: *ADP-Ribosylation of Proteins*. Eds F.R. Althaus, H. Hilz, and S. Shall. Springer-Verlag, Heidelberg. pp. 388–396, 1985.

E. Carafoli. Intracellular calcium homeostasis. In: *Annual Review of Biochemistry*, Ed C.C. Richardson. Annual Reviews Inc., Palo Alto, CA. pp. 395–434, 1987.

B. Chance, J.S. Leigh, B.J. Clark, J. Maris, J. Kent, S. Nioka, and D. Smith. Control of oxidative metabolism and oxygen delivery in human skeletal muscle: a steady state analysis of the work/energy cost transfer function. *Proc. Natl. Acad. Sci. U.S.A.* **82**: 8384–8388 (1985).

B.W. Cherney, R.J. Midura, and A.I. Caplan. Poly(ADP-ribose) synthetase and chick limb mesenchymal cell differentiation. *Dev. Biol.* **112**: 115–125 (1985).

D. Chesnoy-Marchais and J. Fritsch. Voltage-gated sodium and calcium currents in rat osteoblasts. *J. Physiol.* **398:** 291–311 (1988).

D.V. Cohn and B.K. Forscher. Effect of parathyroid extract on the oxidation in vitro of glucose and the production of $^{14}CO_2$ by bone and kidney. *Biochim. Biophys. Acta* **65**: 20–26 (1962).

H. deGroot and T. Noll. The crucial role of hypoxia in halothane-induced lipid peroxidation. *Biochem. Biophys. Res. Commun.* **119**: 139–143 (1984).

H. deGroot and T. Noll. Oxygen gradients: the problem of hypoxia. *Tr. Biochem. Soc.* **15**: 363–365 (1987).

P.K. Dixit. Quantitative histochemistry of cartilage. Alkaline phosphatase and glucose-6-phosphate dehydrogenase activity in different zones of rachitic rat cartilage during healing. *Calcif. Tissue Res.* **10**: 49–57 (1972).

M. Donnelly and I.E. Scheffler. Energy metabolism in respiration-deficient and wild type Chinese hamster fibroblasts in culture. *J. Cell. Physiol.* **89**: 39–52 (1976).

M.R. Duncan, P.R. Rankin, R.L. King, M.K. Jacobson, and R.T. Dell'orco. Stimulation of mono(ADP-ribosyl)ation by reduced extracellular calcium levels in human fibroblasts. *J. Cell. Physiol.* **134**: 161–165 (1988).

J. Dunham, D.R. Shackelton, L. Bitensky, J. Chayen, M.E.J. Billingham, and H. Muir. Enzymic heterogeneity of normal canine articular cartilage. *Cell Biochem. Function* **4**: 43–46 (1986).

B.W. Durkacz, O. Omidisi, D.A. Gray, and S. Shall. (ADP-ribose) participates in DNA excision repair. *Nature (London)* **283**: 593–596 (1980).

H. Eagle, V.I. Oyama, M. Levy, C.L. Horton, and R. Fleischman. The growth response of mammalian cells in tissue culture to L-glutamine and L-glutamic acid. *J. Biol. Chem.* **218**: 607–616 (1956).

H. Eagle. Amino acid metabolism in mammalian cell cultures. *Science* **130**: 432–439 (1959).

K.H. Falchuk, E.J. Goetzl, and J.P. Kulka. Respiratory gases of synovial fluids. *Am. J. Med.* **49**: 223–231 (1970).

R. Felix and H. Fleisch. Increase in fatty acid oxidation in calvarial cells cultured with diphosphonates. *Biochem. J.* **196**: 237–245 (1981).

H.B. Fell. The histogenesis of cartilage and bone in the long bones of the embryonic fowl. *J. Morphol. Physiol.* **40**: 417–459 (1925).

B. Flanagan and G. Nichols. Metabolic studies of bone in vitro. V. Glucose metabolism and collagen biosynthesis *J. Biol. Chem.* **239**: 1261–1265 (1964).

B.D. Gomperts. Involvement of guanine nucleotide binding protein in the gating of Ca^{2+} by receptors. *Nature (London)* **306**: 64–66 (1983).

A.M. Gonzalez, M. Socher, J.S. Hothersall, and P. McLean. Effect of experimental diabetes on the activity of hexokinase in rat lens: an example of glucose overutilization in diabetes. *Biochem. Biophys. Res. Commun.* **84**: 858–864 (1978).

A.S. Greenwald and D.W. Haynes. A pathway for nutrients from the medullary cavity to the articular cartilage of the femoral head. *J. Bone Jt. Surg.* **51B**: 747–753 (1969).

H.L. Guenther, H.E. Guenther, and H. Fleisch. Effects of hydroxyethane-1,1-diphosphonate and dichloromethanediphosphonate on rabbit articular chondrocytes in culture. *Biochem. J.* **184**: 203–214 (1979).

R.W. Haines. Cartilage canals. *J. Anat.* **68**: 45–64 (1934).

M.R. Hammerman, V.M. Corpus, and J.J. Morissey. NAD-induced inhibition of phosphate transport in canine renal brush border membranes. Mediation through a process other than or in addition to NAD hydrolysis. *Biochim. Biophys. Acta* **732**: 110–116 (1983).

C.J. Handley, G. Speight, K.M. Leyden, and D.A. Lowther. Extracellular matrix metabolism by chondrocytes. 7. Evidence that L-glutamine is an essential amino acid for chondrocytes and other connective tissue cells. *Biochim. Biophys. Acta* **627**: 324–331 (1980).

R.B. Heppenstall, G. Grislis, and T.K. Hunt. Tissue gas tensions and oxygen consumption in healing bone defects. *Clin. Orthop. Relat. Res.* **106**: 357–365 (1975).

M.P.M. Hermann-Erlee. The in vitro effect of parathyroid hormone on the amount of NADP in embryonic mouse calvaria. *Biochem. Biophys. Res. Commun.* **23**: 1–4 (1966).

P.W. Hochachka. Defense strategies against hypoxia and hypothermia. *Science* **231**: 234–241 (1986).

S. Hough, J.E. Russell, S.L. Teitelbaum, and L.V. Avioli. Regulation of epiphyseal cartilage metabolism and morphology in the chronic diabetic rat. *Calcif. Tissue Int.* **35**: 115–121 (1983).

D.S. Howell, J.C. Pita, J.F. Marquez, and J.E. Madruga. Partition of calcium, phosphate and protein in the fluid phase aspirated at calcifying sites in epiphyseal cartilage. *J. Clin. Invest.* **47**: 1121–1132 (1968).

C.R. Howlett. The fine structure of the proximal growth plate of the avian tibia. *J. Anat.* **128**: 377–399 (1979).

C.R. Howlett. The fine structure of the proximal growth plate and metaphysis of the avian tibia: endochondral osteogenesis. *J. Anat.* **130**: 745–768 (1980).

C.R. Howlett, M. Dickson, and A.K. Sheridan. The fine structure of the proximal growth plate of the avian tibia: vascular supply. *J. Anat.* **139**: 115–132 (1984).

J.A. Hucul, E.C. Henshaw, and D.A. Young. Nucleoside diphosphate regulation of overall rates of protein biosynthesis acting at the level of initiation. *J. Biol. Chem.* **260**: 15585–15591 (1985).

E.B. Hunziker, R.K. Schenk, and L.-M. Cruz-Orive. Quantitation of chondrocyte performance in growth plate cartilage during longitudinal bone growth. *J. Bone Jt. Surg.* **69-A**: 162–173 (1987).

P.M. Hwang, D.H. Byrne, and P.A. Kitos. Effects of molecular oxygen on chick limb and bud chondrogenesis. *Differentiation* **37**: 14–19 (1988).

J.P. Iannotti, C.T. Brighton, J.L. Stambough, and B.T. Storey. Calcium flux and endogenous calcium content in isolated mammalian growth plate chondrocytes, hyaline cartilage chondrocytes and hepatocytes. *J. Bone Jt. Surg.* **67A**: 113–120 (1985).

W.J. Iglewski, M. Lee, and P. Muller. ADP-ribosyltransferase from beef liver which ADP-ribosylates elongation factor-2. *FEBS Lett.* **173**: 113–118 (1984).

Y. Ishikawa, J.E. Chin, H.L. Hubbard, and R.E. Wuthier. Utilization and formation of amino acids by chicken epiphyseal chondrocytes: comparative studies with cultured cells and native cartilage tissue. *J. Cell. Physiol.* **123**: 79–85 (1985).

L.V. Johnson, M.L. Walsh, J. Bockus, and L.B. Chen. Monitoring of relative mitochondrial membrane potential in living cells by fluorescence microscopy. *J. Cell Biol.* **88**: 526–535 (1981).

D.P. Jones, F.G. Kennedy, B.S. Andersson, T.Y. Aw, and E. Wilson. When is a mammalian cell hypoxic? Insights from studies of cells versus mitochondria *Mol. Physiol.* **8**: 473–482 (1985).

H. Juarez-Salinas, J.L. Sims, and M.K. Jacobson. Poly(ADP-ribose) levels in carcinogen-treated cells. *Nature (London)* **282:** 740–741 (1979).

D.B. Jump, T.R. Butt, and M. Smulson. Nuclear protein modification and chromatin substructure. 3. Relationship between poly(adenosine diphosphate) ribosylation and different functional forms of chromatin. *Biochemistry* **18:** 983–90 (1979).

S. Kakuta, E.E. Golub, J.C. Haselgrove, B. Chance, P. Frasca, and I.M. Shapiro. Redox studies of the epiphyseal growth cartilage: pyridine nucleotide metabolism and the development of mineralization. *J. Bone Miner. Res.* **1**: 433–440 (1986).

G.L.E. Koch. Reticuloplasmin, a novel group of proteins in the endoplasmic reticulum. *J. Cell Sci.* **87**: 491–492 (1987).

S.D. Kokobun and H. Reuter. Dihydropyridine derivatives prolong the open state of Ca channels in cultured cardiac cells. *Proc. Natl. Acad. Sci. U.S.A.* **81**: 4824–4827 (1984).

F.C. Kull, D.A. Brent, I. Parikh, and P. Cuatrecasas. Chemical identification of a tumor-derived angiogenic factor. *Science* **236**: 843–845 (1987).

A.S. Kunin and S.M. Krane. The effect of dietary phosphorus on the intermediary metabolism of epiphyseal cartilage from rachitic rats. *Biochem. Biophys. Acta* **107**: 203–214 (1965).

W.J. Landis and M.J. Glimcher. Electron optical and analytical observations of rat growth plate cartilage prepared by ultracryomicrotomy. *J. Ultrastruct. Res.* **78**: 227–268 (1982).

P.S. Leboy, I.M. Shapiro, B.D. Uschmann, O. Oshima, and D. Lin. Gene expression in mineralizing chick epiphyseal cartilage. *J. Biol. Chem.* **17**: 8515–8520 (1988).

N.H. Lee and I.M. Shapiro. Oxidative phosphorylation by chondrocyte mitochondria. *Calcif. Tissue Res.* **16**: 277–282 (1974).

N.H. Lee and I.M. Shapiro. Ca^{2+} transport by chondrocyte mitochondria of the epiphyseal growth plate. *J. Membr. Biol.* **41**: 349–360 (1978).

A.L. Lehninger. Mitochondria and calcium ion transport. *Biochem. J.* **119**: 129–138 (1970).

C. Levene. The pattern of cartilage canals. *J. Anat.* **98**: 515–538 (1964).

D. Lewison and A.L. Boskey. Calmodulin localization in bone and cartilage. *Cell. Biol. Int. Rep.* **8**: 11–18 (1984).

T.R. Live and E. Kaminskas. Changes in adenylate energy charge in Ehrlich ascites tumor deprived of serum, glucose, or amino acids. *J. Biol. Chem.,* **250**: 1786–1789 (1975).

K. Lund-Oleson. Oxygen tension in synovial fluids. *Arthritis Rheum.* **13**: 769–776 (1970).

R.E. Marcus. The effect of low oxygen concentration on growth, glycolysis and sulfate incorporation by articular chondrocytes in monolayer culture. *Arthritis Rheum.* **16**: 646–656 (1973).

H. Matsumoto, K. DeBolt, and I.M. Shapiro. Adenine, guanine and inosine nucleosides of chick growth cartilage: relationship between energy status and the mineralization process. *J. Bone Miner. Res.* **3**: 347–352 (1988).

J.L Matthews, J.H. Martin, J.A. Lynn, and E.J. Collins. Calcium incorporation in the developing cartilaginous epiphysis. *Calcif. Tissue Res.* **1**: 330–336 (1968).

W.L. Meyer and A.S. Kunin. The inductive effect of rickets on glycolytic enzymes of rat epiphyseal cartilage and its reversal by vitamin D and phosphate. *Arch. Biochem. Biophys.* **129**: 438–446 (1969).

W.L. Meyer and A.S. Kunin. Effects of cortisone, starvation and rickets on oxidative enzyme activities of epiphyseal cartilage from rats. *Arch. Biochem. Biophys.* **156**: 122–133 (1973).

R.J. Midura, B.W. Cherney, and A.I. Caplan. Relationship of nicotinamide adenine dinucleotide to the chondrogenic differentiation of limb bud mesenchymal cells. *Dev. Biol.* **111**: 232–242 (1985).

P.P. Minghetti and A.W. Norman. 1,25$(OH)_2$-vitamin D_3 receptors: gene regulation and genetic circuitry. *FASEB J.* **2**: 3043–3053 (1988).

M. Misawa, J. Doull, P.A. Kitos, and E.M. Uyeki. Teratogenic effects of cholinergic insecticides in chick embryos. 1. Diazinon treatment on acetylcholinesterase and choline acetyl transferase activities. *Toxicol. Appl. Pharmacol.* **57**: 20–29 (1981).

P. Mitchell. Keilins' respiratory chain concept and its chemiosmotic consequences. *Science* **206**: 1148–1159 (1979).

A.M. Nahir. Aerobic glycolysis: a study of human articular cartilage. *Cell Biochem. Funct.* **5**: 109–112 (1987).

W.F. Neuman, M.W. Neuman, and R. Brommage. Aerobic glycolysis in bone: lactate production and gradients in calvaria. *Am. J. Physiol.* **234(1)**: C41–C50 (1978).

A. Nishio, S. Nakanishi, J. Doull, and E.M. Uyeki. Enhanced chondrocytic differentiation in chick limb bud cell cultures by inhibitors of poly(ADP-ribose) synthetase. *Biochem. Biophys. Res. Commun.* **111**: 750–759 (1983).

S.B. Oppenheimer, M. Edidin, C.W. Orr, and S. Roseman. An L-glutamine requirement for intercellular adhesion. *Proc. Natl. Acad. Sci. U.S.A.* **218**: 607–617 (1969).

P. Osdoby and A.I. Caplan. First bone formation in the developing chick limb. *Dev. Biol.* **86**: 147–156 (1981).

V.M. Pain, J.A. Lewis, P. Huvos, E.C. Henshaw, and M.J. Clemens. The effect of amino acid starvation on regulation of polypeptide chain initiation in Ehrlich ascites tumor cells. *J. Biol. Chem.* **255**: 1486–1491 (1980).

J.K. Palmer and D.R. Pfeiffer. The control of Ca^{2+} release from heart mitochondria. *J. Biol. Chem.* **256**: 6742–6750 (1981).

R. Panniers, E.B. Stewart, Merrick, and E.C. Henshaw. Mechanism of inhibition of polypeptide initiation in heat-shocked Ehrlich cells involves reduction of eukaryotic initiation factor 4F activity. *J. Biol. Chem.* **260**: 9648 (1985).

D.G. Pechak, M.J. Kujawa, and A.I. Caplan. Morphological and histochemical events during first bone formation in embryonic chick limbs. *Bone* **7**: 441–458 (1986a).

D.G. Pechak, M.J. Kujawa, and A.L. Caplan. Morphology of bone development and bone remodeling in embryonic chick limbs. *Bone* **7:** 459–472 (1986b).

J. Picard and P. Cartier. La mineralisation du cartilage ossifiable. X. Glycolyse du cartilage ossifiable de rat jaun normal et du rat rachitique. *Bull. Soc. Chim. Biol.* **42**: 1117–1123 (1960).

G.G. Poirer, G. DeMurcia, J., Jongstra-Bilen, C. Neiidergang, and P. Mandel. Poly(ADP-ribosyl)ation of polynucleosomes causes relaxation of chromatin structure. *Proc. Natl. Acad. Sci. U.S.A.* **79**: 3423–3427 (1982).

C.A. Poole, M.H. Flint, and B.W. Beaumont. Chondrons in cartilage: ultrastructural analysis of the pericellular microenvironment in adult human articular cartilages. *J. Orthop. Res.* **5**: 509–522 (1987).

L.J. Reitzer, B.M. Wice, and D. Kennel. Evidence that glutamine not sugar is the major energy source for cultured HeLa cells. *J. Biol. Chem.* **254**: 2669–2676 (1979).

C. Richter, K.H. Winterhalter, S., Baumhuter, H.R. Lotscher, and B. Moser. ADP-ribosylation in inner membrane of rat liver mitochondria. *Proc. Natl. Acad. Sci. U.S.A.* **80**: 3188–3192 (1983).

C. Richter, B. Frei, and J. Schegel. Calcium transport and mono(ADP-ribosylation) in mitochondria. In: *ADP Ribosylation of Proteins,* Eds F.R. Althaus, H. Hilz, and A. Shall. Springer-Verlag, Heidelberg. pp. 530–535, 1985.

J.I. Rodriguez, E. Delgado, and R. Paniagua. Multivacuolated cells in human cartilage canals. *Acta Anat.* **124**: 54–57 (1985).

C.S. Rossi and A.L. Lehninger. Stoichiometric relationships between mitochondrial ion accumulation and oxidative phosphorylation. *Biochem. Biophys. Res. Commun.* **11**: 441–446 (1963a).

C.S. Rossi and A.L. Lehninger. Stoichiometric relationships between accumulation of ions by mitochondria and the energy coupling sites in the respiratory chain. *Biochem. Z.***338**: 698–713 (1963b).

A.G. Ryazanov, E.A. Shestakova, and P.G. Natapov. Phosphorylation of elongation factor 2 by EF-2 kinase affects rate of translation. *Nature (London)* **334**: 170–173 (1988).

H.W. Sampson and M.S. Cannon. Zonal analysis of metabolic profiles or articular-epiphyseal cartilage chondrocytes: a histochemical study. *Histochem. J.* **18**: 233–238 (1986).

J.E. Scott and M. Haigh. Keratan sulphate and the ultrastructure of cornea and cartilage: a "stand-in" for chondroitin sulphate in conditions of oxygen lack? *J. Anat.* **158**: 95–108 (1988).

L. Seveus, D. Brdiczka, and T. Barnard. On the occurence and composition of dense particles in mitochondria in ultrathin frozen dry sections. *Cell Biol. Int. Rep.* **2**: 155–162 (1978).

S. Shall. ADP-ribose in DNA repair. In: *ADP-Ribosylation Reactions,* Eds O. Hayaishi and K. Ueda. Academic Press, New York. pp. 478–520, 1982.

I.M. Shapiro and J.S. Greenspan. Are mitochondria directly involved in biological mineralization? *Calcif. Tissue Res.* **3**: 100–102 (1969).

I.M. Shapiro and N.H. Lee. Effects of Ca^{2+} on the respiratory activity of chondrocyte mitochondria. *Arch. Biochem. Biophys.* **170**: 627–633 (1975).

I.M. Shapiro, A. Burke, and N.H. Lee. Heterogeneity of chondrocyte mitochondria. A study of the Ca^{2+} concentration and density banding characteristics of normal and rachitic cartilage. *Biochim. Biophys. Acta* **451**: 583–591 (1976).

I.M. Shapiro and N.H. Lee. The effects of oxygen, phosphoenolpyruvate, and pH on the release of calcium from chondrocyte mitochondria. *Metab. Bone Dis. Relat. Res.* **1**: 173–177 (1978).

I.M. Shapiro, E.E. Golub, S. Kakuta, J.C. Haselgrove, J. Havery, B. Chance, and P. Frasca. Initiation of endochondral calcification is related to changes in the redox state of hypertrophic chondrocytes. *Science* **217**: 950–952 (1982).

I.M. Shapiro and A. Boyde. Analysis of the mineralizing growth cartilage of the normal and rachitic chick. *Metab. Bone Dis. Relat. Res.* **5**: 317–326 (1984a).

I.M. Shapiro, S. Kakuta, and E.E. Golub. Methods for measurement of energy metabolism in calcified tissues. In: *Methods in Calcified Tissue Preparation,* Ed G.R. Dickson. Elsevier, pp. 587–622, 1984b.

S.F. Silverton, L.C. Wagerle, M.E. Robiolo, J.C. Haselgrove, and R.E. Forster. Oxygen supply to the cartilage growth plate: an estimation of oxygen gradients in two regions of the epiphyseal growth plate. In: *Oxygen Transport to Tissue XI,* Ed K. Rakusan et al. Plenum Press, New York. pp. 809–815, 1989.

S.F. Silverton, H. Matsumoto, K. DeBolt, A. Reginato, and I.M. Shapiro. Pentose phosphate shunt metabolism by cells of the chick growth cartilage. *Bone* **10**: 45-51 (1989).

M. Smulson, P. Stark, M. Gazzoli, and J. Roberts. Release of template restriction for DNA synthesis by poly (ADP ribosyl)ation polymerase during the HeLa cell cycle. *Exp. Cell Res.* **90**: 175–182 (1975).

M. Socher, N.Z. Baquer, and P. McLean. Glucose overutilization in diabetes: evidence from studies on the changes in hexokinase, the pentose phosphate pathway and glucuronate-xylulose pathway in rat kidney cortex in diabetes. *Biochem. Biophys. Res. Commun.* **86**: 32–39 (1979).

A.P. Somlyo, M. Bond, and A.V. Somlyo. Calcium content of mitochondria and endoplasmic reticulum in liver frozen rapidly in vivo. *Nature (London)* **314**: 622–625 (1985).

A.M. Star, J.P. Iannotti, C.T. Brighton, and P.F. Armstrong. Cytosolic calcium concentration in bovine growth plate chondrocytes. *J. Orthop. Res.* **5**: 122–127 (1987).

R.A. Stockwell. The cell density of human articular cartilage. *J. Anat.* **101**: 753–763 (1967).

R.A. Stockwell. Changes in the acid glycosoaminoglycan (mucopolysaccharide) content of aging human articular cartilage. *Ann. Rheum. Dis.* **29**: 509–515 (1970).

R.A. Stockwell. Metabolism of cartilage. In: *Cartilage,* Ed B.K. Hall. Academic Press, New York. pp. 81–123, 1983.

R.S. Thomas and J.W. Greenwalt. Microincineration, electron microscopy and electron diffraction of calcium phosphate-loaded mitochondria. *J. Cell Biol.* **39**: 55–76 (1968).

C.A. Towle, H.J. Mankin, J. Avruch, and B.V. Treadwell. Insulin promoted decrease in the phosphorylation of protein synthesis initiation factor EIF-2. *Biochem. Biophys. Res. Commun.* **121**: 134–140 (1984).

P. Treuhaft and D. McCarty. Synovial fluid pH, lactate, pO_2 and pCO_2 in various joint fluids. *Arthritis Rheum.* **14**: 475–484 (1971).

J. Trueta and M.H.M. Harrison. The normal vascular anatomy of the femoral head in adult man. *J. Bone Jt. Surg.* **35B**: 442–461 (1953).

J. Trueta and V.P Amato. The vascular contribution to osteogenesis. III. Changes in the growth cartilage. *J. Bone Jt. Surg.* **42B**: 571–587 (1960).

K. Ueda and O. Hayaishi. ADP-Ribosylation. In: *Annual Reviews of Biochemistry,* Ed C.C. Richardson. Annual Reviews Inc., Palo Alto, CA. pp. 73–100, 1985.

G.M. Walton and G.N. Gill. Preferential regulation of protein synthesis initiation complex formation by purine nucleotides. *Biochim. Biophys. Acta* **447**: 11–19 (1976).

E.C. Weinbach and T. von Brand. The isolation and composition of dense granules from Ca^{2+} loaded mitochondria. *Biochem. Biophys. Res. Commun.* **19**: 133–137 (1965).

E.C. Weinbach and T. von Brand. Formation, isolation and composition of dense granules from mitochondria. *Biochim. Biophys. Acta* **148**: 256–266 (1967).

R.G. Whitehead and S.M. Weidmann. Oxidative enzyme systems in ossifying cartilage. *Biochem. J.* **72**: 56–69 (1959).

N.J. Wilsman and D.C. Van Sickle. Cartilage canals, their morphology and distribution. *Anat. Rec.* **173**: 79–84 (1972).

R.E. Wuthier, G.S. Rice, J.E.B. Wallace, R.L. Weaver, R.Z. LeGeros, and E.D. Eanes. In vitro precipitation of calcium phosphate under intracellular conditions: formation of brushite from an amorphous precursor in the absence of ATP. *Calcif. Tissue Int.* **347**: 401–410 (1985).

T. Yamamoto and C.V. Gay. Ultrastructural analysis of cytochrome oxidase in chick epiphyseal growth plate cartilage. *J. Histochem. Cytochem.* **36**: 1161–1166 (1988).

H.R. Zielke, P.T. Ozand, J.T. Tildon, D.A. Sevdalian, and M. Cornblath. Growth of human fibroblasts in the absence of glucose utilization. *Proc. Natl. Acad. Sci. U.S.A.* **73**: (4110–4114 (1976).

H.R. Zielke, C.L. Zielke, and P.T. Ozand. Glutamine: a major energy source for cultured mammalian cells. *Fed. Proc.* **43**: 121–125 (1984).

4

Cartilage Growth and Differentiation Factors

SAEID M. SEYEDIN
Metra Biosystems
3141 Porter Drive
Palo Alto, California

DAVID M. ROSEN
Celtrix Laboratories
2500 Faber Place
Palo Alto, California

Introduction

The proliferation and differentiation of normal animal cells is a well controlled and highly conserved process. The control mechanisms are influenced by a variety of local substances, including a group of protein growth and differentiation factors which are hormone-like in both structure and function. When compared to true endocrine hormones, their synthesis, release, mechanism of action, and regulation are still largely unknown.

The specificity of the signal induced by most protein factors is mediated through specific cell surface receptor(s). Protein factors bind to their receptors with dissociation constants in the range of 10^{-9} to 10^{-10} M or tighter. The receptor is believed to be a signal transducing moiety. The role of the protein factor is to activate a molecular program initiating a wide range of biochemical and cellular responses which may include mitosis and the induction or inhibition of genes for new function and activity. In many systems, changes in cellular activities are associated with the cytoskeleton and therefore with morphological changes of the cell.

Protein factors have been implicated in growth and maintenance of the cartilage phenotype, and induction of uncommitted mesenchymal-like cells into cartilage. These factors are likely to be participants in embryological as well as adult skeletal development. In this chapter protein factors have been divided into two classes: growth factors and inducing or differentiation factors. Growth factors stimulate cell replication and metabolism in the absence of exogenous serum factors, whereas inducing factors promote phenotypic change or differentiation of cells and are not normally found in an active form in serum. In addition, inducing factors may require supplemental factors to elicit their full biological activity (e.g., transforming growth factor beta [TGF-β] and fibroblast growth factor (FGF), see Kimelman and Kirschner, 1987).

Growth Factors

In the past 15 years, a number of growth factors have been identified, and in many instances their primary structures and their gene sequences have been determined. The signal transduction mechanisms of these factors via their receptors have also been extensively studied. One major process of signal transduction involves the stimulation of a tyrosine-specific protein kinase. Several growth factors, such as platelet-derived growth factor (PDGF) and epidermal growth factor (EGF), are known to stimulate tyrosine kinase activity. Other factors, such as transforming growth factor-beta (TGF-β), apparently do not. Another mechanism of signal transduction involves the hydrolysis of specific inositol containing lipids to give two second messengers, diacylglycerol (DAG) and inositol 1,4,5-trisphosphate (IP_3), (see Berridge, 1986, for review). PDGF activates this cascade of events as does 12-O-tetradecanoyl phorbol-13-acetate (TPA). Several oncogene products have also been found to be involved in, or act via, both or either of these two transduction mechanisms.

Over the years, a number of studies have been published describing the effects of factors or crude tissue extracts on cartilage and cartilage cells. Most of these investigations have focused on factors that either promote cell replication or increase production of cartilage matrix molecules. Other activities, such as the induction and differentiation of cartilage, are discussed in the next section. A major effort has been made to isolate such growth factors from cartilage itself. It is becoming increasingly clear that many of these factors isolated from cartilage are either the same or closely related to other known factors. Discussion of such activities will be consolidated whenever appropriate. This section will concentrate on what are thought to be locally acting growth factors and will focus on three major areas. First, some of the factors known to have effects on cartilage and cartilage cells will be briefly described. Second, the activities of these factors on isolated chondrocytes and cartilage *in vitro* will be discussed. Reference will be made to mechanisms of action (including receptors) and synergistic activities of growth factors. Finally, and probably most importantly, we will discuss what is known about the activity of these factors *in vivo*. Here, we will be concerned with some of the proposed roles of factors in the normal growth of cartilage as well as

Table 1
Growth Factors That Affect Cartilage or Chondroblasts

Factor	Mr	Subunits	Carbohydrate	Tyr-kinase receptor	Related or synonymous protein
IGF-I	~7.5K	Single chain	No	Yes	Sulfation factor, SmC, CDF
IGF-II	~7.5K	Single chain	No	Yes	MSA, SGF
bFGF	~16K	Single chain	No	Yes	CDGF, MDGF
EGF	~16K	Single chain	No	Yes	TGF-α
PDGF	28—32K	Hetero- and homodimers 16K, 18K	Yes	Yes	c-sis, ODGF
TGF-β1	26K	Homodimer, 13K	No	No	CIF-A, BDGF-I
TGF-β2	26K	Homodimer, 13K	No	No	CIF-B
IL-1	~17.5K	Single chain	No	No?	Catabolin, OAF

the potential therapeutic application of such factors. A summary of the factors known to effect cartilage or isolated chondroblasts is presented in Table 1.

Insulin-Like Growth Factors (Somatomedins)

It has been known for some time that there are circulating factors that can affect the growth of cartilage. Serum from hypophysectomized rats has been shown to have diminished "sulfation factor" activity *in vitro* (Salmon and Daughaday, 1957). Subsequent studies revealed that this was not directly due to a lack of growth hormone as initially thought, but rather was due to another substance apparently induced by growth hormone. This substance, originally called sulfation factor, was renamed somatomedin (Daughaday *et al.,* 1972). Several molecules have since been purified which possess somatomedin-like characteristics. These include insulin-like growth factors I and II (IGF-I, IGF-II) (Rinderknecht and Humbel, 1976), and the homologous rat protein, multiplication stimulating activity (MSA) (Moses *et al.,* 1980). These molecules are low molecular weight proteins that circulate in plasma bound to high molecular weight carrier proteins (Zapf *et al.,* 1975). To date, two general classes of binding proteins have been identified. There is a large protein with a molecular weight of about 150,000 and a smaller group of proteins with molecular weights around 50,000. The exact role of the binding proteins is still unknown, but it may be to deliver factors to a desired site and to control the amount of biologically active material (for review, see Baxter, 1987). In their bound form, the IGFs do not appear to have significant biological activity (Hintz and Live, 1977; Mevli *et al.,* 1978; Zapf *et al.,* 1979). *In vitro,* they can be released from their carrier proteins by dissociating agents (e.g., guanidine) or by acid treatment. Recently, there has been significant progress in the identification and characterization of these binding proteins (Lyons and Smith, 1986; Hey *et al.,* 1987). It was originally thought that the liver was the only source of IGFs. There is now emerging evidence that both factors can also be made by a variety of cells and tissues, including chondroblasts (Burch *et al.*, 1986), and thus probably act locally.

Several factors have been isolated from cartilage and bone that have IGF-like activity. These include cartilage derived factor (CDF) (Kato *et al.*, 1980, 1981a) and skeletal growth factor (SGF) (Kato *et al.*, 1982). Recent data suggest that CDF is most probably IGF-I (Kato *et al.*, 1981b), whereas SGF is identical, or closely related, to IGF-II (Mohan *et al.*, 1988). The two major activities used to define or identify CDF are its ability to stimulate ^{3}H-thymidine incorporation into DNA and ^{35}S-sulfate incorporation into proteoglycans by chondrocytes. These effects on primary cultures of rabbit costal chondrocytes were extensively described by Kato *et al.* (1980, 1981a). In these studies, CDF was partially purified from bovine cartilage using a combination of gel filtration and ion-exchange chromatographies. In this system, CDF caused four- to fivefold and twofold increases in ^{3}H-thymidine and $^{35}SO_4$ incorporation, respectively, at concentrations of 1 to 5 μg/ml. Similar studies done with MSA, isolated from the conditioned medium of BRL cells, yielded virtually identical results (Kato *et al.*, 1981c), as did studies done with purified IGF-I using bovine articular cartilage (Osborn *et al.*, 1989).

Cellular receptors for IGF-I and IGF-II have been identified and characterized (Trippel *et al.*, 1983). Recent data suggest that chondrocytes contain two classes of IGF receptors (Trippel *et al.*, 1988). As seen in Figure 1, the two classes of receptors differ in their relative affinities for IGF-I and IGF-II. Functional differences between each class of receptor remain to be determined.

More recently, the interactions of CDF and MSA with other growth factors has been evaluated. It was found that both factors have synergistic activities with either EGF or FGF on DNA synthesis (Osborn *et al.*, 1989; Trippel *et al.*, 1983, 1988; Kato *et al.*, 1983; Hiraki *et al.*, 1987). Synergistic effects of growth factors have been reported with several systems (e.g., Kimelman and Kirschner, 1987). Since cells do not see only one growth factor at a time *in vivo,* it is not surprising that several factors would act together to exert a concerted biological activity. This point will be a consistent theme emphasized throughout this chapter.

The limited availability of purified growth factors has precluded any significant amount of *in vivo* testing. As recombinant forms of these factors become more readily available, there will undoubtedly be a rapid expansion in number and types of studies done to determine the true biological activities of these factors. However, there have been several studies done *in vivo* with somatomedin C (IGF-I). Rats injected with IGF-I (100 ng/day) into the growth plate showed a significant increase in epiphyseal width (Russell and Spencer, 1985). In another study, IGF-I or IGF-II was administered using an osmotic minipump (Schoenle *et al.*, 1985). In this study, 43 or 103 μg/day of IGF-I or 131 μg/day IGF-II were administered over a 6-day period into hypophy-

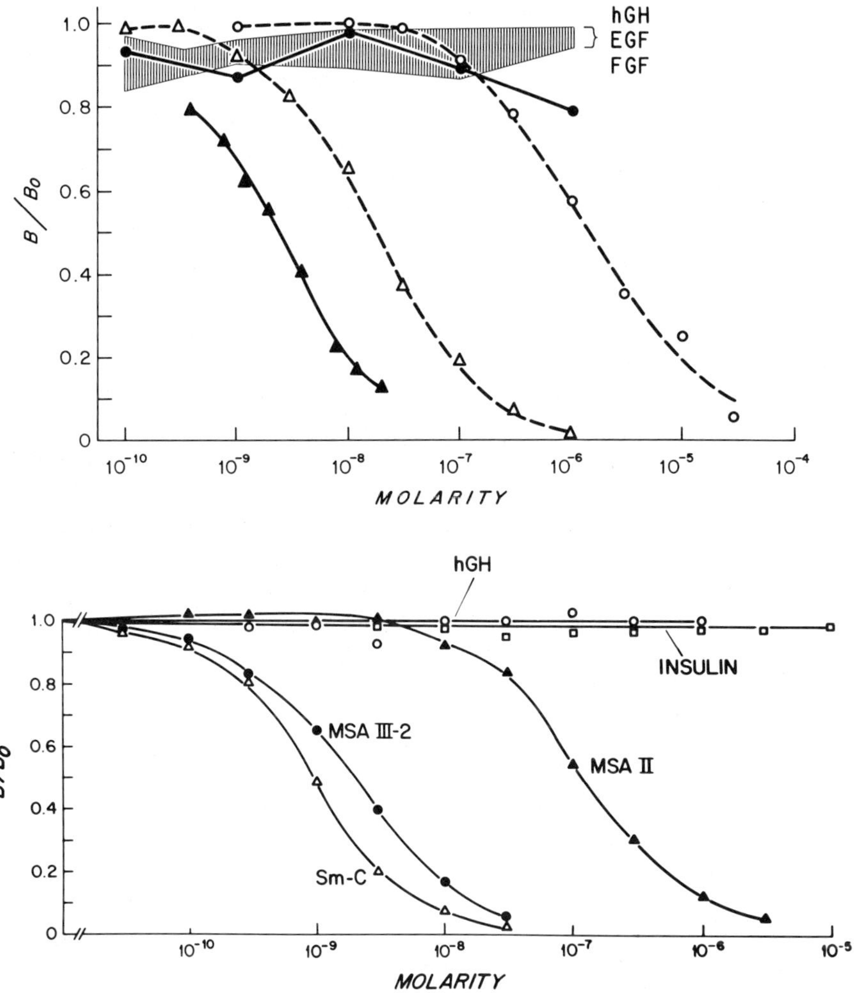

Figure 1. Competition for ^{125}I-IGF-I and IGF-II binding to bovine growth plate chondrocytes. *Upper:* The effects of increasing concentrations of IGF-I (▲), rat IGF-II (△), NGF (●), and insulin (○) on ^{125}I-IGF-I specific binding are expressed as competition curves. The entire range of B/B_0 for EGF, bFGF, and hGH are indicated by brackets. Note that both insulin and rIGF-II competed for the IGF-I receptor, but at concentrations which were 8- and 600-fold higher, respectively, than the unlabeled IGF-I. *Lower:* The effects of increasing concentrations of IGF-I (△-△-△-), MSA-III-2 (-●-●-●-), MSA-II (-▲-▲-▲-), insulin (-□-□-□-), and hGH (-○-○-○-) on ^{125}I-rIGF-II (MSA-III) specific binding to bovine growth plate chondrocytes. MSA-III-2 and MSA-II were isolated as described by Moses *et al.* Note that IGF-I was slightly more potent than MSA-III (rIGF-II) and approximately 30 times more potent than MSA-II in competing for ^{125}I-rIGF-II. Growth hormone and insulin did not compete for binding. Data supplied by Dr. S. Trippel *et al.* (1988).

sectomized rats. This study also demonstrated a dose-dependent increase in epiphysial width with IGF-I. Here, IGF-I was found to be more potent than IGF-II, a situation that has been observed in some, but not all, *in vitro* systems (Schmid *et al.*, 1983). In this situation, there appears to be a strong correlation between activities observed *in vitro* and *in vivo*.

In addition to their direct effects on bone and cartilage *in vivo,* modulation of IGFs *in vivo* during development has been studied by several investigators. Early studies of IGF-I and IGF-II showed a positive correlation between the circulating concentration of IGF-I and birthweight (Bennett *et al.*, 1983). Similar studies have shown that levels of IGF-II (MSA) are much higher in fetal rat serum than in adult serum (Moses *et al.*, 1980). These findings have led to the speculation that IGF-II may be the fetal form of IGF, whereas the predominant adult form would be IGF-I. More recently, immunolocalization studies have shown a strong correlation between the intensity of anti IGF-I staining and the degree of chondrocyte differentiation. Cells in the proliferative zone stained very intensely, whereas cells in the hypertrophic zone stained much weaker (Nilsson *et al.*, 1986). Furthermore, hypophysectomized rats showed significantly reduced staining for IGF-I, which could be restored upon administration of growth hormone (GH) (D'Ercole *et al.*, 1988). Thus, this study not only supports the role of IGF-I in directly promoting cartilage growth *in vivo,* but also suggests that GH may be acting directly on chondrocytes to produce their own IGF-I. Nilsson *et al.* (1986) have observed more intense staining of cells in the proliferative region as compared to germinal and hypertrophic cells. Whether IGF is acting through an autocrine or paracrine mechanism is not clear. In addition to immunostaining, recent *in situ* hybridization studies have revealed that mRNAs for IGF-I and IGF-II are abundant in many mesenchymal cells, including perichondrial cells (Han *et al.*, 1987; Stylianopoulou *et al.*, 1988). These studies also suggest that the IGFs made by mesenchymal cells could act in a paracrine fashion. For example, since extensive hybridization was found in perichondral cells, but not in mature chondrocytes, production of IGFs by perichondrial cells could act locally on potential neighboring target cells. Consistent with this hypothesis is the fact that a gradient of IGF-II mRNA was found in cartilage. Early cartilage had the highest levels of message and the amount declined as the chondrocytes matured (Beck *et al.*, 1987).

Fibroblast Growth Factor and Related Peptides

As with the IGFs, several activities have been isolated from cartilage which probably represent the same or very similar factors as the fibroblast growth factors (FGFs). A major factor in this category is cartilage-derived growth factor (CDGF), which is distinct from the IGF-like cartilage-derived factors (CDF) which was discussed in the previous section. The principal features of CDGF that relate it to FGF are its ability to bind tightly to heparin-Sepharose (Sullivan and Klagsbrun, 1985) and its other physical characteristics (e.g., basic pI and molecular weight of around 18,000). Its strong mitogenic effects on chondroblasts and capillary endothelial cells are also reminiscent of FGF-like activity.

As originally described in 1977 by Klagsbrun *et al.*, CDGF is mitogenic for BALB/C 3T3 cells. Subsequent studies showed it to be mitogenic for chondrocytes as well (Klagsbrun and Smith, 1980). Similar studies with pituitary FGF (basic FGF) showed that it also promoted an increase in ^{3}H-thymidine incorporation into chondroblasts (Sachs *et al.*, 1982). Early reports describing the effects of FGF and CDGF on glycosaminoglycan (GAG) production appear to be somewhat different. Hamermon *et al.*, (1986) demonstrated that CDGF enhanced hyaluronic acid synthesis by isolated chondrocytes, but actually slightly inhibited sulfated GAG synthesis. In contrast, others (Kato and Gospodarowicz, 1984) have rather convincingly demonstrated that FGF enhances sulfated GAG production by chondrocytes. The most likely explanation for this apparent discrepancy is that each group is using slightly different culture conditions. As Kato and Gospodarowicz point out, there is a profound but variable effect of FGF on GAG synthesis by chondroblasts, depending on culture conditions. Sparse cultures of cells cultured with FGF resulted in well differentiated confluent cultures producing more GAG than untreated cultures. Sacks *et al.* (1982) found that FGF inhibited $^{35}SO_4$ incorporation in 48-h cultures, but stimulated incorporation by (72 h) cultures. Thus, these studies suggest that treatment with FGF results in a short-term decrease, but long-term increase in proteoglycan synthesis by isolated chondroblasts. The reason for the initial decrease is not clear. It may be tied to the rapid proliferation of cells occurring at early times.

Other molecules described as being identical or related to the FGFs are endothelial cell growth factor (ECGF) and macrophage-derived growth factor (MDGF). Work by Rifas *et al.* (1984) demonstrated that rat peritoneal macrophages secrete a factor which is mitogenic for osteoblasts and chondroblasts. Subsequently, Baird *et al.* (1985) have shown similar cultures of macrophages to secrete a factor that is cross reactive with basic FGF antisera. The work done to date with FGF is almost exclusively with basic FGF. In several *in vitro* systems, the homologous acidic FGF was found to be less potent (Gospodarowicz *et al.*, 1986). Whether this is true for its activity on chondroblasts is not well documented.

The heparin binding property of FGF is interesting, yet poorly understood. Several groups have found that FGF is deposited in the extracellular matrix (ECM) by endothelial cells. The binding to the ECM is abolished by pretreatment with heparinase or heparitinase, but not other GAG degradative enzymes (Baird and Ling, 1987). Whether or not this feature is important for the biology of FGF *in vivo* remains to be determined. This feature may be an important storage mechanism for FGF *in vivo*.

Although there has been a rather large volume of work done with FGF *in vitro*, the true test of a factor's function and utility ultimately have to come from *in vivo* studies. Although to date there are few reports evaluating the effects of FGF on cartilage *in vivo*, Cuevas *et al.* (1988) demonstrated that FGF has a potent effect on cartilage when administered locally. In this study, basic FGF was administered to lesions in rabbit knees. In animals with no lesions, there was no apparent change in the cartilage morphology. In contrast, when FGF was administered via minipumps to knees with intrachondrial lesions (approximately 200 ng over the course of 2 weeks), there was a dramatic effect on cartilage healing. After 20 days, the lesions were filled in with clusters of chondroblasts surrounded by extensive extracellular matrix (Figure 2). Thus, FGF does have a profound effect on cartilage *in vivo* as well as *in vitro*. The mechanism by which FGF exerts this healing response is not clear. *In vitro*, FGF stimulates both cell replication and matrix production. The pattern of FGF-induced cartilage healing *in vivo* seems to indicate that both processes are stimulated *in vivo* as well. It is interesting to note that cartilage is an avascular tissue, yet FGF is a potent angiogenic factor. In this model, FGF did not elicit an angiogenic response. Thus, cartilage may contain a potent and specific anti-angiogenic activity as suggested (Eisenstein *et al.*, 1975). Thus, FGF may be an attractive molecule to utilize in cartilage repair.

Transforming Growth Factor-β (TGF-β) Family

Transforming growth factor beta (TGF-β), a 25 kDa homodimer, has been found in several naturally occurring forms, the best studied of which are TGF-β1 and TGF-β2 (see Sporn *et al.*, 1987, for review). The initial studies leading to the discovery that TGF-β may play a role in cartilage and bone development were made by Seyedin *et al.* (1985). Here, two bone-derived factors were purified based on their ability to induce rat mesenchymal cells to assume a chondroblastic phenotype. The factors, originally called cartilage-inducing factors A and B (CIF-A, CIF-B), are now known to be identical to TGF-β1 and TGF-β2, respectively (Seyedin *et al.*, 1986, 1987). Work by several groups showed that TGF-β inhibits the expression of a differentiated phenotype by mature chondroblasts. Skantze *et al.* (1985) demonstrated a decrease in both collagen and GAG production by chondroblasts cultured with TGF-β in agarose cultures. Rosen *et al.* (1988) showed that TGF-β induces a ''dedifferentiated'' state in chondroblasts grown in monolayer. These cells assume a fibroblastic morphology, exhibit decreased production of proteoglycans, and switch collagen phenotype from type II to type I. In contrast, Redini *et al.* (1988) found that TGF-β stimulated the synthesis of both collagen and GAGs. The reason for this discrepancy is not clear. One possible explanation is that there may be differences in culture conditions, i.e., cell density or the serum content of the media. For example, O'Keefe *et al.* (1988) reported that TGF-β decreased collagen synthesis by chick chondroblasts, but much more so in cultures containing 10% fetal bovine serum (FBS) as compared to cultures with no serum. Interestingly, they also observed an increase in GAG production after culturing with TGF-β. In the case of GAGs, there was

A) CONTROL

B) basic FGF (1ng/ul/hr)

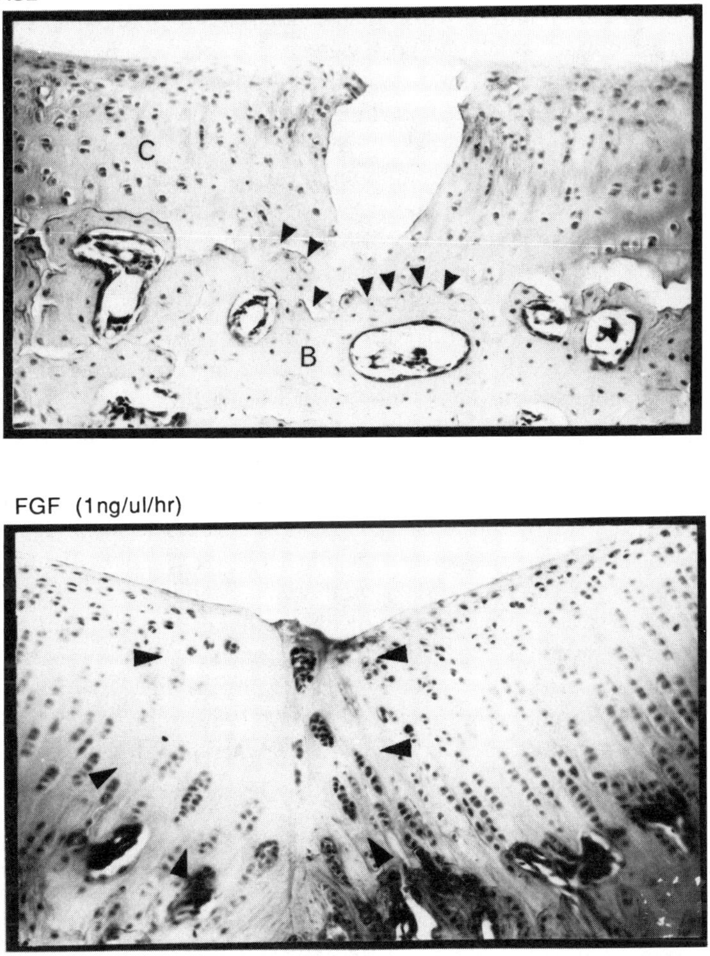

Figure 2. Effects of saline and basic FGF on the healing of an intrachondral lesion. A 2-mm lesion was made in the patella of New Zealand white rabbits. Basic FGF (1 μg/ml) or saline was administered via osmotic minipumps at a rate of 0.5 μl/h for 2 weeks. Lesions were evaluated histologically 20 days postsurgery. (A) Saline-treated lesion. (B) FGF-treated lesion. Note the boundary (indicated by small arrowheads) between calcified articular cartilage (labeled C) and underlying bone (labeled B) for the saline treated lesion (Panel 1A). Note the presence of large quantities of extracellular matrix containing chondrocytes (indicated by large arrowheads) in bFGF-treated lesion (Panel B). Reprinted with permission. Data supplied by Dr. A. Baird.

a dose-dependent increase in GAG production in the absence of serum, but a sharply biphasic response in the presence of serum. Still different responses were observed when dialysed FBS was used. These results underscore the need to carefully interpret *in vitro* results and to consider the potential for multifactor interactions (i.e., from other serum factors). There has been a considerable amount of work done evaluating the mitogenic effects of TGF-β on osteoblasts which will be discussed in another chapter (see also Centrella *et al.*, 1988, for a review). With chondroblasts, Skantze *et al.* (1985) saw no mitogenic effect of TGF-β alone, but observed a synergism between TGF-β and EGF in agarose cultures. We have observed variable mitogenic effects of TGF-β on chondrocytes in monolayer culture depending on culture conditions. Cells cultured in Hams F-12 media responded very poorly to TGF-β whereas cells cultured in DMEM exhibited a strong mitogenic response to TGF-β.* The reason for this response is under investigation. This difference does not appear to be related to variations in the phenol red content of each media, which has been linked to possible estrogen-like activity (Berthios *et al.*, 1986). O'Keefe *et al.* (1988) noted a biphasic mitogenic response of chick chondroblasts grown in monolayer when assays were done using DMEM. Similar to their results with collagen synthesis, the mitogenic activity of TGF-β was much more pronounced in the presence of FBS than in its absence.

Recently, the effects of TGF-β on organ cultures of bovine cartilage have been evaluated by Morales and Roberts (1986). In these studies, bovine articular cartilage was used from joints of young calves. In organ culture, TGF-β increased proteoglycan synthesis and inhibited catabolism of proteoglycans. There were no observed changes in DNA or collagen content of the cultures. Thus, there are varied reports on the effects of TGF-β on chondroblasts *in vitro*. The effects of TGF-β on cartilage *in vivo* have yet to be published. Recent work by Nemeth *et al.* (1989) showed that TGF-β administered at a fracture site can stimulate the generation of new cartilage. Further studies evaluating TGF-β and cartilage repair will surely be conducted in the near future.

Other Cytokines (PDGF, EGF, IL-1)

To date, the majority of investigations on cartilage has focused on the IGFs and FGFs and their related family members. More recently, there has been a surge in studies concerning TGF-βs. There has been little work describing the effects of other growth factors on cartilage and chondrocytes. Studies by Prins *et al.* (1982a, b) investigated the effects of PDGF and EGF, both of which elicited a mitogenic response. Hiraki *et al.* (1987) also found EGF to be a potent mitogen for chondroblasts. In contrast, while PDGF increased $^{35}SO_4$ incorporation, EGF caused either no increase or a slight decrease (Prins *et al.*, 1982b). Hiraki *et al.* (1987) and Kato *et al.* (1983) also showed that EGF and FGF act synergistically to stimulate DNA synthesis, as did EGF and the IGFs (MSA and CDF). In contrast, it was found that EGF reduced the stimulation of $^{35}SO_4$ incorporation by both FGF and the IGFs. Thus, EGF appears to have a very distinct profile of biologic activity on chondroblasts.

* D.M. Rosen and S.A. Stempien, unpublished observations.

While EGF is a potent mitogen, it also appears to inhibit the production of matrix. Similar observations have been made when evaluating the effects of EGF on fetal rat calvaria (Canalis *et al.*, 1979).

The involvement of an immune component in degenerative cartilage disease has been known for some time. The presence of macrophages and macrophage products has been observed in joints of patients with rheumatoid arthritis. Specifically, the role of IL-1 has drawn recent attention. IL-1 has been associated with chronic inflammation and has been found in the synovial fluid of diseased joints (Nowi *et al.*, 1984). *In vitro* it has been shown to cause a decrease in ^{35}S-sulfate incorporation and type II collagen synthesis while increasing the release of proteoglycan from cartilage explants (Benton and Tyler, 1988; Tyler and Benton 1986). On isolated chondrocytes, IL-1 has been shown to stimulate prostaglandin production and secretion of proteases, including collagenase and elastase (Gowan *et al.*, 1984; Chin and Lin, 1988; Redini *et al.*, 1988). Thus several investigators have attempted to correlate the production of macrophage products, including IL-1, with cartilage degeneration.

Differentiation Factors

Interest in differentiative factors began with the early work of Urist (1965) who observed ectopic endochondral bone formation with allogeneic demineralized bone matrix (DBM). Urist reported that fresh allogeneic bone matrix that is demineralized in 0.5 *N* HCl induces cartilage formation at an ectopic site. Subsequent studies were directed to delineate the mechanism of endochondral bone development with DBM implants at ectopic sites. One of the earliest responses to DBM implants is the recruitment and attachment of mesenchymal cells to the implant. This activity is then followed by proliferation and differentiation of these cells into chondrocytes and the formation of a cartilagenous matrix. This cartilage then undergoes calcification and resorption followed by formation and remodeling of bone (Reddi and Huggins, 1972; Reddi, 1981). This localized response is similar to the sequence of events that occur during normal embryonic endochondral bone formation as well as fracture repair in adults.

Several factors, including surface charge and the geometry of DBM particles, seem to influence the induction process (Reddi and Huggins, 1973, 1974). This model system has been extremely useful in exploring the biochemistry of distinct phases of endochondral bone development. Incorporation of $^{35}SO_4$ into proteoglycan and induction and expression of type II collagen have served as indices of cartilage formation. ^{45}Ca incorporation and increased alkaline phosphatase activity have been used as markers of calcification and bone formation (Reddi, 1981). When DBM was implanted in diffusion chambers with 0.2 μm filters to allow diffusion of proteins but not cells across the filter, endochondral bone formation occurred outside the chamber (Urist *et al.*, 1973), suggesting that induction is under the influence of a diffusible substance which has been called bone morphogenetic protein (BMP). Clearly one of the most challenging tasks in bone biology has been to identify and isolate factor(s) in DBM responsible for triggering this biological cascade. In an attempt to solubilize inducing factors from DBM, Urist and associates digested the DBM with bacterial collagenase and were able to solubilize and recover activity (Urist *et al.*, 1979). Subsequently, Hanamura *et al.* (1980) reported that inductive activity can be extracted from DBM using 4 *M* guanidine-HCl. Since then, many laboratories have used 4 *M* guanidine-HCl extracts of DBM as a starting material for the isolation and purification of cartilage/bone inducing factor(s). This general extraction procedure has also been used to solubilize endochondral bone inducing factor(s) from a murine osteosarcoma (Hanamura et al., 1980).

In vitro or *in vivo* assays have been used to isolate cartilage/bone inducing factor(s) from complex mixtures of proteins obtained from DBM or osteosarcoma extracts. There are several advantages to using *in vitro* assays; they are more sensitive and more rapid. They can, however, be risky and may not correlate with the true physiological activity of an isolated molecule. Therefore, many laboratories have relied on *in vivo* assay systems for the purification of cartilage/bone inducing factor(s).

Cartilage-Inducing and Stimulating Factors (CIFs)

Seyedin *et al.* (1985) followed the protein factors in DBM that were responsible for differentiation of mesenchymal cells into chondrocytes using culture conditions favorable for chondroblast growth. An assay was developed to allow expression of a chondrogenic phenotype using cartilage-specific proteoglycan and type II collagen expression as markers of chondrogenic induction. Embryonic rat muscle mesenchymal cells were embedded in agarose gels and treated with DBM extract. These cells undergo morphological changes and begin to assume a chondroblastic phenotype, including the production of cartilage-specific macromolecules. Using this assay, two cartilage-inducing factors, CIF-A and CIF-B, were isolated and characterized (Seyedin *et al.*, 1985). Both molecules were active at 1 to 10 ng/ml in this chondrogenic assay, as was determined using an ELISA assay for rat type II collagen and cartilage proteoglycan.

Evidence suggests that bone is a major source of TGF-β1 and TGF-β2 in the body (Seyedin *et al.*, 1985). These proteins are known to have multifunctional activities and have been implicated in wound healing and bone remodeling, as well as being an important growth modulators for many cells in the immune system. The *in vivo* response to subcutaneous injection of TGF-β1 or TGF-β2 is the induction and formation of granulation tissue, but not cartilage (Roberts *et al.*, 1986), suggesting a role in the wound healing process. Similar responses have been seen with TGF-β1 and TGF-β2 using collagen carriers in an injectable or implantable form.* Collagen has also been used to deliver partially purified cartilage/bone inducing factor(s) (Bentz *et al.*, 1987). Although TGF-β1 or TGF-β2 alone do not produce cartilage *in vivo*, as we will see later, they appear to play an important role in this process.

Recent studies looking at cell surface binding proteins to TGF-β indicate that there is a high molecular weight component that is found only on primary cells of mesenchymal origin, including chondroblasts and osteoblasts. The high molecular weight binding component has been characterized and found to be proteoglycan in nature (Segarini *et al.*, 1989). Cell surface proteoglycans are known to interact with extracellular matrix components (Rapraeger *et al.*, 1987). The proteoglycan binding component may provide cells with a unique matrix recognition tool. TGF-β binding may modulate this interaction and affect the way in which cells attach to matrix and differentiate.

Isolated chick limb bud mesenchymal cells will form cartilage, bone, or nonskeletal connective tissue (fibroblastic cells) depending upon culture conditions. These conditions include initial plating density, nutrient supply, and growth matrices that allow variable cell adhesion (Caplan, 1970; Finch et al., 1978; Solursh, 1984). Low density culture discourages differentiation of mesenchymal cells into chondrocytes and promotes a fibroblastic phenotype. This system has been used to investigate the effects of chondrogenic stimulating factors. The positive staining of cultures with toluidine blue and $^{35}SO_4$ uptake have been used as indices of cartilage formation. A water soluble fraction from a guanidine-HCl extract of demineralized bone has been shown to stimulate low density chick limb bud to differentiate into chondrocytes (Syftestad *et al.*, 1985). Fractionation of 4 *M* guanidine-HCl extracts of DBM revealed that the soluble fraction contains several proteins, all less than 31,000 in molecular weight with varying degrees of activity in this assay. Currently, it is not known whether *in vivo* chondrogenic activity observed during endochondral bone formation by DBM is related to the cartilage stimulating activity (CSA) described by Syftested *et al.* (1987). Furthermore, it is not known whether or not these proteins are related to other known growth factors. However, TGF-β at relatively high doses seems to have chondrogenic activity in this system (Syftestad *et al.*, personal communication).

Endochondral Bone Inducing Factors

In 1965 Urist reported that bone-inducing activity, ascribed to a specific BMP, is retained in collagenous bone matrix that had been decalcified by hydrochloric acid. In the ectopic model of bone formation described by Urist, the formation of cartilage preceded bone formation. This model has been used by many investigators as an assay system to isolate and purify cartilage/bone inducing factors. However, the method of introducing factor(s) to the ectopic

* Seyedin, Bentz, Armstrong *et al.*, unpublished observations.

site is quite varied. Urist *et al.* reported a 18,500 Da protein with BMP activity. BMP implanted alone was rapidly resorbed; however, an aggregate of BMP with other bone proteins induced endochondral bone in mouse thigh muscle. Therefore, the slow release of BMP from the implantation site appears to be an important component in eliciting its response. This BMP was defined as an acidic protein with an isoelectric point between 4.8 and 5.2 (Urist *et al.*, 1984).

Using a very similar assay, but a different delivery system, Sampath et al. (1987) isolated a bone inducing protein designated ''osteogenin'' from bovine DBM. Osteogenin was assayed in the rat subcutaneum using allogenic rat demineralized bone matrix residue as a carrier. Residue was prepared by extracting allogenic DBM with 4 *M* guanidine-HCl to remove endogenous bone inducing factor(s). Residue prepared by this method was shown to be inactive. In this model, the species-specific residue appears to play an important role in generating osteogenic activity. In the rat subcutaneous model, 1 to 5 μg of osteogenin induced endochondral bone formation as determined by alkaline phosphatase activity and histological evaluation of the implants. The purified osteogenin exhibited a molecular weight of 22,000 on SDS-polyacrylamide gels and appears to be different than other known protein factors. However, there are no amino acid sequence data yet available for osteogenin.

Using the reconstitution method and the model described by Sampath and Reddi (1987), Wang *et al.* (1988) reported the isolation of several bone morphogcnetic proteins with molecular weights of approximately 30,000. Limited proteolytic digests of a mixture of these proteins yielded sequence information which was used to clone their respective genes. Using oligonucleotide probes based on these sequences, Wozney *et al.* (1989) reported cloning four BMPs; BMP-1, BMP-2a, BMP-2b, and BMP-3. Genes for each BMP were inserted into mammalian expression systems, and all of the recombinantly expressed proteins apparently induced cartilage, but not bone. BMP-2a and BMP-2b have strong homology to each other and share some sequence homology to TGF-β gene family as does BMP-3. BMP-1 appears to be a unique protein and not related to the other BMPs. The nonreduced BMPs migrated as 30,000 Da proteins on SDS-polyacrylamide gels. Upon reduction, bands were observed migrating at 30,000, 18,000, and 16,000 Da. The 30 Kd band has been termed BMP-1, the 18 Kd band BMP-2 (BMP-2a), and the 16 Kd band BMP-3. Unlike the acidic BMP described by Urist *et al.* (1984), these BMPs are basic proteins.

In our own laboratory we have been using purified collagen or a composite of collagen and ceramic to deliver osteoinductive factor(s) (Bentz *et al.*, 1987; Nathan *et al.*, 1988). As indicated earlier, subcutaneous implantation of TGF-β (β1 or β2) resulted in an induction of granulation tissue with no evidence of any cartilage formation (Roberts *et al.*, 1986; Bentz *et al.*, 1987). However, several studies from this laboratory have suggested that TGF-β may play a key role in the process of endochondral bone formation (Bentz *et al.*, 1987). A partially purified osteoinductive factor (OIF) with a molecular weight ranging between 20,000 and 30,000, when further fractionated on reverse-phase HPLC and implanted in rats, showed little or no activity. However, when these fractions were spiked with TGF-β, large amounts of cartilage formed (Figure 3). These data strongly suggested to us that bone induction requires a complex mixture of polypeptides, one of which appears to be TGF-β. Recently, we have further purified this OIF and, using a collagen/ceramic carrier, confirmed the fact that TGF-β (β1 or β2) enhances bone induction at a subcutaneous site (Bentz *et al.*, 1987). Current data indicate that OIF migrates over a broad region, between 22 and 28 Kd, on SDS-PAGE. Furthermore, OIF elutes in a very broad peak on HPLC suggesting posttranslational modifications of OIF. Deglycosylation with glycopeptidase F yields a single sharp band with a molecular weight of 12,000, suggesting that OIF exists as a single core protein with multiple glycosylations. The relatedness of the glycoforms and the deglycosylated core protein has been confirmed immunologically using a monoclonal antibody to OIF (Rosen *et al.*, 1989). More recent information revealed that the purified OIF does not possess osteoinductive activity. Partially purified preparations appeared to contain a mixture of BMP-2 and BMP-3 whose activity was substantially enhanced by TGF-β (Bentz *et al.*, submitted). The protein termed OIF belongs to the leucine-rich family of proteins with an as yet unknown biological activity.

Summary

In summary, even though the model systems used to detect *in vivo* cartilage/bone inducing factors appear very similar, there are subtle but very significant differences in the methods used to evaluate the biological activity of isolated protein factors. These critical differences may account for the varied identity of factors isolated by many laboratories. The carrier is a seemingly inert but quite important component of the system and can influence the purification and identification of cartilage/bone inducing factors. Carriers themselves may elicit a specific biological response, such as recruitment of specific or non-specific inflammatory cells. These cells and their products can have a significant impact on the process of cartilage or bone formation *in vivo*. The site of implantation, e.g., subcutaneous or intramuscular, as well as the species, age, and sex of the animal used may also be important variables. Although there has been significant progress made on the isolation of novel cartilage growth and inducing factors, the biology and mechanisms of action of these factors, e.g., receptor interactions, which specific cell types are involved,

A. Fractionation of CM-1 on C18 Reversed-Phase HPLC

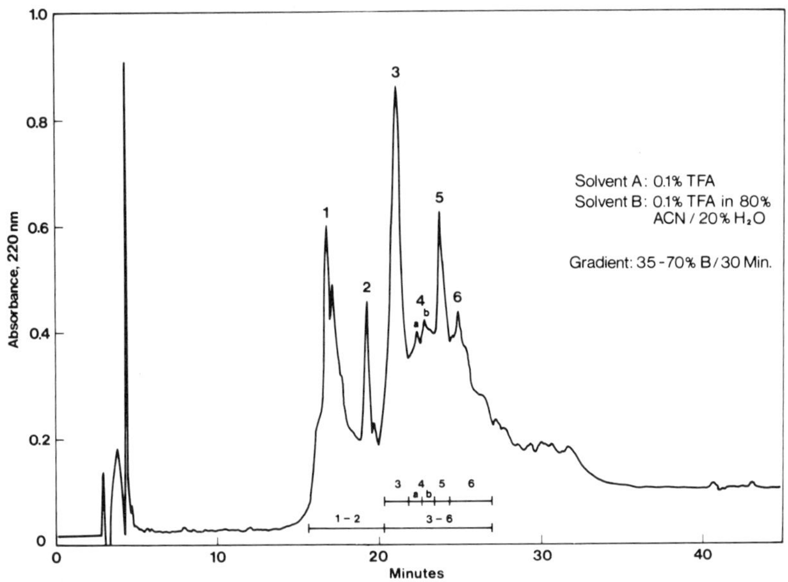

B. Osteogenic Activity of CM-1 Fractionated on C18 Reversed-Phase HPLC

Fraction	Histology Score (0-5+)			
	− TGF-β		+ TGF-β	
	Cartilage	Bone	Cartilage	Bone
Combined	1+	5+	ND	ND
1,2	0	0	0	0
3	0	0	0	0
4a	trace	trace	3+	trace
4b	trace	trace	trace	trace
5	trace	trace	4+	trace
6	trace	1+	5+	trace

Figure 3. Role of TGF-β in osteoinduction. Semipurified bovine bone extract (CM-1) was fractionated by C18-reversed phase HPLC chromatography (A). Fractions were implanted subcutaneously in Sprague-Dawley rats as described (Ref. 74) with or without exogenous TGF-β1 as indicated. New bone and cartilage formation was determined histologically. Implants were scored from 0 to 5+, with 5+ indicating that 80 to 100% of the implant was occupied by newly formed bone or cartilage. Note the substantial enhancement of osteoinductive activity when samples were spiked with exogenous TGF-β1 (B). Data supplied by Dr. H. Bentz (1987).

etc., are poorly understood. However, with molecular biological and cell biological tools applied to this field, the future will continue to bring many exciting breakthroughs and discoveries.

References

A. Baird and N. Ling. Fibroblast growth factors are present in the extracellular matrix produced by endothelial cells in vitro: implications for a role of heparinase-like enzymes in the neovascular response. *Biochem. Biophys. Res. Commun.* **142**: 428–435 (1987).

A. Baird, P. Mormede, and P. Bohlen. Immunoreactive fibroblast growth factor in cells of peritoneal exudate suggests its identity with macrophage-derived growth factor. *Biochim. Biophys. Res. Commun.* **126**: 358–364 (1985).

R.C. Baxter. The somatomedins: insulin-like growth factors. *Adv. Clin. Chem.* **25**: 49–115 (1987).

F. Beck, N.J. Samani, J.D. Penschow, B. Thorley, G.W. Tregear, and J.P. Coghlan. Histochemical localization of IGF-I and -II mRNA in the developing rat embryo. *Development* **101**: 175–184 (1987).

A. Bennett, D.M. Wilson, F. Liu, R. Nagashima, R.G. Rosenfeld, and R.L. Hintz. Levels of insulin-like growth factors I and II in human cord blood. *J. Clin. Endocrinol. Metab.* **57**: 609–612 (1983).

H.P. Benton and J.A. Tyler. Inhibition of cartilage proteoglycan synthesis by interleukin I. *Biochem. Biophys. Res. Commun.* **154**: 421–428 (1988).

H. Bentz, R. Armstrong, and S.M. Seyedin. Cartilage induction and differentiation: the role of bone derived cartilage inducing factor (CIF-A). *Development and Diseases of Cartilage and Bone Matrix.* Alan R. Liss, Inc., New York. pp. 137–147, 1987.

H. Bentz, R. Nathan, D. Rosen, R. Armstrong, A. Thompson, P. Segarini, M. Mathews, J. Dasch, K. Piez, and S. Seyedin. Purification of an osteoinductive factor from bovine demineralized bone. *J. Cell Biol.* **107**: 162 (abst.) (1989).

H. Bentz, A. Thompson, R. Armstrong, R. Chang, K. Piez, and D. Rosen (submitted).

M.J. Berridge. Growth factors, oncogenes, and inositol lipids. *Cancer Supv.* **5**: 413–430 (1986).

Y. Berthios, J.A. Katzenellenbogen, and B.S. Katzenellenbogen. Phenol red in tissue culture media is a weak estrogen: implications concerning the study of estrogen-responsive cells in culture. *Proc. Natl. Acad. Sci. U.S.A.* **83**: 2496–2500 (1986).

W.M. Burch, S. Weir, and J.J. Van Wyk. Embryonic chick cartilage produces its own somatomedin-like peptide to stimulate cartilage growth in vitro. *Endocrinology* **119**: 1370–1376 (1986).

E. Canalis and L.G. Raisz. Effect of epidermal growth factor on bone formation in vitro. *Endocrinology* **104**: 862–869 (1979).

A.I. Caplan. Effect of nicotimamide sensitive teratogen 3-acetypryidine on chick limb cells in culture. *Exp. Cell Res.* **62**: 341–348 (1970).

M. Centrella, T.L. McCarthy, and E. Canalis. Skeletal tissue and transforming growth factor β. *FASEB J.* **2**: 3066–3073 (1988).

J.E. Chin and Y. Lin. Effects of recombinant human interleukin-1β on rabbit articular chondrocytes. Stimulation of prostanoid release and inhibition of cell growth. *Arthritis Rheum.* **31**: 1290–1296 (1988).

P. Cuevas, J. Burgos, and A. Baird. Basic fibroblast growth factor (FGF) promotes cartilage repair in vivo. *Biochem. Biophys. Res. Commun.* **156**: 611–618 (1988).

W.H. Daughaday, K. Hall, M.S. Raben, W.D. Salmon, J.L. Van den Brande, and J.J. Van Wyk. Somatomedins: proposed designation for sulphation factor. *Nature (London)* **235**: 107–109 (1972).

A.J. D'Ercole, L.E. Stiles, and L.E. Underwood. Tissue concentrations of somatomedin C: further evidence for multiple sites of synthesis and paracrine or autocrine mechanisms of action. *Proc. Natl. Acad. Sci. U.S.A.* **81**: 935–939 (1988).

R. Eisenstein, K. Kuettner, C. Neapolitan, L.W. Soble, and N. Sorgente. The resistance of certain tissues to invasion. III. Cartilage extracts inhibit the growth of fibroblasts and endothelial cells in culture. *Am. J. Pathol.* **81**: 337–348 (1975).

R.A. Finch, C.L. Parker, and S.T. Walton. The lack of an inhibitory effect of hyaluronate on chondrogenesis in chick limb bud mesoderm cells grown in culture. *Cell Diff.* **7**: 283–293 (1978).

D. Gospodarowicz, G. Neufeld, and L. Schweigerer. Fibroblast growth factor. *Mol. Cell. Endocrinol.* **46**: 187–206 (1986).

M. Gowan, D.D. Wood, E.J. Ihrie, J.E. Meats, and R.G.G. Russell. Stimulation by interleukin-1 of cartilage breakdown and production of collagenase and proteoglycanase by human chondrocytes but not by human osteoblasts in vitro. *Biochim. Biophys. Acta* **797**: 186–193 (1984).

D. Hamermon, J. Sasse, and M. Klagsbrun. A cartilage-derived growth factor enhances hyaluronate synthesis and diminishes sulfated glycosaminoglycan synthesis in chondrocytes. *J. Cell. Physiol.* **127**: 317–322 (1986).

V.K.M. Han, A. D'Ercole, and P.K. Lund. Cellular localization of somatomedin (insulin-like growth factor) messenger RNA in the human fetus. *Science* **236**: 193–197 (1987).

H. Hanamura, Y. Higuchi, M. Nakagawa, H. Iwata, H. Nogami, and M.R. Urist: Solubilized bone morphogenetic protein (BMP) from mouse osteosarcoma and rat demineralized bone matrix. *Clin. Orthop. Res.* **148**: 281–290 (1980).

W. Hey, C.A. Browne, and G.D. Thorburn. Fetal sheep serum contains a high molecular weight insulin-like growth factor (IGF) binding protein that is acid stable and specific for IGF-II. *Endocrinology* **121**: 1975–1984 (1987).

R.L. Hintz and F. Live. Demonstration of specific plasma protein binding sites for somatomedin. *J. Clin. Endocrinol. Metab.* **45**: 988–995 (1977).

Y. Hiraki, H. Inoue, Y. Kato, M. Fukuya, and F. Suzuki. Combined effects of somatomedin-like growth factors with fibroblast growth factor and epidermal growth factor in DNA synthesis in rabbit chondrocytes. *Mol. Cell. Biochem.* **76**: 185–193 (1987).

Y. Kato and D. Gospodarowicz. Sulfated proteoglycan synthesis by confluent cultures of rabbit costal chondrocytes grown in the presence of fibroblast growth factor. *J. Cell. Biol.* **100**: 477–485 (1984).

Y. Kato, Y. Hiraki, H. Inoue, M. Kinoshita, Y. Yutani, and F. Suzuki. Differential and synergistic actions of somatomedin-like growth factors, fibroblast growth factor, and epidermal growth factor in rabbit costal chondrocytes. *Eur. J. Biochem.* **129**: 685–690 (1983).

Y. Kato, Y. Nomura, Y. Daikuhara, N. Nasu, M. Tsuji, A. Asuda, and F. Suzuki. Cartilage-derived factor (CDF). I. Stimulation of proteoglycan synthesis in rat and rabbit costal chondrocytes in culture. *Exp. Cell Res.* **130**: 73–81 (1980).

Y. Kato, Y. Nomura, M. Tsuji, H. Ohmae, M. Kinoshita, S. Hamamoto, and F. Suzuki. Cartilage-derived factor (CDF). II. Somatomedin-like action on cultured chondrocytes. *Exp. Cell Res.* **132**: 339–347 (1981a).

Y. Kato, Y. Nomura, M. Tsuji, H. Ohmae, and F. Suzuki. Somatomedin-like peptides isolated from bovine cartilage (cartilage-derived factor): isolation and some properties. *Proc. Natl. Acad. Sci. U.S.A.* **78**: 6831–6835 (1981b).

Y. Kato, Y. Nomura, M. Tsuji, H. Ohmae, T. Nakazawa, and F. Suzuki. Multiplication-stimulating activity (MSA) and cartilage-derived factor (CDF): biological actions in cultured chondrocytes. *J. Biochem.* **90**: 1377–1386 (1981c).

Y. Kato, R. Watanabe, Y. Nomura, M. Tsuji, F. Suzuki, L.G. Raisz, and E. Canalis. Effect of bone-derived growth factor on DNA, RNA, and proteoglycan synthesis in cultures of rabbit costal chondrocytes. *Metabolism* **31**: 812–815 (1982).

D. Kimelman and M. Kirschner. Synergistic induction of mesoderm by FGF and TGF-β and the identification of an mRNA coding for FGF in the early Xenopus embryo. *Cell* **51**: 869–877 (1987).

M. Klagsbrun and S. Smith. Purification of a cartilage-derived growth factor. *J. Biol. Chem.* **255**: 10859–10866 (1980).

M. Klagsbrun, R. Langer, R. Levenson, S. Smith, and C. Lillihei. The stimulation of DNA synthesis and cell division in chondrocytes and 3T3 cells by a growth factor isolated from cartilage. *Exp. Cell Res.* **105**: 99–108 (1977).

R.M. Lyons and G.L. Smith. Characterization of multiplication-stimulating activity (MSA) carrier protein. *Mol. Cell. Endocrinol.* **45**: 263–270 (1986).

C. Mevli, J. Zapf, and E.R. Froesch. NSILA-carrier protein abolishes the activity of nonsuppressible insulin-like activity (NSILAs) on perfused rat heart. *Diabetologia* **14**: 255–259 (1978).

S. Mohan, J.C. Jennings, T.A. Linkhart, and D.J. Baylink. Primary structure of human skeletal growth factor: homology with insulin-like growth factor-II. *Biochim. Biophys. Acta* **966**: 44–55 (1988).

T.I. Morales and A.B. Roberts. Transforming growth factor β regulates the metabolism of proteoglycans in bovine cartilage organ cultures. *J. Biol. Chem.* **263**: 12828–12831 (1988).

A.C. Moses, S.P. Nissley, P.A. Short, M.M. Rechler, and J.M. Podskalny. Purification and characterization of multiplication stimulating activity. *Eur. J. Biochem.* **103**: 387–400 (1980).

A.C. Moses, S.P. Nissley, P.A. Short, M.M. Rechler, R.M. White, A.B. Knight, and O.Z. High. Elevated levels of multiplication-stimulating activity, an insulin-like growth factor, in fetal rat serum. *Proc. Natl. Acad. Sci. U.S.A.* **77**: 3649–3653 (1980).

R.M. Nathan, H. Bentz, R.A. Armstrong, K.A. Piez, T.L. Smestad, L.R. Ellingsworth, J.M. McPherson, and S.M. Seyedin. Osteogenesis in rats with an inductive bovine composite. *J. Orthop. Res.* **6**: 324–334 (1988).

G.G. Nemeth, A. Heydemann, M.E. Joyce, M.E. Bolander. Expression and localization of transforming growth factor-β in a model of fracture healing. *Trans. J. Orthop. Res.* **14**: 587 (Abst.) (1989).

A. Nilsson, J. Isgaard, A. Lindhal, A. Dahlstrom, A. Skottner, and O.G.P. Isaksson. Regulation by growth hormone of number of chondrocytes containing IGF-I in rat growth plate. *Science* **233**: 571–574 (1986).

M.E. Nowi, G.S. Panayi, and S.M. Goodman. Cytokines and the chronic inflammation rheumatic disease. I. The presence of interleukin-1 in synovial fluids. *Clin. Exp. Immunol.* **55**: 295–302 (1984).

R.J. O'Keefe, J.E. Puzas, J.S. Brand, and R.N. Rosier. Effects of transforming growth factor β on matrix synthesis by chick growth plate chondrocytes. *Endocrinology* **122**: 2953–2961 (1988).

K.D. Osborn, S.B. Trippel, and H.J. Mankin. Growth factor stimulation of adult articular cartilage. *J. Orthop. Res.* **7**: 35–42 (1989).

A.P.A. Prins, J.M. Lipman, and L. Sokoloff. Effect of purified growth factors on rabbit articular chondrocytes in monolayer culture. I. DNA Synthesis. *Arthritis Rheum.* **25**: 1217–1227 (1982a).

A.P.A. Prins, J.M. Lipman, C.A. McDevitt, and L. Sokoloff. Effect of purified growth factors on rabbit articular chondrocytes in monolayer culture. II. Sulfated proteoglycan synthesis. *Arthritis Rheum.* **25**: 1228–1238 (1982b).

A. Rapraeger, J. Koda, and M. Bernfield. Matrix interactions of cell surface proteoglycans. In: *Connective Tissue: Molecular Pathology of the Extracellular Matrix*, Ed J. Vitto and A.J. Perejda. Marcel Dekker, New York. pp. 213–231, 1987.

A.H. Reddi. Cell biology and biochemistry of endochondral bone development. *Collagen Relat. Res.* **1**: 209–229 (1981).

A.H. Reddi and C.B. Huggins. Biochemical sequences in the transformation of normal fibroblasts in adolescent rats. *Proc. Natl. Acad. Sci. U.S.A.* **69**: 1601–1605 (1972).

A.H. Reddi and C.B. Huggins. Influence of geometry of transplanted tooth and bone on transformation of fibroblasts. *Proc. Soc. Exp. Biol. Med.* **143**: 634–637 (1973).

A.H. Reddi and C.B. Huggins. Cyclic electrochemical inactivation and restoration of competence of bone matrix to transform fibroblasts. *Proc. Natl. Acad. Sci. U.S.A.* **71**: 1648–1652 (1974).

F. Redini, P. Galera, A. Mauriel, G. Loyau, and J.P. Pujol. Transforming growth factor-β stimulates collagen and glycosaminoglycan biosynthesis in cultured rabbit articular chondrocytes. *FEBS Lett.* **234**: 172–176 (1988).

F. Redini, C. Lafuma, J.-P. Pujol, L. Robert, and W. Hornbeck. Effects of cytokines and growth factors on the expression of elastase activity by human synoviocytes, dermal fibroblasts, and rabbit articular chondrocytes. *Biochem. Biophys. Res. Commun.* **155**: 786–693 (1988).

L. Rifas, V. Shen, K. Mitchell, and W.A. Peck. Macrophage-derived growth factor for osteoblast-like cells and chondrocytes. *Proc. Natl. Acad. Sci. U.S.A.* **81**: 4558–4562 (1984).

E. Rinderknecht and R.E. Humbel. Amino-terminal sequences of two polypeptides from human serum with non-suppressible insulin-like and cell-growth-promoting activities: evidence for structural homology with insulin B chain. *Proc. Natl. Acad. Sci. U.S.A.* **73**: 4379–4381 (1976).

A.B. Roberts, M.B. Sporn, R.K. Assoian, L.M. Smith, N.S. Roche, L.M. Wakefield, U.I. Heine, L.A. Liotta, V. Falanga, J.H. Kehrl, and A.S. Fauci. Transforming growth factor type-beta: rapid induction of fibrosis and angiogenesis *in vivo* and stimulation of collagen formation *in vitro*. *Proc. Natl. Acad. Sci. U.S.A.* **83**: 4167–4171 (1986).

D.M. Rosen, S.A. Stempien, L. Bonewald, G. Mundy, H. Bentz, and S.M. Seyedin. Effects of a novel osteoinductive factor on osteoblastic cells in vitro. *J. Cell Biol.* **107**: 162a (Abst.) (1989).

D.M. Rosen, S.A. Stempien, A.Y. Thompson, and S.M. Seyedin. Transforming growth factor-β modulates the expression of osteoblast and chondroblast phenotypes in vitro. *J. Cell. Physiol.* **134**: 337–346 (1988).

S.M. Russell and E.M. Spencer, Local injections of human or rat growth hormone or of purified Somatomedin-C stimulate unilaterial tibial epiphyseal growth in hypophysectomized rats. *Endocrinology* **116**: 2563–2567 (1985).

B.L. Sachs, V.M. Goldberg, R.W. Moskowitz, and C.J. Malemud. Response of articular chondrocytes to pituitary fibroblast growth factor (FGF). *J. Cell. Physiol.* **112**: 51–59 (1982).

W.D. Salmon and W.H. Daughaday. A hormonally controlled serum factor which stimulates sulfate incorporation by cartilage in vitro. *J. Lab. Clin. Med.* **49**: 825–836 (1957).

T.K. Sampath, N. Muthukumaran, and A.H. Reddi. Isolation of osteogenin, an extracellular matrix-associated bone-inductive protein, by heparin affinity chromatography. *Proc. Natl. Acad. Sci.* **84**: 7109–7113 (1987).

C. Schmid, T. Steiner, and E.R. Froesch. Insulin-like growth factors stimulate the synthesis of nucleic acids and glycogen in cultured calvaria cells. *Calcif. Tissue Int.* **35**: 578–585 (1983).

E. Schoenle, J. Zapf, C. Hauri, T. Steiner, and E.R. Froesch. Comparison of in vivo effects of insulin-like growth factors I and II and of growth hormone in hypophysectomized rats. *Acta Endocrinol.* **108**: 167–174 (1985).

P.R. Segarini, D.M. Rosen, and S.M. Seyedin. Binding of transforming growth factor-β to cell surface proteins varies with cell type. *Mol. Endocrinol.* **3**: 261–272 (1989).

S.M. Seyedin, P.R. Segarini, D.M. Rosen, A.Y. Thompson, H. Bentz, and J. Graycar. Cartilage-inducing factor B is a unique protein structurally and functionally related to transforming growth factor β. *J. Biol. Chem.* **262**: 1946–1949 (1987).

S.M. Seyedin, T.C. Thomas, A.Y. Thompson, D.M. Rosen, and K.A. Piez. Purification and characterization of two-cartilage-inducing factors from bovine demineralized bone. *Proc.*

Natl. Acad. Sci. U.S.A. **82**: 2267–2271 (1985).

S.M. Seyedin, A.Y. Thompson, H. Bentz, D.M. Rosen, J.M. McPherson, A. Conti, N.R. Siegel, J.R. Gallupi, and K.A. Piez. Cartilage-inducing factor A: apparent identity to transforming growth factor β. *J. Biol. Chem.* **261**: 5693–5695 (1986).

K.A. Skantze, C.E. Brinckerhoff, and J.P. Collier. Use of agarose culture to measure the effect of transforming growth factor β and epidermal growth factor on rabbit articular chondrocytes. *Cancer Res.* **45**: 4416–4421 (1985).

M. Solursh. Cell and matrix interactions during limb chondrogenesis *in vitro*. In: *The Role of Extracellular Matrix in Development,* Ed R. Trelstad. Alan R. Liss, New York. p. 277, 1984.

M.B. Sporn, A.B. Roberts, L.M. Wakefield, and B. deCrombrugghe. Some recent advances in the chemistry and biology of transforming growth factor-beta. *J. Cell Biol.* **105**: 1039–1045 (1987).

F. Stylianopoulou, A. Efstratiadis, J. Herbert, and J. Pintar. Pattern of the insulin-like growth factor II gene expression during rat embryogenesis. *Development* **103**: 497–506 (1988).

R. Sullivan and M. Klagsbrun. Purification of cartilage-derived growth factor by heparin affinity chromatography. *J. Biol. Chem.* **260**: 2399–2403 (1985).

G.T. Syftestad, P.A. Lucas, and A.I. Caplan. The *in vitro* chondrogenic response of limb-bud mesenchyme to a water-soluble fraction prepared from demineralized bone matrix. *Differentiation* **29**: 230–237 (1985).

G.T. Syftestad, P.A. Lucas, H. Ohgushi, and A.I. Caplan. Chondrogenesis as an *in vitro* response to bioactive factors extracted from adult bone and non-skeletal tissues. *Development and Diseases of Cartilage and Bone Matrix.* Alan R. Liss, New York. pp. 137–147, 1987.

S.B. Trippel, S.D. Chernausek, J.J. Van Wyk, A.C. Moses, and H.J. Mankin. Demonstration of type I and type II somatomedin receptors on bovine growth plate chondrocytes. *J. Orthop. Res.* **6**: 817–826 (1988).

S.B. Trippel, J.J. Van Wyk, M.G. Foster, and M.E. Svoboda. Characterization of a specific somatomedin-C receptor on isolated bovine growth plate chondrocytes. *Endocrinology* **112**: 2128–2136 (1983).

J.A. Tyler and H.P. Benton. Synthesis of type II collagen is decreased in cartilage cultured with interlukin I while the rate of intracellular degradation remains unchanged. *Collagen Relat. Res.* **8**: 393–405 (1988).

M.R. Urist. Bone formation by autoinduction. *Science* **150**: 893–899 (1965).

M.R. Urist, Y.K. Huo, A.G. Brownell, W.M. Hohl, J. Buyske, A. Lietze, P. Tempst, M. Hunkapiller, and R.J. DeLange. Purification of bovine bone morphogenetic protein by hydroxyapatite chromatography. *Proc. Natl. Acad. Sci.* **81**: 371–375 (1984).

M.R. Urist, H. Iwata, P.L. Ceccotti, R.C. Dorfman, S.D. Boyd, R.M. McDowell, and C. Chien. Bone morphogenesis in implants of insoluble bone gelatin. *Proc. Natl. Acad. Sci. U.S.A.* **79**: 3511–3515 (1973).

M.R. Urist, A. Mikulski, and A. Lietze. Solubilized and insolubilized bone morphogenetic protein (BMP). *Proc. Natl. Acad. Sci. U.S.A.* **76**: 1828 (1979).

E.A. Wang, V. Rosen, P. Cordes, R.M. Hewick, M.J. Kriz, D.P. Luxenberg, B.S. Sibley, and J.M. Wozney. Purification and characterization of other distinct bone-inducing factors. *Proc. Natl. Acad. Sci. U.S.A.* **85** (1988).

J.M. Wozney, V. Rosen, A.J. Celeste, L.M. Mitsock, M.J. Whitters, R.W. Kriz, R.M. Hewick, and E.A. Wang. Novel regulators of bone formation: molecular clones and activities. *Science* **242**: 1528–1534 (1988).

J. Zapf, E. Schoenle, G. Jagars, I. Sand, J. Gurnwald, and E.R. Froesch. Inhibition of the action of nonsuppressible insulin-like activity on isolated rat fat cells by binding to its carrier protein. *J. Clin. Invest.* **63**: 1077–1084 (1979).

J. Zapf, M. Waldvogel, and E.R. Froesch. Binding of nonsuppressible insulin-like activity to human serum: Evidence for a carrier protein. *Arch. Biochem. Biophys.* **168**: 638–645 (1975).

5

Ionic Regulation of Chondrogenesis

ROCKY S. TUAN
Department of Orthopaedic Surgery
Thomas Jefferson University
Philadelphia, Pennsylvania

Introduction

Chondrogenesis is the process whereby mesenchymal cells differentiate into chondrocytes with the histogenesis of cartilage. Histologically, chondrogenesis involves dramatic changes in cell shape as well as in the nature of the extracellular matrix. The mature cartilage consists of polygonal or round cells actively synthesizing a prodigious amount of extracellular matrix macromolecules, including type II collagen and high-M_r, sulfated proteoglycans. Thus, the conversion of a mesenchymal cell into a chondrocyte is necessarily a complex process, involving changes in gene expression and biochemical and cellular events. Additional macromolecular organizational steps eventually lead to the assembly of chondrocytes and their extracellular matrix to form a cartilaginous tissue. The aim of this review is to examine how ionic factors and their cellular interactions may regulate chondrogenesis.

Chondrogenesis *In Vivo* and Cellular Condensation

During embryonic development, many chondrogenic events take place leading to the formation of the skeletal structure. Chondrogenesis has long been an active area of research in embryology and developmental biology. In particular, limb chondrogenesis and somitic chondrogenesis have been studied extensively by many workers. The focus of this review is primarily on limb chondrogenesis as an example to illustrate the importance of ionic regulation.

Limb chondrogenesis is a histologically well defined phenomenon. A brief description of this developmental process in the chick embryo is as follows (Amprino, 1984; Gilbert, 1988; Hunter and Caplan, 1983; Saunders, 1972; Solursh, 1984): During early development, at the primitive streak stage, limb buds are initially derived from a single zone of cells of the embryonic mesodermal mesenchyme layer. Shortly afterwards the cells of this flank region segregate and become committed to either fore or hind limb morphology.

These cells then form bud structures which contain both ectoderm and mesoderm, and enter the ''morphogenetic phase'' (Zwilling, 1968). Throughout this period, the mesenchymal cells of the pre-limb exert a specific inductive effect on the overlying ectoderm which causes it to form the apical ectodermal ridge (AER), a structure which is subsequently involved in dictating the proximo-distal polarity of limb outgrowth (Saunders, 1977; Saunders and Gasseling, 1968). According to Zwilling (1968), the central core of mesenchyme represents the pluripotent cell population of the bud at least up to Hamilton-Hamburger stage 21-22 (Ahrens *et al.*, 1977), although work of Christ *et al.* (1977) and Chevallier *et al.* (1977) strongly indicate that the myogenic and chondrogenic cells have distinct lineages from the earliest emergence of the limb bud. Between stages 21/22 to 23/24, the cells of the core of the limb bud undergo definitive phenotypic ''commitment'' to give rise to the various tissue-cell types of the limb, including muscle and cartilage. Morphologically, this is accompanied by a phenomenon known as cellular condensation, i.e., an increase in cell-packing density (Ede, 1983; Fell and Canti, 1935; Mohammed, 1978; Thorogood and Hinchliffe, 1975), in the limb core. It is generally believed that condensation is a result of relative cell movement, either direct or matrix-mediated, rather than increased cell division (Ede, 1983; Ede *et al.*, 1977; Newman *et al.*, 1985; Oster *et al.*, 1984, 1985). Phenotypic changes do not occur until after stage 23-24, when overt chondrogenesis takes place in the condensed zones where cells begin to display typical chondrocyte morphology and produce prodigious quantities of extracellular matrix (Shinomura *et al.*, 1984; Toole and Linsenmayer, 1977; Vasan and Lash, 1977, 1979; von der Mark *et al.*, 1976), and thereby gradually separate from one another. Formation of the cartilage anlage then ensues.

The temporal correlation between the abrupt increase in cell density in the central core of the limb mesenchyme and the cellular commitment to chondrogenesis suggests that the two events may be causally related, perhaps involving direct interactions between these similar cells. In support of this possibility, one may also consider the concept of prepatterning as formulated by Newman and Frisch (1979), which would entail that the nature or extent of cell interactions defines the local microenvironments conducive to chondrogenesis in the developing limb bud.

Chondrogenesis *In Vitro*

The process of chondrogenesis has been observed *in vitro* using mesenchymal cells dissociated from the embryonic limb bud (Ahrens *et al.*, 1977; Osdoby and Caplan, 1979; Solursh *et al.*, 1978). For example, limb mesenchymal cells obtained from stage 23-24 chick embryos plated as high-density micromass cultures (Ahrens *et al.*, 1977) exhibit the following typical morphological changes (Ahrens *et al.*, 1977; San Antonio and Tuan, 1986; Solursh *et al.*, 1978): (1) cells begin to form microscopically visible aggregates by 24 h (end of day 1); (2) during day 2 of culture, these aggregates enlarge by apparently recruiting neighboring cells; and (3) during day 3 of culture, distinct, metachromatic nodules are formed; these nodules are cartilaginous based on their positive staining with alcian blue, and presence of type II collagen and sulfated proteoglycans. Interestingly, the chondrogenic potential of mesenchymal cells *in vitro* is highly dependent on the initial cell plating density (Ahrens *et al.*, 1977; Osdoby and Caplan, 1979). Thus, a threshold value of cell plating density exists, such that cultures plated below this value do not yield any nodules. In fact, early experiments by Moscona (1956) and Umansky (1966) also demonstrated that chondrogenic potential *in vitro* was totally suppressed when limb bud explants were dissected into pieces smaller than a threshold size or diluted by mixing with non-bud tissues, findings which are also observed with isolated mesenchymal cells *in vitro* (Solursh and Reiter, 1980). This density dependence characteristics of chondrogenesis *in vitro* serves as further support for the functional requirements of cell-cell interactions as described above.

Cell-Cell Interactions in Chondrogenesis

If cell-cell interactions, observed histologically as cellular condensation, are indeed crucial to the induction of chondrogenesis, it would be of great interest to address the following experimental questions: (1) what is the mechanism(s) mediating cellular condensation; (2) what is the nature of these interactions and cellular condensation; and finally, (3) what kind of signals, if any, are generated by this process to result in, first, the cellular commitment to, and then, the expression of the chondrogenic phenotype. Indeed, these questions have been addressed by many developmental biologists who have studied chondrogenesis. In this review, specific attention will be paid to the possibility that ions and polyions may play important regulatory roles in chondrogenesis.

At a cellular and molecular level, cell-cell interactions may be classified into at least the following. First, the interactions may operate at a membrane-to-membrane level, i.e., membrane-associated components (e.g., the CAM family of proteins; see review by Edelman, 1986) could serve as the interactive molecular species in a homotypic or heterotypic manner. Second, the interactions could be cell-to-matrix or matrix-to-cell; in this manner, a cell-surface receptor(s) interacts with a specific extracellular matrix component(s) which acts as a molecular signal (Buck and Horwitz, 1987; Hynes, 1987; Ruoslahti, 1988). (Note: cells may also interact by utilizing their extracellular matrix as a mechanical transmitter of physical forces in a vectorial manner [Harris *et al.*, 1984; Stopak and Harris, 1982]). Third, the interaction could be through the short range delivery of humoral factors in a paracrine- and/or autocrine-like manner; thus, various growth factors could play the role of the molecular messenger (Deuel, 1987). Fourth, cells may communicate directly via juxtaposed membrane channels, i.e., gap junctions (Bennett and Spray, 1985), which allow the transmittal of low-M_r molecules to perhaps act as signaling messengers. A central requirement inherent in all of these modes of cell-cell interactions is molecular specificity, namely that both components of the interacting system, i.e., the ligand/receptor or the donor/recipient, have to show specificity toward each other. Furthermore, to function as part of a developmental program, the interactions need to be activated (and deactivated) in a time- and position-dependent manner.

How may specificity be dictated in biomolecular interactions? For noncovalent molecular interactions in general, ionic forces and hydrophobic and hydrogen bonding are the most important elements, all of which are strictly dependent on the structure and conformation of the macromolecules involved. In particular, if these biomacromolecules and their ligands are in fact expressed or made available in a site-specific and temporally defined manner in the limb bud, a three-dimensional developmental programming of limb chondrogenesis may indeed take place.

Calcium Ions and Chondrogenesis *In Vitro*

Divalent cations, specifically Ca^{2+}, have been strongly implicated in mediating cell-cell interactions (e.g., Brakenbury *et al.*, 1981; Damluji and Riley, 1979; Hennings and Holbrook, 1983; Magnani *et al.*, 1981; Takeichi, 1987; Takeichi *et al.*, 1981). Several recent investigations have shown that Ca^{2+} is indeed intimately involved in cell-cell interactions of the embryonic limb mesenchyme. Bee and von der Mark (1984) reported that dissociated chick limb bud mesenchyme undergoes two temporal phases of aggregation when placed in suspension culture. The first occurs shortly after dissociation of the bud and, based on several criteria, is likely to be a nonspecific, cell membrane mediated association. The second aggregation phase, detected 16 h after dissociation of the bud, is Ca^{2+}-dependent and is sensitive to cycloheximide. It was found that antiserum directed against the late aggregating cells inhibited the late, but not the early, aggregation, and that the antiserum stained specifically the chondrogenic core of the limb bud. These workers concluded that the late aggregation is mediated by Ca^{2+}-dependent cell surface protein interactions. A preliminary identification and characterization of the Ca^{2+}-binding surface moiety has also been reported by these workers (Bee and von der Mark, 1990).

The functional aspects of Ca^{2+}-mediated events during chondrogenesis have been recently examined by San Antonio and Tuan (1986). These workers observed that $CaCl_2$ supplementation to high-density limb mesenchyme micromass cultures enhances chondrogenesis in a concentration-dependent manner (0.6 to 3.3 m*M*). This effect is not mimicked by other divalent cations, including Cd, Co, Ni, Mg, Mn, and Sr, and is not correlated with any increase in DNA synthesis. Interestingly, the enhancement of chondrogenesis is optimal when $CaCl_2$ is supplemented during day 1 of culture, the time period during which mesenchymal cells actively aggregate. Most importantly, Ca^{2+} promotes chondrogenesis even in normally nonchondrogenic, low-density, limb mesenchyme cultures. More recently, Bee and Jeffries (1987) reported that addition of the Ca^{2+} ionophore, A23187, to limb mesenchyme micromass cultures enhances chondrogenesis at 0.1 or 0.01 μ*M* concentrations of the ionophore, but inhibits chondrogenesis at 1 μ*M*, suggesting that modulation of intracellular Ca^{2+} level may regulate the chondrogenic differentiation of limb mesenchymal cells. These authors have also estimated that the intracellular Ca^{2+} level in mesenchyme cultures transiently increases between days 1 and 2 in culture, falls, then increases as a function of overt chondrogenesis until day 6 of culture. On the other hand, cultures at low density, which do not undergo chondrogenesis, do not exhibit the early rise in intracellular Ca^{2+} but still maintain the subsequent Ca^{2+} increase as a function of time in culture, albeit at lower levels. However, since only ^{45}Ca uptake is used as the parameter in these studies, it is not possible to conclude whether the measurements actually reflect "intracellular Ca^{2+}", since ^{45}Ca uptake represents the net of Ca^{2+} influx/efflux, Ca^{2+} binding and absorption, the activities of intracellular Ca^{2+} reservoirs, and cytosolic Ca^{2+}. The most recent observations from the author's laboratory (Tuan and DiBello, unpublished results) indicate that the enhancement of chondrogenesis by Ca^{2+} exhibits two interesting characteristics: (1) only a short exposure to exogenous Ca^{2+}, for the first 12 h or more on day 1 of culture, is sufficient; and (2) a brief treatment (2 to 5 min) of the Ca^{2+}-treated cells with 2 m*M* EGTA, a specific chelator of Ca^{2+}, which by itself does not affect the chondrogenic potential of control cultures, is able to completely abolish the enhancement of chondrogenesis by Ca^{2+}. Most importantly, it is found that the EGTA treatment is effective only when the cells have been exposed to Ca^{2+} for $\leq$12 h on day 1 of culture, whereas after a 24-h exposure to Ca^{2+} (i.e., for the complete day 1), EGTA treatment is ineffective in blocking the enhancement effect. These findings therefore suggest that the Ca^{2+}-mediated event is complete by 24 h, at which point the treated cells have acquired a "differentiation memory" of the stimulation. In addition, these findings also strongly suggest that the primary action of Ca^{2+} on enhancing mesenchymal chondrogenesis may be exerted on the cell surface, probably to influence cell-cell interactions. Finally, it was found that Ca^{2+} ionophores (e.g., A23187), as well as Ca^{2+} channel blockers (including verapamil, nifedipine, and nitrendipine), when supplemented at various doses in the presence of varying exogenous [Ca^{2+}], do not alter the effect of Ca^{2+} enhancement of chondrogenesis. These observations indicate that Ca^{2+} ion channel formers or blockers are irrelevant to the Ca^{2+} effect, and are therefore consistent with the Ca^{2+}-mediated effect being exerted primarily at the cell surface.

How does Ca^{2+} promote mesenchymal chondrogenesis *in vitro*? The available evidence as described above strongly suggests that Ca^{2+} is primarily required for the formation or stabilization of interactions occurring on the cell surface of mesenchymal cells which are undergoing aggregation *in vitro*, i.e., during the first day of culture (San Antonio and Tuan, 1986; Ahrens *et al.*, 1977; Solursh *et al.*, 1978). It is thus reasonable to postulate that the molecular interactions mediating cell aggregation may involve components which specifically bind Ca^{2+}, and that the binding of Ca^{2+} is essential for their ability to interact effectively. Since cellular aggregation is quantitatively correlated with the formation of cartilaginous nodules in limb mesenchyme cultures (Ahrens *et al.*, 1977; Evans and Tuan, 1988), such an enhancement or stabilization of aggregate formation should result in the promotion of chondrogenesis. Thus, understanding the mechanism of action of Ca^{2+} or chondrogenesis requires a detailed examination of the nature of the Ca^{2+}-dependent or Ca^{2+}-binding cell surface components which are responsible for cell aggregation, including their biochemistry and topography, and the pattern and regulation of their expression. Indeed a family of Ca^{2+}-dependent cell-cell adhesion proteins, the cadherins (Takeichi, 1987) has been identified in vertebrate tissues, and has been postulated to play important functional or regulatory roles in development and tissue morphogenesis (Ekblom *et al.*, 1986). It should be of great interest to evaluate whether cadherins and other Ca^{2+}-dependent molecules are involved in limb chondrogenesis.

In addition to the functional role discussed above, there is another aspect of chondrogenesis which may involve Ca^{2+}. This aspect concerns the stabilization or maintenance of the chondrogenic phenotype. Several studies (Deshmukh *et al.*, 1976, 1977; Maor *et al.*, 1985) have suggested that chondrocytes cultured *in vitro* are modulated by Ca^{2+} supplementation such that collagen type synthesis and mitotic rates are significantly altered. In addition, growth plate chondrocytes at various stages of their life cycle must also respond differently to graded changes in environmental calcium as they advance from the resting and proliferative stages to hypertrophy and mineralization (Brighton, 1978); for example, recent data suggest that matrix vesicle biogenesis, a property of hypertrophic chondrocytes, may be regulated by Ca^{2+} (Iannotti *et al.*, 1989). Taken together, these properties of chondrocytes, the cell type derived from chondrogenic differentiation of mesenchymal cells, suggest that they are likely to be continuously influenced by their calcium milieu throughout their life cycle. In this manner, the role of calcium in chondrogenesis may include a postdifferentiation phase. This is in fact an important aspect to consider, since histogenesis of the cartilage does extend beyond the initial cytodifferentiation event. At the present, little is known about how calcium may influence the activities of the chondrocyte during its life cycle.

Matrix Calcium and Chondrogenesis

During skeletogenesis cartilage generally precedes bone formation, particularly during endochondral ossification, where the cartilage anlage is progressively replaced by invading osteoblasts (Fell, 1925; Caplan and Pechak, 1987; Jackson and Randall, 1956). In particular, in the epiphyseal cartilage growth plate, chondrocytes undergo a discrete series of histomorphological changes that include proliferation, hypertrophy, and mineralization, all prerequisites for subsequent bone replacement (Ali, 1983). In contrast, the skeletal tissues which undergo intramembranous ossification, e.g., the calvarial head bones, develop without the intermediate cartilaginous phase (Bernard and Pease, 1969; Jackson and Randall, 1956), i.e., mesenchymal cells are observed to condense and differentiate directly into osteoblasts, which later become matrix-embedded osteocytes. What is the basis for the different developmental pathways taken by the cranial neural crest cells, which become the membrane bones of the head, and the dorsal paraxial mesodermal cells, which give rise to the axial and the appendicular skeleton? Is there an intrinsic nonequivalence between these mesenchymal populations which dictates their different development pathways? Recent experiments in several laboratories (see below) have in fact shown that cells of the embryonic calvaria actually resemble cells of the endochondral tissues in that they possess the potential to undergo chondrogenesis and, furthermore, that this potential is regulated as a function of calcium in the mineralized matrix.

The experiments by Tuan and Lynch (1983) and Jacenko and Tuan (1986a, b) exploit a unique feature of calcium metabolism of the developing chick embryo, i.e., that its calcium is supplied by the eggshell (Tuan, 1987). Consequently, by maintaining chick embryos in long-term, shell-less culture, the effect of systemic calcium deficiency on matrix calcification and cytodifferentiation in the skeletal tissues can be observed. Tuan and Lynch (1983) observed that the calvaria of the calcium-deficient chick embryo synthesize collagen type II, a characteristic component of the cartilage matrix. By means of immunohistochemistry, Jacenko and Tuan (1986a) revealed that the cells expressing collagen type II reside in the undermineralized central core of the calvaria. The cells occupy isolated islands which are substantially deficient in mineral, and are also stained positively with alcian blue indicating the presence of sulfated glycosaminoglycans. Recent studies (Jacenko, 1988; Jacenko, Mattessich, and Tuan, unpublished observations) indicate that the calcium-deficient calvaria synthesize high-M_r, chondroitin-sulfate-rich proteoglycans, similar to those characteristic of cartilage (Hascall *et al.*, 1976). These findings therefore clearly indicate that chondrocytes, which in effect represent an aberrant cell type for normal intramembranous ossification, have become part of the phenotype of the embryonic calvarium as a result of systemic calcium deficiency. Interestingly, several other studies (Aubin *et al.*, 1982; Nefussi *et al.*, 1985; Rifas *et al.*, 1982; Villanueva and Nimni, 1989) have further revealed and confirmed that the embryonic calvarium indeed contains cells capable of undertaking the chondrogenic pathway.

What then regulates the selection of differentiation pathway by the mesenchyme of the embryonic calvarium? The undermineralized matrix of the calvarium of the shell-less chick embryo has been shown to be a permissive factor, based on recent work in the author's laboratory (Jacenko, 1988; Jacenko and Tuan, manuscript in preparation). These workers have further examined the nature of the aberrant chondrogenesis in the calvarium by (1) placing the tissue in various culture conditions, such as organ cultures *in vitro* and as tissue grafts on the chorioallantoic membrane of normal or shell-less chick embryos; and (2) culturing isolated calvarial cells *in vitro* at varying Ca^{2+} concentrations. In summary, the results from these experiments have shown that: (1) the chondroprogenitor cells are an endogenous cell population of the calvarium, and are present even in the tissue of the normal embryo; (2) expression of the cartilaginous phenotype requires an intact matrix, and is accentuated as a result of induced calcium deficiency and undermineralization of the matrix; and (3) the calcium effect is not mediated by changes in the level of Ca^{2+} in the extracellular milieu, for example in the culture medium of isolated cells maintained *in vitro*.

From these studies, it is proposed that embryonic intramembranous ossification involves the participation of a population of cells, at least some of which are inherently capable of switching to chondrogenesis. However, this latter step takes place only when a substantial malformation of the matrix, e.g., as a result of systemic calcium deficiency and resultant undermineralization of the tissue, has occurred. Perhaps the loose, undermineralized matrix permits ready cellular condensation and the requisite cell-cell interactions, which then lead to chondrogenesis. Recent data consistent with this supposition have been obtained by McDonald and Tuan (1989), who have used *in situ* hybridization to collagen type II cDNA to identify a subpopulation of cells in the calvaria of both normal and shell-less chick embryos which express collagen type II mRNA. Interestingly, the distribution of these cells within the calvarium is closely correlated with the extent of matrix mineralization, i.e., fewer collagen type II positive cells are found in the more mineralized areas.

The chondrogenic differentiation by cells of the calvarium (Jacenko and Tuan, 1986a, b, 1987; Nefussi *et al.*, 1985; Rifas *et al.*, 1982; Tuan and Lynch, 1983; Villanueva and Nimni, 1989) and the apparent effect of calcium (Jacenko and Tuan, 1986a, b; McDonald and Tuan, 1989; Tuan and Lynch, 1983) are thus distinct from chondrogenesis of the limb bud mesenchyme, since the former takes place within a tissue already undergoing programmed osteogenesis. Moreover, for the calvarium, the calcium effect is most likely mediated primarily via the matrix. Thus, matrix "loosening" due to undermineralization may in fact be the permissive condition for the cell-cell interactions, i.e., cellular condensation, important for chondrogenic differentiation. It should be pointed out that the relative abundance of calcium in the calvarium, even in the calcium-deficient shell-less embryo, should adequately ensure that the Ca^{2+}-mediated cellular aggregation events, similar to those in limb development described above, may be properly carried out.

Polyionic Molecules

In biomacromolecular interactions, specificity and kinetics are often the result of charge and conformation parameters (Cantor and Schimmel, 1980). Cell surfaces are in general negatively charged under physiological conditions, largely because of the presence of sialic acid residues (Alberts *et al.*, 1989). Therefore unless specifically interacting moieties, either homotypic or heterotypic, are present, cells are unlikely to interact directly and readily. On the other hand, cells may interact with one another by means of their extracellular matrix; in this manner, specific extracellular matrix components may serve as a "bridge" between the cells (Ekblom *et al.*, 1986). A large number of molecular components have been identified which may play direct and indirect functional roles in cell-cell interactions (for reviews see Ekblom *et al.*, 1986; Höök *et al.*, 1984). Some of these molecules have in fact been suggested to be involved in regulating chondrogenesis, e.g., N-CAM (Crossin *et al.*, 1985), fibronectin (Swalla and Solursh, 1984; Frenz *et al.*, 1989a, b), proteoglycans (Kosher and Lash, 1975; Kosher *et al.*, 1973; Lash and Vasan, 1978), collagen (Dessau *et al.*, 1980; Kosher and Church, 1975), hyaluronic acid (Knudson and Toole, 1985; Kujawa and Caplan, 1986; Kujawa *et al.*, 1986; Toole, 1972), a blastema-specific peanut agglutinin binding protein (Aulthouse and Solursh, 1987; Zimmerman and Theis, 1984), etc. Although these molecules, if indeed functional in regulating chondrogenesis, may act differently, a likely common theme is that they probably serve as an informational link by structurally interacting with the mesenchymal cells.

Polycations

A series of recent experiments in the author's laboratory have explored the consequence of ionically cross-linking limb mesenchymal cells on their potential for subsequent chondrogenic differentiation *in vitro*. The strategy is based on the principle of charged, polyionic interactions. In the first set of experiments (San Antonio and Tuan, 1986), the cationic polymer, poly-L-lysine (PL) is administered in the culture medium to stage 23-24 chick limb mesenchyme in micromass culture *in vitro* (Ahrens *et al.*, 1977). Chondrogenesis, as measured by alcian blue staining, and type II collagen and sulfated proteoglycan synthesis, is significantly stimulated by PL in a dose-dependent manner. The effect of PL is dependent on its M_r (only species $\geq$8 kDa are active) and charge, and is mimicked by poly-D-lysine and poly-D,L-lysine, but not by lysine or other analogs of PL or lysine. Lysine-containing heteropolypeptides are effective only when their lysine content exceeds 75%. The stimulation of chondrogenesis by PL does not appear to be correlated with increased DNA synthesis. Interestingly, PL exerts optimal stimulation of chondrogenesis when administered on day 2 of culture, the time period during which active recruitment and assimilation of neighboring cells into existent cellular aggregates are believed to be taking place in the mesenchyme micromass culture. Finally, it was observed that PL elicits nodule formation even in low-density cultures that are normally non-chondrogenic.

How does PL enhance chondrogenesis by the limb mesenchyme? A most likely explanation, based on its time of action and molecular size requirement, is that PL serves to promote the growth of the cellular aggregates, which is brought about by PL interacting with and crosslinking negatively charged molecular domains on the surface of cells. In this manner, growing networks of cells are formed that can give rise to cellular aggregates, which in turn are capable of eventually forming cartilaginous nodules. The results from the heteropolypeptide studies, which show a minimum of 75% lysine being required for the effect on chondrogenesis, indicate the chemical specificity of these charge-based interactions. On the other hand, the fact that the PL stereoisomers, poly-D-lysine and poly-D,L-lysine, are also active indicates that the specificity does not depend on the a particular conformation of the molecule. This is a reasonable finding since, at physiological pH, PL and its stereoisomers should exist mostly as random-coiled structures because of the uniformly distributed positive charge along the molecule (Cantor and Schimmel, 1980).

A recent report by Takeda and co-workers (1989) indicates that PL is also active on mature cartilage. Adult bovine articular cartilage explants showed a marked increase in proteoglycan synthesis (measured on the basis of ^{35}S-sulfate incorporation and core protein synthesis) upon treatment with high-M_r PL at 1 to 20 μg/ml *in vitro*. The PL effect is highly specific for proteoglycans since glycoprotein production is not affected. Thus the action of PL may be interpreted as the promotion of chondrogenesis as well as the further stabilization of the chondrogenic phenotype.

What is the action of PL (or other high-M_r cationic polypeptides and proteins) on pre-chondrogenic mesenchyme and mature chondrocytes? It is reasonable to speculate that, by virtue of its high positive charge density, PL should interact with polyanionic molecular moieties. Gelman and co-workers (1973) have characterized the interaction between PL and the glycosaminoglycan (GAG), chondroitin-6-sulfate, in aqueous solution, and reported that PL is converted from a charged coil to an α-helical conformation upon high affinity binding to the GAG. That the chondrogenesis stimulating effect of PL is both charge and size dependent (San Antonio and Tuan, 1986) would in fact support the hypothesis that the molecule(s) interacting with PL is a high-M_r, negatively charged moiety. In this manner, PL may act by binding ionically (charge dependent) and thereby cross-linking (length dependent) certain anionic component(s) of the cell surface or the extracellular matrix to elicit its chondrogenic effect.

The potent stimulation of chondrogenesis by PL *in vitro* suggests that PL-like, cationic polymers may be present and actually functional *in vivo*. Kuettner *et al.* (1975) had found previously that the cartilaginous matrix contains a substantial amount of cationic molecules, including lysozyme. Recently,

Fassbender *et al.* (1989) have reported that a variety of cationic proteins and polypeptides, including PL, are able to bind with high affinity to the cartilaginous matrix. The extractability of the bound polycations is inversely correlated to their molecular size, such that low-M_r (3.6 kD) PL is fully extractable with 1 *M* NaCl, whereas high-M_r (260 kD) PL is essentially nonextractable ($<5\%$). These authors have also shown that binding of the polycations results in the displacement of monovalent cations, such as Na^+. These studies therefore indicate that polycationic molecules are present and are accumulated in the extracellular matrix of mature cartilage, thus permitting their cellular action(s), such as promoting proteoglycan synthesis (Takeda *et al.,* 1989). Although similar information is presently unavailable for the embryonic limb bud, it should be pointed out that the requirement for polycationic molecules may be satisfied provided specific matrix components of the limb bud actually contain localized stretches of highly positively charged molecular domains. Indeed, type I collagen (Obrink, 1973; Obrink *et al.,* 1975) and fibronectin (Hynes and Yamada, 1982) both contain such domains. The relevance and significance of these structural domains in regulating chondrogenesis remain to be studied.

Interestingly, recent work from the author's laboratory (Gawande and Tuan, 1987, 1990) has demonstrated that PL promotes chondrogenesis via a mechanism which is most likely independent of that involved in the stimulation of chondrogenesis by bone-matrix derived factors (Syftestad and Caplan, 1981; Syftestad *et al.,* 1984). Specifically, the latter family of factors act differently in that the density dependence of chondrogenic differentiation (Ahrens, 1977; San Antonio and Tuan, 1986; Osdoby and Caplan, 1979) is not overcome. Thus, PL is not likely to act by simply mimicking bone-derived chondrogenesis-stimulating activities. It should also be pointed out that PL has been shown to exhibit other biological activities, including stimulation of endocytosis (Morad *et al.,* 1984; Ryser and Hancock, 1965; Ryser *et al.,* 1978), surfactant-like action in stimulation of mitosis (Whitfield *et al.,* 1968), and cell agglutination by destabilization of membrane interactions (Coakley *et al.,* 1985). Whether these activities of PL are also functional and/or responsible for its stimulation of chondrogenesis *in vitro* remains to be investigated.

Polyanions

The dramatic effect of PL on chondrogenesis as described above strongly suggests the possible involvement of a polyanionic molecular counterpart, which is present either on the cell surface or associated with the extracellular matrix, to execute the polyionic regulatory function. The most likely polyanionic candidates, particularly for cells destined to become cartilage, are the proteoglycans. Proteoglycans are polymers of highly negatively charged GAG chains covalently linked to a protein core (Ruoslahti, 1988; Hassell *et al.,* 1986). Extracellular and membranous forms of proteoglycans are found in various tissues and cells (Höök *et al.,* 1984). Among the various types of proteoglycans, the cell surface heparan sulfate proteoglycans have recently received much attention concerning their possible functional role as cellular receptors for various extracellular matrix molecules (Cole and Glaser, 1986; Fransson, 1985; Koda *et al.*, 1985; Ratner *et al.*, 1985). In particular, cell surface proteoglycans have been invoked as participants in cell surface-extracellular matrix interactions during various developmental processes (e.g., see Hay, 1984), perhaps serving as an interface or informational link between the cell and the matrix. These postulates are based in part on the findings that several extracellular matrix molecules (such as collagen, fibronectin, laminin, and others), which interact with the cell surface and in some instances co-localize with heparan sulfate proteoglycan, also contain specific GAG-binding sites (Hayman *et al.*, 1982; Laterra *et al.*, 1983; Sakashita *et al.*, 1980).

As discussed above, cell-extracellular matrix interactions have long been thought to be important for chondrogenesis, including those thought to involve proteoglycans and their components. For example, in the limb core the concentration of hyaluronic acid decreases before the mesenchymal condensation process (Toole, 1972), and the concentrations of type I collagen and fibronectin both reach a maximum during condensation and decrease at the onset of chondrocyte differentiation (Dessau *et al.*, 1980). Results from *in vitro* studies demonstrate that some of these extracellular matrix molecules significantly influence chondrogenesis (Hadhazy *et al.*, 1983; Kosher and Church, 1975; Kujawa and Caplan, 1986; Swalla and Solursh, 1984). In particular, Kosher and Lash (1975) had proposed earlier that proteoglycans may regulate chondrogenesis based on studies on somite chondrogenesis *in vitro*. These workers found that the chondrogenic effect of the notochord on somites is considerably impaired if, prior to this interaction, notochords are pre-exposed to hydrolytic enzyme treatments which remove cell surface-associated proteoglycans. The identity of the cell surface-associated factor(s) required for chondrogenesis remains unknown. Interestingly, Hadhazy *et al.* (1983) have reported that high doses of heparin and other polyanions inhibit chondrogenesis of limb bud mesenchyme *in vitro*.

Recent studies in the author's laboratory have addressed the possible involvement of proteoglycans in limb chondrogenesis by studying the effect of their GAG components on micromass cultures of chick stage 23/24 embryonic limb mesenchyme *in vitro* (San Antonio *et al.*, 1987). Heparan sulfate and several of its analogs (heparin, dermatan sulfate, and dextran sulfate) are found to significantly stimulate cartilage nodule formation; in addition, heparan sulfate and heparin also promote nodule growth. Chondroitin sulfate, on the other hand, does not stimulate chondrogenesis. The activity of the GAGs is dependent on their dose, molecular size, charge, and chemical structure. Similar effects are also observed in micromass cultures derived from stage 25 embryonic wing-tip subridge mesoderm, which represents an enriched population of chondroprogenitor cells (Kosher *et al.*, 1979; Newman,

1977; Newman *et al.*, 1981). Interestingly, the time of action of the GAGs on chondrogenesis is also observed to be maximal on culture day 2, reminiscent of the effect of the polycation, PL (San Antonio and Tuan, 1986), suggesting that the action of the GAGs may be involved in the postcellular aggregation events of chondrogenesis, such as initial expression of the chondrocyte phenotype or growth of cartilage nodules.

The highly specific nature of the chondrogenesis stimulatory activity of heparan sulfate (and other *L*-iduronic acid containing GAGs) is analogous to its effect on several other biological processes, such as endothelial cell migration (Azizkhan *et al.*, 1980), salivary gland morphogenesis (Bernfield *et al.*, 1984), and smooth muscle cell metabolism (Majack and Bornstein, 1985). Furthermore, for these processes, as well as for chondrogenesis, dextran sulfate has been found to mimic the activity of heparan sulfate. It is also interesting to note that although the effect is charge-dependent, it is also highly structure-specific, since the chondroitin sulfate used is of higher sulfation density but is nevertheless ineffective. A large polymeric structure is also needed since monomers, such as glucose-6-sulfate, N-acetylglucosamine, D-glucuronic acid, are all ineffective.

How may heparan sulfate and structurally related GAGs regulate chondrogenesis? Many experimental observations have implicated the important role of GAGs in modulating the phenotypic expression of various cell types. For example, synthesis of heparan sulfate and chondroitin sulfate by the corneal epithelium *in vitro* is significantly enhanced by the supplementation of these GAGs to the culture medium (Meier and Hay, 1974); these effects are highly specific since GAG expression is stimulated only in cultures supplemented with the GAG species normally produced by the epithelium. Cartilage matrix synthesis by chondrocytes in suspension cultures is stimulated in the presence of a variety of polyanions, including chondroitin sulfate, dermatan sulfate, heparin, heparan sulfate, and dextran sulfate (Nevo and Dorfman, 1972), and inhibited by hyaluronate (Wiebkin and Muir, 1973). The stimulation of chondrogenesis by heparan sulfate-like GAGs could therefore be a consequence of their effects on extracellular matrix production by the limb mesenchyme.

The time of action of heparan sulfate, culture day 2, and its continuous action through day 3, suggests that it is unlikely to act solely on mesenchymal cell aggregation, which occurs primarily on day 0.5 to 1.25 (Ahrens *et al.*, 1977; San Antonio and Tuan, 1986). Thus, postaggregation events, such as the initial expression of the cartilage phenotype and the growth of cartilage nodules, are most likely affected. It is possible that these actions are mediated via enhanced cell division, since the passage of limb or somitic mesenchyme through a quantal cell cycle is thought to be a prerequisite for the emergence of the differentiated state (Holtzer *et al.*, 1975). Indeed, there is much evidence implicating heparan sulfate-like molecules in the regulation of mitosis in a variety of cell types (Castellot *et al.*, 1981; Fritze *et al.*, 1985; Ratner *et al.*, 1985) and in limb mesenchyme (Hadhazy *et al.*, 1983).

Alternatively, heparan sulfate may act by (1) interacting with specific cell surface receptors to initiate chondrogenesis, or (2) disrupting certain cell surface heparan sulfate proteoglycan-extracellular matrix interactions which are inhibitory to chondrogenesis. Consistent with the former postulate are the observations that heparan sulfate and structurally related molecules bind specifically to the surfaces of various cell types, including hepatocytes (Kjellen *et al.,* 1980) and smooth muscle cells (Castellot *et al.,* 1985). On the other hand, the latter postulate would assume that limb mesenchyme contains a membrane-associated heparan sulfate proteoglycan that functions as an extracellular matrix receptor, as has been shown for certain epithelia (Koda *et al.,* 1985). In this manner, the exogenous GAGs may compete with the endogenous heparan sulfate proteoglycans for the binding of specific matrix molecules. A consequence of such competition may be the disruption of specific cell-matrix interactions, leading to cell rounding. Indeed, it has been shown that agents or conditions that promote cell rounding of chondrogenic mesenchymal cells or chondrocytes also promote the expression of the cartilage phenotype (Holtzer and Abbott, 1968; Solursh *et al.,* 1982; Zanetti and Solursh, 1984). Perhaps the effects of other matrix molecules, such as collagen and fibronectin, on chondrogenesis may also be explained in terms of their interactions with the heparan sulfate proteoglycan of the mesenchyme cell surface to affect cell shape. Interestingly, recent studies by Frenz *et al.,* (1989a, b) suggest that limb mesenchymal condensation *in vitro* is indeed dependent on interactions between the cell surface via the amino-terminal heparin-binding domain of fibronectin, consistent with the hypothetical scheme described above. These *in vitro* effects of heparan sulfate and related GAGs on chondrogenesis are, in fact, suggestive of their function *in vivo*. Indeed, the presence of heparan sulfate has been demonstrated in the prechondrogenic limb bud of the chick (Kimata *et al.,* 1986) and the mouse (Matsui *et al.,* 1986), as well as in confluent cultures derived from chick limb buds (Pennypacker *et al.,* 1978). Recently Vasan (1986) has demonstrated that only during the condensation phase does the prechondrogenic mesenchyme of the chick limb bud produce heparan sulfate in greater quantities than the other GAGs. These observations, taken together, are therefore strongly supportive of the postulate that a heparan sulfate-like molecule may play an important regulatory role in chondrogenesis *in vivo*.

Many components of the extracellular matrix have been proposed to play development-specific functional roles in limb bud chondrogenesis. For example, in the early limb bud, hyaluronate is produced and is then removed before the cellular condensation phase (Toole, 1972). It has been postulated that hyaluronate allows cell movement and migration, but prohibits cell-cell interaction, and that its dissolution thus permits or enhances direct cell-cell interaction (Oster *et al.,* 1985; Toole, 1982). Furthermore, Dessau *et al.* (1980) have shown that the levels of type I collagen and fibronectin increase in the condensing limb core, coincident with the expression of a high-M_r

chondroitin sulfate proteoglycan (designated PG-M) which contains binding sites for hyaluronate, type I collagen, and fibronectin (Yamagata *et al.*, 1986). These authors have proposed that these extracellular matrix molecules are functionally involved in the condensation process. Recently, Newman *et al.* (1985) have shown *in vitro* that cells are translocated mechanically along an artificial concentration gradient of fibronectin, a phenomenon termed "matrix-driven translocation". The authors have suggested that perhaps the establishment of such a gradient of fibronectin serves to bring about cellular condensation in the developing limb bud. Recent studies from these workers (Frenz *et al.*, 1989a, b) implicate the heparin-binding domain of fibronectin as functionally important. It is possible that some of the effects of these extracellular matrix molecules on chondrogenesis are mediated via specific heparan sulfate proteoglycans which are expressed in a development-specific manner in the limb bud.

Interactions Between Polycations and Polyanions

In the discussions above concerning the functional roles of the polycation (PL) and the polyanion (heparan sulfate-like GAGs) in regulating and stimulating chondrogenesis, an inherent assumption is that they act by each interacting with a high-M_r component(s) of opposite charge. In fact, the latest observations from this author's laboratory (San Antonio, 1988; San Antonio and Tuan, 1987; San Antonio *et al.*, submitted for publication), as described below, are consistent with the hypothesis that PL promotes chondrogenesis by interacting electrostatically with the sulfated GAGs, and furthermore indicate that PL significantly influences the production and distribution of the sulfated components of the extracellular matrix.

In the first set of experiments, it is found that GAGs, added exogenously to the PL-supplemented cultures of limb mesenchyme *in vitro*, are able to modify the PL stimulation of chondrogenesis in a dose-dependent manner. Interestingly, each of the GAGs tested exhibits a characteristic dose profile: (1) heparin and heparan sulfate—an initial dose-dependent inhibition of the PL effect until an "equivalence" concentration, following which a gradual dose-dependent increase of chondrogenesis is observed; and (2) chondroitin sulfate and dermatan sulfate—a steady dose-dependent inhibition of the PL effect. The most ready explanation of these findings is that PL and the sulfated GAGs are interacting electrostatistically in a "neutralizing" manner, so that only at concentrations greater than equivalence are heparin and heparan sulfate able to exert its chondrogenesis stimulatory effect (San Antonio *et al.*, 1987), whereas chondroitin sulfate continues to inhibit. However, it should be noted that heparan sulfate only inhibits the PL effect very slightly compared to heparin, suggesting that it probably does not interact with PL in a totally "neutralizing" manner.

In a second set of experiments, it is found that stimulation of chondrogenesis by PL is accompanied by significant changes in the metabolism and distribution of the sulfated extracellular matrix. This is studied by metabolically radiolabeling high density cultures of limb mesenchyme with ^{35}S-sulfate and ^{3}H-leucine on day 2 of culture, either in the presence of PL or with subsequent exposure to PL. It is found that the release into the culture medium of sulfated macromolecules, but not of proteins in general, is significantly inhibited within 1 h of PL treatment, and by 24 h a net increase in total sulfated GAG content of the PL-treated cultures is observed. This increase is a result of stimulated biosynthesis and not of inhibited degradation of sulfated materials. Furthermore, the composition of the sulfated materials synthesized in the presence of PL is similar to that of control cultures. Interestingly, dissociative chromatography of the released GAGs reveals a significant, relative enrichment of low-M_r species of GAGs in the media of PL-treated cultures. Finally, the specificity of these effects is indicated by the fact that treatment of the cultures with a low-M_r (4 kD), nonstimulatory form of PL is totally ineffective.

In interpreting these findings to gain insight into the mechanism of action of PL, it should be pointed out first that PL does not act on limb mesenchyme by solely influencing GAG release and metabolism, since another cartilage marker, type II collagen, co-localizes with all of the alcian blue positive foci generated by PL treatment (San Antonio and Tuan, 1986). Thus, PL exposure results in the apparent stimulation of true cartilage nodule formation. This point is critical in that it lends itself to two interpretations. First, PL treatment could result merely in the enhanced expression (and thus facilitated detection) of preexisting nodules, in that PL may not only inhibit the release of GAGs from the cell layer, but also of other matrix components, including type II collagen. Alternatively, the primary action of PL may actually be the retention and/or metabolism of GAGs, which in turn affects the generation or expression of cartilage nodules. For example, the inhibition of GAG release by PL may somehow trigger enhanced biosynthesis of the cartilage matrix. It has been previously observed that the PL effect is maximal when administered on day 2 of culture; that day 3, a time period of rapid nodule formation and matrix synthesis, is actually not the optimal time for PL action would argue against the notion that PL enhances nodule formation merely by the simple retention of matrix components. On the other hand, it is possible that many of the cartilage nodules in control cultures which begin to emerge on day 2 are actually selected against by nonoptimal environmental conditions, thus causing their disappearance by day 3. Indeed, culture conditions have been shown to exert profound, cell-type specific influences on the survival and/or phenotype of chondrogenic mesenchymal cell and other interfering cell types (Coon, 1966; Hunter and Caplan, 1983; Thal *et al.*, 1986). Thus, the activity of PL, whether mediated via matrix trapping or other mechanisms, could function to stabilize or magnify the expression of existing nodules which are transiently expressed on day 2. In this manner, PL treatment on day 3 would be less effective simply because fewer candidate nodules remain. Consistent with this idea are the following findings: (1) at their optimal concentration, both PL (San Antonio and Tuan, 1986) and the polyanion, heparin (San Antonio *et al.*, 1987), stimulate the formation of the same (perhaps *maximal*) number of nodules; and (2) both agents are maximally stimulatory when applied on day 2.

These effects of PL on limb mesenchymal cells are reminiscent of the findings of Kosher and Church (1975), who showed that supplementation of procollagen to the media of somite explants results in an increased production of cartilage matrix, as well as an increased retention of labeled GAGs and a lesser retention of noncollagenous proteins. It is tempting to speculate that PL may mimic the chondrogenesis stimulatory activity of procollagen; indeed, collagen contains cationic domains which have been postulated to be involved in polyionic interactions with proteoglycans during the formation of the cartilage matrix (Obrink, 1973; Obrink *et al.*, 1975).

Model for Polyionic Regulation of Chondrogenesis: Matrix Trapping

The experimental findings described above strongly suggest that polyionic molecules are functional in regulating chondrogenesis *in vitro*. Although the exact mechanism of these molecules remains to be elucidated, a ''matrix trapping'' model (Figure 1) is proposed here to explain their activities, and draws on various lines of evidence gained from observations on chondrogenesis *in vitro* and *in vivo* as discussed above in this review.

The central theme of this model proposes that exogenous polyions stabilize cartilage nodule formation and extracellular matrix expression by promoting the trapping of cartilage matrix components. The process is believed to proceed via the following steps:

1. Polyionic substances interact with an endogenous counterpolyionic complement molecule produced by the mesenchymal cell. Thus, PL may bind the anionic proteoglycan(s), and heparin or heparan sulfate may bind the cationic domains of type II collagen, or other matrix components.
2. The polyionic complex formed in this manner may promote or stabilize chondrogenesis by creating a pericellular ''trap'' to encapsulate the emerging cartilage nodule, and thus allow undisturbed assembly of the extracellular matrix, or the progression of cell-cell interactions required for cartilage matrix biosynthesis.

A corollary of this model is that it also provides a relatively simple explanation for the cell density dependence of chondrogenesis *in vitro*. When limb mesenchymal cells are cultured at greater than confluent densities, the cartilage matrix secreted by some of the chondroblasts may be trapped within the surrounding cells, which provide the partitioned environment for matrix assembly and the emergence of the nodule. Similarly, cells cultured clonally

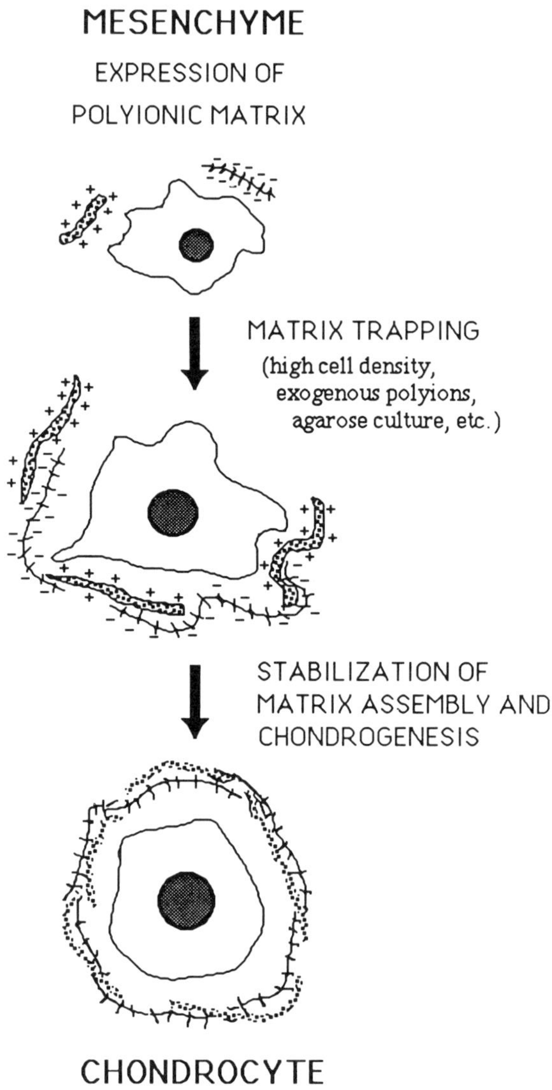

Figure 1. Matrix trapping model of limb chondrogenesis.

within an artificial matrix of collagen gel or agarose would also experience such matrix trapping, and are thus capable of forming cartilaginous colonies. On the other hand, cells in cultures seeded at less than confluent densities may also produce cartilage matrix, but in this case the matrix is not stabilized in its assembly, and the nodule subsequently regresses or is never formed. Indeed, a cartilage nodule may simply be a manifestation of the undisturbed, cooperative assembly of cartilage matrix by a group of chondroblasts.

Conclusion and Perspectives

This review has summarized some of the salient findings that directly implicate ions and polyions as important regulators of chondrogenesis. It is proposed that during chondrogenesis Ca^{2+} is involved in the regulation of initial cellular aggregation events, whereas polycations and polyanions are involved in postaggregational processes, such as matrix synthesis and cell recruitment for nodule formation. At the present, the available evidence is too incomplete to indicate the exact basic mechanism by which these substances regulate chondrogenesis. Some of the key issues that certainly need to be addressed in future studies include:

1. *Calcium.* If Ca^{2+} is interacting with the cell surface, what is the identity of the Ca^{2+}-binding "receptor" molecule? What is the nature of the immediate and long-term cellular signalling and response to the exogenous calcium? What cellular and molecular events mediate cell aggregation? What are the molecular consequences, e.g., change in gene expression, of cell aggregation? What is the relationship between Ca^{2+}-mediated cell aggregation and cell shape change? Is Ca^{2+} level of the limb core actually modulated during chondrogenesis *in vivo*: if so, how is this modulation brought about?

2. *Polyions.* What are the natural analogs of PL and heparan sulfate? What is the biochemical basis for the molecular specificity governing the interactions between the polycation(s) and the polyanion(s)? How does PL modify the metabolism of sulfated GAGs? If the "matrix trapping" model for polyionic action on chondrogenesis is correct, what is the cellular and molecular response(s) to "matrix trapping"? Is it via cell shape changes, or perhaps the entrapment of autocrine growth factors?

3. *Chondrogenesis and Development.* At what stage of commitment to chondrogenesis do Ca^{2+} and polyions exert their regulatory effects? How do these regulators contribute to the overall morphogenesis of the embryonic limb *in vivo*? How do these regulators act in concert with other putative effectors and modulators of chondrogenesis?

A common theme that has emerged from the work reviewed here is that the action of Ca^{2+} and the polyionic factors is to transmit extracellular signals to the mesenchymal cell, resulting in the specific cellular response of chondrogenesis. Consequently, a most relevant approach to studying the mechanism of their action should be an in-depth examination of the components of the cell membrane, the living link between the "extracellular" and the "intracellular". Answers to the experimental questions posed here should eventually lead to a fuller understanding of the basic mechanism of chondrogenesis.

Acknowledgment

The author is grateful to J. San Antonio and O. Jacenko who have made some of the seminal observations described here; M. Lynch, S. Gawande, and S. McDonald for other experimental studies; and Jennie Platt for excellent assistance in preparing the manuscript. Work done in the author's laboratory has been supported in part by the National Institutes of Health, March of Dimes Birth Defects Foundation, and the U.S. Department of Agriculture.

References

P. Ahrens, M. Solursh, and R. Reiter. Stage-related capacity for limb chondrogenesis in cell culture. *Dev. Biol.* **60**: 69–72 (1977).

B. Alberts, D. Bray, J. Lewis, M. Raff, K. Robert, and J. Watson. *Molecular Biology of the Cell,* 2nd ed. Garland, New York. chap. 6, 1989.

S. Ali. Calcification of cartilage. In: *Cartilage: Structure, Function, and Biochemistry,* Vol. 1, Ed B.K. Hall. Academic Press, New York. pp. 343–378, 1983.

R. Amprino. The development of the vertebrate limb. *Clin. Orthop. Relat. Res.* **188**: 263–284 (1984).

J. Aubin, J. Heersche, M. Merriless, and J. Sodek. Isolation of bone cell clones with differences in growth, hormone responses and extracellular matrix production. *J. Cell Biol.* **92**: 452–461 (1982).

A. Aulthouse and M. Solursh. The detection of a precartilage, blastema-specific marker. *Dev. Biol.* **120**: 377–384 (1987).

R. Azizkhan, J. Azizkhan, B. Zetter, and J. Folkman. Mast cell heparin stimulates migration of capillary endothelial cells *in vitro*. *J. Exp. Med.* **152**: 931–944 (1980).

J. Bee and R. Jeffries. The relationship between intracellular calcium levels and limb bud chondrogenesis *in vitro*. *Development* **100**: 73–81 (1987).

J. Bee and K. von der Mark. A molecular approach to analysis of prechondrogenic condensation in the avian limb. *J. Embryol. Exp. Morphol.* (Suppl.) **82**: 80 (1984).

J. Bee and K. von der Mark. An analysis of chick limb bud intercellular adhesion underlying the establishment of cartilage aggregates in suspension culture. *J. Cell Sci.* **96**: 527–536 (1990).

M. Bennett and D. Spray. *Gap Junctions,* Cold Spring Harbor Laboratory, Cold Spring Harbor, NY (1985).

G. Bernard and D. Pease. An electron microscopic study of initial intramembranous osteogenesis. *Am. J. Anat.* **125**: 271–290 (1969).

M. Bernfield, S. Banerjee, J. Koda, and A. Rapraeger. Remodelling of the basement membrane: morphogenesis and maturation. *Ciba Found. Symp.* **108**: 179–196 (1984).

R. Brakenbury, U. Rutishauser, and G. Edelman. Distinct calcium-independent and calcium-dependent adhesion systems of chick embryo cells. *Proc. Natl. Acad. Sci. U.S.A.* **78**: 387–391 (1981).

C. Brighton. Structure and function of the growth plate. *Clin. Orthop. Relat. Res.* **136**: 22–32 (1978).

C. Buck and A. Horwitz. Cell surface receptors for extracellular matrix molecules. *Annu. Rev. Cell Biol.* **3**: 179–205 (1987).

A. Caplan and D. Pechak. The cellular and molecular embryology of bone formation. In: *Bone and Mineral Research,* Ed W. Peck. Elsevier Press, New York. pp. 117–183, 1987.

C. Cantor and P. Schimmel. *Biophysical Chemistry*. W.H. Freeman, San Francisco. 1980.

J. Castellot, K. Wong, B. Herman, R. Hoover, D. Albertini, T. Wright, B. Caleb, and M. Karnovsky. Binding and internalization of heparin by vascular smooth muscle cells. *J. Cell. Physiol.* **124**: 13–20 (1985).

A. Chevallier, M. Kieny, and A. Mauger. Limb-somite relationship: origin of the limb musculature. *J. Embryol. Exp. Morphol.* **41**: 245–258 (1976).

B. Christ, H. Jacob, and M. Jacob. Experimental analysis of the origin of the wing musculature in avian embryos. *Anat. Embryol.* **150**: 171–186 (1977).

W. Coakley, L. Hewison, and D. Tilley. Interfacial instability and the agglutination of erythrocytes by polylysine. *Eur. Biophys. J.* **13**: 123–130 (1985).

G. Cole and L. Glaser. A heparin-binding domain from N-CAM is involved in neural cell-substratum adhesion. *J. Cell Biol.* **102**: 403–412 (1986).

K. Crossin, C. Chuong, and G. Edelman. Expression sequence of cell adhesion molecules. *Proc. Natl. Acad. Sci. U.S.A.* **82**: 6942–6946 (1985).

R. Damluji and P. Riley. On the role of calcium in adhesion of cells to solid substrates. *Exp. Cell Biol.* **47**: 226–237 (1979).

H. Coon. Clonal stability and phenotype expression of chick cartilage cells *in vitro*. *Proc. Natl. Acad. Sci. U.S.A.* **55**: 66–73 (1966).

T. Deuel. Polypeptide growth factors: roles in normal and abnormal growth. *Annu. Rev. Cell Biol.* **3**: 443–492 (1987).

K. Deshmukh, W. Kline, and B. Sawyer. Role of calcium in the phenotypic expression of rabbit articular chondrocytes and culture. *FEBS Lett.* **67**: 48–51 (1976).

K. Deshmukh, W. Kline, and B. Sawyer. Synthesis of collagen by chondrocytes in suspension culture: modulation by calcium, 3′,5′-cyclic AMP, and prostaglandins. *Proc. Natl. Acad. Sci. U.S.A.* **74**: 3864–3868 (1977).

W. Dessau, H. von der Mark, K. von der Mark, and S. Fischer. Changes in the patterns of collagens and fibronectin during limb bud chondrogenesis. *J. Embryol. Exp. Morphol.* **57**: 51–60 (1980).

D. Ede. Cellular condensations and chondrogenesis. In: *Cartilage: Development, Differentiation, and Growth,* Vol. 2, Ed B.K. Hall. Academic Press, New York. pp. 143–185, 1983.

D. Ede, O. Flint, O. Wilby, and P. Colquhoun. The development of precartilage condensations in limb-bud mesenchyme of normal and mutant embryos *in vivo* and *in vitro*. In: *Vertebrate Limb and Somite Morphogenesis,* Eds D. Ede, J. Hinchliffe, and M. Balls. Cambridge University Press, London. pp. 161–179, 1977.

G. Edelman. Cell adhesion molecules in the regulation of animal form and pattern. *Annu. Rev. Cell Biol.* **2**: 81–116 (1986).

P. Ekblom, D. Vestweber, and R. Kemler. Cell-matrix interactions and cell adhesion during development. *Annu. Rev. Cell Biol.* **2**: 27–48 (1986).

M. Evans and R. Tuan. Cellular condensation and collagen type II expression during chondrogenesis *in vitro*. *J. Cell Biol.* 107, 163a (1988).

K. Fassbender, D. Pietryla, M. DiMuzio, M. Lenz, T. Schwitzer, J. Williams, K. Kuettner, and E. Thonar. Affinity of cationic proteins of small and high molecular weights for anionic components of the extracellular matrix of articular cartilage. *Trans. Orthop. Res. Soc.* **14**: 156 (1989).

H. Fell. The histogenesis of cartilage and bone in the long bones of the embryonic fowl. *J. Morphol. Physiol.* **40**: 417–459 (1925).

H. Fell and R. Canti. Experiments on the development *in vitro* of the avian knee joint. *Proc. R. Soc. London Ser. B* **116**: 316–351 (1934).

L. Fransson. Mammalian glycosaminoglycans. In: *The Polysaccharides,* Vol. 3, Ed G.O. Aspinall. Academic Press, Orlando. pp. 337–415, 1985.

D. Frenz, S. Akiyama, D. Paulsen, and S. Newman. Latex beads as probes of cell surface-extracellular matrix interactions during chondrogenesis: evidence for a role for amino-terminal heparin-binding domain of fibronectin. *Dev. Biol.* **136**: 87–96 (1989).

D. Frenz, N. Jaikaria, and S. Newman. The mechanism of precartilage mesenchymal condensation: a major role for interaction of the cell surface with the amino-terminal heparin-binding domain of fibronectin. *Dev. Biol.* **136**: 97–103 (1989).

L. Fritze, C. Reilly, and R. Rosenberg. An antiproliferative heparan sulfate species produced by postconfluent smooth muscle cells. *J. Cell Biol.* **100**: 1041–1049 (1985).

S. Gawande and R. Tuan. Stimulation of limb mesenchyme chondrogenesis *in vitro* by bovine bone extract. *J. Cell Biol.* **105**: 292a (1987).

S. Gawande and R. Tuan. Characterization of bone-derived chondrogenesis-stimulating activity on embryonic limb mesenchymal cells *in vitro*. *Cell Tiss. Kinet.* **23**: 375–390 (1990).

R. Gelman, W. Rippon, and J. Blackwell. Interactions between chondroitin-6-sulfate and poly-L-lysine in aqueous solution: circular dichroism studies. *Biopolymers* **12**: 541–558 (1973).

S. Gilbert. *Developmental Biology,* 2nd ed. Sinauer Associates, Sunderland, MA, chap. 16, 1988.

Cs. Hadhazy, J. Szollosi, and M. Laszlo. *In vitro* cartilage differentiation: modification of division cycle and effect of some polyanions. *Acta Histochem.* (Suppl.) **27**: S285–S288 (1983).

A. Harris. D. Stopak, and P. Warner. Generation of spatially periodic patterns by a mechanical instability: a mechanical alternative to the Turing model. *J. Embryol. Exp. Morphol.* **80**: 1–20 (1984).

V. Hascall, T. Oegema, M. Brown, and A. Caplan. Isolation and characterization of proteoglycans from chick limb bud chondrocytes grown *in vitro*. *J. Biol. Chem.* **251**: 3511–3519 (1976).

J. Hassell, J. Kimura, and V. Hascall. Proteoglycan core protein families. *Annu. Rev. Biochem.* **55**: 539–567 (1986).

E. Hay. Cell-matrix interaction in the embryo: cell shape, cell surface, cell skeletons, and their role in differentiation. In: *The Role of Extracellular Matrix in Development,* Ed R.L. Trelstad. Alan R. Liss, New York. pp. 1–31, 1984.

E. Hayman, A. Oldberg, G. Martin, and E. Ruoslahti. Codistribution of heparan sulfate proteoglycan, laminin, and fibronectin in the extracellular matrix of normal rat kidney cells and their coordinate absence in transformed cells. *J. Cell Biol.* **94**: 28–35 (1982).

H. Hennings and K. Holbrook. Calcium regulation of cell-cell contact and differentiation of epidermal cells in culture. An ultrastructure study. *Exp. Cell Res.* **143**: 127–142 (1983).

H. Holtzer and J. Abbott. Oscillations of the chondrogenic phenotype *in vitro*. In: *The Stability of the Differentiated State,* Ed H. Ursprung. Springer-Verlag, New York. pp. 1–16, 1968.

H. Holtzer, N. Rubinstein, S. Fellini, G. Yeoh, J. Chi, J. Birnhaum, and M. Okayama. Lineages, quantal cell cycles, and generation of cell diversity. *Rev. Biophys.* **8**: 523–557 (1975).

M. Höök, L. Kjellen, S. Johansson, and J. Robinson. Cell-surface glycosaminoglycans. *Annu. Rev. Biochem.* **53**: 847–869 (1984).

S. Hunter and A. Caplan. Control of cartilage differentiation. In: *Cartilage: Development, Differentiation, and Growth,* Vol. 2, Ed B.K. Hall. Academic Press, New York. pp. 87–119, 1983.

R. Hynes. Integrins: a family of cell surface receptors. *Cell* **48**: 549–554 (1987).

R. Hynes and K. Yamada. Fibronectins: multifunctional modular proteins. *J. Cell Biol.* **95**: 369–377 (1982).

J. Ianotti, S. Naidu, Y. Noguchi, R. Hunt, C. Brighton. Calcium-induced matrix vesicle biogenesis. *Trans. Orthop. Res. Soc.* **14**: 125 (1989).

O. Jacenko. Calcium and Cell Differentiation During Chick Embryonic Skeletogenesis, Ph.D. thesis, University of Pennsylvania, Philadelphia.

O. Jacenko and R. Tuan. Calcium deficiency induces expression of cartilage-like phenotype in chick embryonic calvaria. *Dev. Biol.* **115**: 215–232 (1986a).

O. Jacenko and R. Tuan. Changes in the extracellular matrix of chick embryonic bone during induced calcium deficiency. *Prog. Clin. Relat. Res.* **217B**: 401–404 (1986b).

O. Jacenko and R. Tuan. Chondrogenic potential of chick embryonic calvaria. *J. Cell Biol.* **105**: 141a (1987).

S. Jackson and J. Randall. Fibrogenesis and the formation of matrix in developing bone. In: *Ciba Foundation Symposium of Bone Structure and Metabolism,* Eds G. Wolstenholme and H. O'Connor. Little, Brown, Boston. pp. 47–62, 1956.

K. Kimata, Y. Oike, K. Tani, T. Shinomura, M. Yamagata, M. Uritani, and S. Suzuki. A large chondroitin sulfate proteoglycan (PG-M) synthesized before chondrogenesis in the limb bud of chick embryo. *J. Biol. Chem.* **261**: 13517–13525 (1986).

L. Kjellen, A. Oldberg, and M. Höök. Cell surface heparan sulfate. Mechanisms of proteoglycan-cell association. *J. Biol. Chem.* **255**: 10407–10413 (1980).

C. Knudson and B. Toole. Changes in the pericellular matrix during differentiation of limb bud mesoderm. *Dev. Biol.* **112**: 308–318 (1985).

J. Koda, A. Rapraeger, and M. Bernfield. Heparan sulfate proteoglycans from mouse mammary epithelial cells. Cell surface proteoglycan as a receptor for interstitial collagens. *J. Biol. Chem.* **260**: 8157–8162 (1985).

R. Kosher and R. Church. Stimulation of *in vitro* somite chondrogenesis by procollagen and collagen. *Nature (London)* **258**: 327–329 (1975).

R. Kosher and R. Lash. Notochordal stimulation of *in vitro* somite chondrogenesis before and after enzymatic removal of perinotochordal materials. *Dev. Biol.* **42**: 362–378 (1975).

R. Kosher, J. Lash, and R. Minor. Environmental enhancement of *in vitro* chondrogenesis. IV. Stimulation of somite chondrogenesis by exogenous chondromucoprotein. *Dev. Biol.* **35**: 210–220 (1973).

R. Kosher, M. Savage, and S. Chan. *In vitro* studies on the morphogenesis and differentiation of the mesoderm subjacent to the apical ectodermal ridge of the embryonic chick limb bud. *J. Embryol. Exp. Morphol.* **50**: 75–97 (1979).

K. Kuettner, N. Sorgente, and R. Eisenstein. Lysozyme in calcifying tissues. *Clin. Orthop. Relat. Res.* **112**: 316–339 (1975).

M. Kujawa and A. Caplan. Hyaluronic acid bonded to cell culture surfaces stimulates chondrogenesis in stage 24 limb mesenchyme cell cultures. *Dev. Biol.* **114**: 504–518 (1986).

M. Kujawa, D. Carrino, and A. Caplan. Substrate-bonded hyaluronic acid exhibits a size dependent stimulation of chondrogenic differentiation of stage 24 limb mesenchymal cells in culture. *Dev. Biol.* **114**: 519–528 (1986).

J. Lash and N. Vasan. Somite chondrogenesis *in vitro*: stimulation by exogenous extracellular matrix components. *Dev. Biol.* **66**: 151–171 (1978).

J. Laterra, J. Silbert, and L. Culp. Cell surface heparan sulfate mediates some adhesive responses to glycosaminoglycan-binding matrices, including fibronectin. *J. Cell Biol.* **96**: 112–113 (1983).

S. McDonald and R. Tuan. Expression of collagen type II transcripts in chick embryonic bone detected by *in situ* cDNA-mRNA hybridization. *Dev. Biol.* **133**: 221–234 (1989).

J. Magnani, W. Thomas, and M. Steinberg. Two distinct adhesion mechanisms in embryonic neural retina cells. I. A kinetic analysis. *Dev. Biol.* **81**: 96–105 (1981).

R. Majack and P. Bornstein. Heparin regulates the collagen phenotype of vascular smooth muscle cells: induced synthesis of an M_r 60,000 collagen. *J. Cell Biol.* **100**: 613–619 (1985).

G. Maor, D. Lewinson, and M. Silverman. The possibility of calcium involvement in the inhibitory effect of dexamethasone on DNA synthesis in cartilage. In: *Current Advances in Skeletogenesis,* Eds A. Ornoy, A. Harrell, and J. Sela. Elsevier, Amsterdam. pp. 105–112, 1985.

F. Matsui, O. Atsuhiko, R. Shoji, and H. Nogami. Glycosaminoglycans and proteoglycans synthesized by rat limb buds during prechondrogenic and chondrogenic stages. *Differentiation* **31**: 29–34 (1986).

S. Meier and E. Hay. Stimulation of extracellular matrix synthesis in the developing cornea by glycosaminoglycans. *Proc. Natl. Acad. Sci. U.S.A.* **71**: 2310–2313 (1974).

M. Mohammed. Cell Division in Normal and Mutant Chick Embryos, Ph.D. thesis, University of Glasgow, United Kingdom, 1978.

N. Morad, H. Ryser, and W. Shen. Binding sites and endocytosis of heparin and polylysine are changed when the two molecules are given as a complex to chinese hamster ovary cells. *Biochim. Biophys. Acta* **801**: 117–126 (1984).

A. Moscona. Development of heterotypic combinations of dissociated embryonic chick cells. *Proc. Soc. Exp. Biol. Med.* **92**: 410–416 (1956).

J. Nefussi, M. Boy-Lefevre, H. Boulekbache, and N. Forest. Mineralization *in vitro* of matrix formed by osteoblasts isolated by collagenase digestion. *Differentiation* **29**: 160–168 (1985).

Z. Nevo and A. Dorfman. Stimulation of chondromucoprotein synthesis in chondrocytes by extracellular chondromucoprotein. *Proc. Natl. Acad. Sci. U.S.A.*, **69**: 2069–2072 (1972).

S. Newman. Lineage and pattern in the developing wing bud. In: *Vertebrate Limb and Somite Morphogenesis,* Eds D. Ede, J. Hinchliffe, and M. Balls. Cambridge University Press, Cambridge. pp. 181–197, 1977.

S. Newman and H. Frisch. Dynamics of skeletal pattern formation in developing chick limb. *Science* **205**: 662–668 (1979).

S. Newman, M.-P. Pautou, and M. Kieny. The distal boundary of myogenic primordia in chimeric avian limb buds and its relation to an accessible population of cartilage progenitor cells. *Dev. Biol.* **84**: 440–448 (1981).

S. Newman, D. Frenz, J. Tomasek, and D. Rabuzzi. Matrix-driven translocation of cells and non-living particles. *Science* **228**: 885–889 (1985).

B. Obrink. The influence of glycosaminoglycans on the formation of fibers from monomeric tropocollagen *in vitro*. *Eur. J. Biochem.* **34**: 129–137 (1973).

B. Obrink, T. Laurent, and B. Carlsson. The binding of chondroitin sulfate to collagen. *FEBS Lett.* **56**: 166–169 (1975).

G. Oster, J. Murray, and P. Maini. Forces and patterns and limb morphogenesis. *J. Embryol. Exp. Morphol.* **82** (Suppl.): 98a (1984).

G. Oster, J. Murray, and P. Maini. A model for chondrogenic condensations in the developing limb: the role of extracellular matrix and cell tractions. *J. Embryol. Exp. Morphol.* **89**: 93–112 (1985).

P. Osdoby and A. Caplan. Osteogenesis in cultures of limb mesenchymal cells. *Dev. Biol.* **73**: 84–102 (1979).

J. Pennypacker, C. Lewis, and J. Hassell. Altered proteoglycan metabolism in mouse limb mesenchyme cell cultures treated with vitamin A. *Arch. Biochem. Biophys.* **186**: 351–358 (1978).

N. Ratner, R. Bunge, and L. Glaser. A neuronal cell surface heparan sulfate proteoglycan is required for dorsal root ganglion neuron stimulation of Schwann cell proliferation. *J. Cell Biol.* **101**: 744–754 (1985).

L. Rifas, J. Uitto, V. Memoli, K. Kuettner, R. Henry, and W. Peck. Selective emergence of differentiated chondrocytes during serum-free culture of cells derived from fetal rat calvaria. *J. Cell Biol.* **92**: 493–504 (1982).

E. Ruoslahti. Fibronectin and its receptors. *Annu. Rev. Biochem.* **57**: 375–414 (1988).

E. Ruoslahti. Structure and biology of proteoglycans. *Annu. Rev. Cell Biol.* **4**: 229–255 (1988).

H. Ryser and R. Hancock. Histones and basic polyamino acids stimulate the uptake of albumin by tumor cells in culture. *Science* **150**: 501–503 (1965).

H. Ryser, W. Shen, and F. Merk. Membrane transport of macromolecules: new carrier functions of proteins and poly(amino acids). *Life Sci.* **22**: 1253–1260 (1978).

S. Sakashita, E. Engvall, and E. Ruoslahti. Basement membrane glycoprotein laminin binds to heparin. *FEBS Lett.* **116**: 243–246 (1980).

J. San Antonio. Polyionic Regulation of Chondrogenesis, Ph.D. thesis, University of Pennsylvania, Philadelphia, 1988.

J. San Antonio and R. Tuan. Chondrogenesis of limb bud mesenchyme *in vitro:* stimulation by cations. *Dev. Biol.* **115**: 313–324 (1986).

J. San Antonio and R. Tuan. Polyionic regulation of chondrogenesis. *J. Cell Biol.* **105**: 293a (1987).

J. San Antonio and R. Tuan. Polyionic regulation of chondrogenesis. II. Promotion of chondrogenesis by polylysine is associated with an enhanced retention of the extracellular matrix. Submitted for publication.

J. San Antonio, O. Jacenko, and R. Tuan. Polyionic regulation of chondrogenesis. I. Interaction between polylysine and sulfated glycosaminoglycans of limb mesenchyme and other embryonic cell types. Submitted for publication.

J. San Antonio, B. Winston, and R. Tuan. Regulation of chondrogenesis by heparan sulfate and structurally related glycosaminoglycans. *Dev. Biol.* **123**: 17–24 (1987).

J. Saunders. Developmental Control of three-dimensional polarity in the avian limb. *Ann. N.Y. Acad. Sci.* **193**: 29–42 (1972).

J. Saunders and M. Gasseling. Ectodermal-mesenchymal interactions in the origin of limb symmetry. In: *Epithelial-Mesenchymal Interactions,* Eds R. Fleishmajer and R. Billingham. Williams and Williams, Baltimore. pp. 78–97, 1968.

T. Shinomura, K. Kimata, Y. Oikc, N. Maeda, S. Yano, and S. Suzuki. Appearance of distinct types of proteoglycan in a well-defined temporal and spatial pattern during early cartilage formation in the chick limb. *Dev. Biol.* **103**: 211–220 (1984).

M. Solursh. Cell and matrix interactions during limb chondrogenesis *in vitro*. In: *The Role of Extracellular Matrix in Development,* Ed R. Trelstad. Alan R. Liss, New York. pp. 277–303, 1983.

M. Solursh. Cell-cell interactions and chondrogenesis. In: *Cartilage: Development, Differentiation, and Growth,* Vol. 2, Ed B.K. Hall. Academic Press, New York. pp. 121–141, 1984.

M. Solursh, P. Ahrens, and R. Reiter. A tissue culture analysis of the steps in limb bud chondrogenesis. *In Vitro* **14**: 51–61 (1978).

M. Solursh and R. Reiter. Evidence of histogenic interactions during *in vitro* limb chondrogenesis. *Dev. Biol.* **78**: 141–150 (1980).

M. Solursh. T. Linsenmayer, and K. Jensen. Chondrogenesis from single limb bud mesenchyme cells. *Dev. Biol.* **94**: 259–264 (1982).

D. Stopak and A. Harris. Connective tissue morphogenesis by fibroblast traction. I. Tissue culture observations. *Dev. Biol.* **90**: 383–398 (1982).

B. Swalla and M. Solursh. Inhibition of limb chondrogenesis by fibronectin. *Differentiation* **26**: 42–48 (1984).

G. Syftestad and A. Caplan. A fraction from extracts of demineralized adult bone stimulates the conversion of mesenchymal cells into chondrocytes. *Dev. Biol.* **104**: 348–356 (1981).

G. Syftestad, J. Triffitt, M. Urist, and A. Caplan. An osteo-inductive bone matrix extract stimulates the *in vitro* conversion of mesenchyme into chondrocytes. *Calcif. Tissue Int.* **30**: 625–627 (1984).

M. Takeda, T. Schwitzer, B. Matijevitch, G. Campion, M. Aydelotte, K. Kuettner, and E. Thonar. Stimulation of cartilage proteoglycan synthesis by high and low molecular weight poly-L-lysine. *Trans. Orthop. Res. Soc.* **14**: 172 (1989).

M. Takeichi. Cadherins: a molecular family essential for selective cell-cell adhesion and animal morphogenesis. *Trends Genet.* **3**: 213–217 (1987).

M. Takeichi, T. Atsumi, C. Yoshida, K. Uno, and T. Okada. Selective adhesion of embryonal cells by Ca^{2+}-dependent sites. *Dev. Biol.* **87**: 340–350 (1981).

P. Thorogood and J. Hinchliffe. An analysis of the condensation process during chondrogenesis in the embryonic chick bud limb. *J. Embryol. Exp. Morphol.* **33**: 581–606 (1975).

B. Toole. Hyaluronate turnover during chondrogenesis in the developing chick limb and axial skeleton. *Dev. Biol.* **29**: 321–329 (1972).

B. Toole. Developmental role of hyaluronate. *Connect. Tissue Res.* **10**: 83–100 (1982).

B. Toole and T. Linsenmayer. Newer knowledge of skeletogenesis: macromolecular transitions in the extracellular matrix. *Clin. Orthop. Relat. Res.* **129**: 258–278 (1977).

R. Tuan. Mechanism and regulation of calcium transport by the chick embryonic chorioallantoic membrane. *J. Exp. Zool.* (Suppl.) **1**: 1–13 (1987).

R. Tuan and M. Lynch. Effect of experimentally induced calcium deficiency on the developmental expression of collagen types in chick embryonic skeleton. *Dev. Biol.* **100**: 374–386 (1983).

R. Umansky. The effect of cell population density on the developmental fate of reaggregating mouse limb mesenchyme. *Dev. Biol.* **13**: 31–56 (1966).

N. Vasan. Analysis of extracellular matrix material during mesenchymal cell condensation in limb development. *J. Cell Biol.* **103**: 97a (1986).

N. Vasan and J. Lash. Heterogeneity of proteoglycans in developing chick limb cartilage. *Biochem. J.* **164**: 179–183 (1974).

N. Vasan and J. Lash. Monomer and aggregate proteoglycans in the chondrogenic differentiation of embryonic chick limb buds. *J. Embryol. Exp. Morphol.* **49**: 47–59 (1979).

J. Villanueva and M. Nimni. Cells isolated from fetal calvaria by isopycnic separation express different collagen phenotypes. *Trans. Orthop. Res. Soc.* **14**: 123 (1989).

K. von der Mark, H. von der Mark, and S. Gay. Study of differential collagen synthesis during development of the chick embryo by immunofluorescence. II. Localization of type I and type II collagen during long bone development. *Dev. Biol.* **53**: 153–170 (1976).

J. Whitfield, A. Perris, and T. Youdale. The role of calcium in the mitotic stimulation of rat thymocytes by detergents, agmatine, and poly-lysine. *Exp. Cell Res.* **53**: 155–165 (1968).

O. Wiebkin and H. Muir. The inhibition of sulfate incorporation in isolated adult chondrocytes by hyaluronic acid. *FEBS Lett.* **37**: 42–46 (1973).

M. Yamagata, K. Yamada, M. Yoneda, S. Suzuki, and K. Kimata. Chondroitin sulfate proteoglycan (PG-M-like chondroitin sulfate proteoglycan) is involved in the binding of hyaluronic acid to cellular fibronectin. *J. Biol. Chem.* **261**: 13526–13535 (1986).

N. Zanetti and M. Solursh. Induction of chondrogenesis in limb mesenchymal cultures by disruption of the actin cytoskeleton. *J. Cell Biol.* **99**: 115–123 (1984).

B. Zimmerman and M. Theis. Alterations of lectin binding during chondrogenesis in mouse limb buds. *Histochemistry* **81**: 353–361 (1984).

E. Zwilling. Morphogenetic phases in development. *Dev. Biol.* (Suppl.) **2**: 184–207 (1968).

6

The Growth Plate: Cellular Physiology, Cartilage Assembly and Mineralization

A. ROBIN POOLE
Joint Diseases Laboratory
Shriners Hospital for Crippled Children
Division of Surgical Research
Department of Surgery
McGill University
Montreal, Quebec, Canada

Structure and Organization of the Growth Plate

Overview

The primary growth plates of developing mammalian long bones offer the best insights into growth plate structure and organization (Figure 1a). This is because the cells of the different zones are well defined about a longitudinal axis stretching from the cartilage or bone of the epiphysis (depending upon the age of the individual) to the metaphyseal woven bone. The primary growth plate is responsible for longitudinal growth, whereas the secondary growth plate or ossification center which forms within the epiphysis (Figure 1b) is responsible for epiphyseal endochondral ossification and is more "spherical" in shape. In this chapter discussions will be restricted to that part of the growth plate which is composed of cartilaginous structures, namely the resting, proliferative, maturing, and hypertrophic zones (Figure 2).

The reserve zone, which abuts the proliferating zone, contains approximately spherical cells which exhibit little or no cell division (Kember, 1960). In the proliferative zone, where oxygen tension is highest (Brighton and Heppenstall, 1971), cells divide to give rise to long columns of flattened cells which secrete a hyaline extracellular matrix containing type II collagen. This is composed of transverse septa and longitudinal septa elaborated by these chondrocytes (Figure 1). The flattened nature of the cells suggests that they are under a longitudinal compressive force. Their appearance is reminiscent of the flattened chondrocytes of the superficial zone of articular cartilage, which are exposed to the compressive and sheer forces of articulation. Also, flattened cells are found in the perichondria of other cartilages where cell proliferation is centered. Thus the appearance of the proliferative zone suggests that it results from axial loading accompanying active cell division.

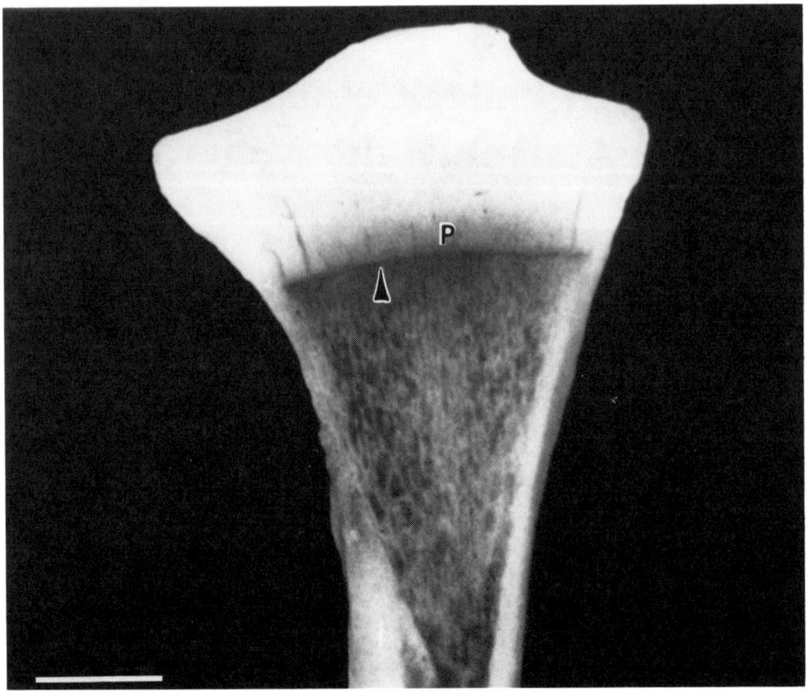

A

Figure 1. Sagittal sections of (A) the primary growth plate (P) of the proximal fetal bovine tibia and (B) the primary (P) and secondary (S) growth plates of the distal bovine fetal femur. The junction with the metaphysis is indicated by an arrow. Scale = 1 cm.

The cells of the proliferative zone suddenly start to enlarge and round up to form the zone of maturation in which the chondrocytes mature into hypertrophic chondrocytes. The zone is clearly identifiable in that it is preceded by the flattened proliferating cells and followed by the much enlarged hypertrophic cells of the upper hypertrophic zone. The hypertrophic cells enlarge five- to tenfold (Buckwalter *et al.*, 1986; Hunziker *et al.*, 1987). At the same time there is a reduction in matrix volume per unit volume of tissue. In addition, there is an increase in the volume of the territorial matrix (Buckwalter *et al.*, 1986). It is in this zone that the mineralization or calcification of cartilage takes place: it starts suddenly in the lower hypertrophic zone and is mainly concentrated in the longitudinal septa.

Calcification starts in discrete focal sites between collagen fibrils (Figure 3). It rapidly radiates out from these centers. Decalcification of such cartilages reveals the presence of "mineralization nodules" (see Figure 10f in Poole *et al.*, 1984). Calcification is initiated in sites close to cells in the territorial

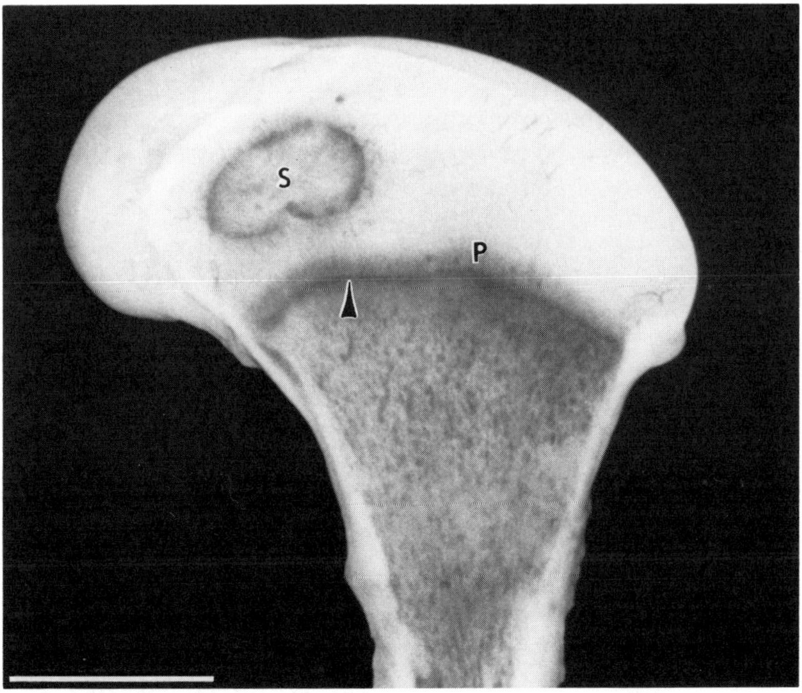

FIGURE 1B.

matrix. Eventually calcification spreads throughout the cartilage matrix of the longitudinal septa. The transverse septa usually remain uncalcified. The pericellular zone immediately around chondrocytes, which is approximately 2 μm wide, also does not calcify, even in the longitudinal septa.

The hypertrophic cells in the lowermost layer of the hypertrophic zone eventually die, probably as a result of the extensive mineralization of matrix around them leading to a loss of nutrition. The condensed cells can be observed on the last transverse septa (Farnum and Wilsman, 1987).

At this time, capillary sprouts erode the last transverse septa separating the hypertrophic zone from the metaphysis (Figure 4) (Schenk *et al.*, 1967, 1968; Arsenault, 1987). Eventually, vessels break through into the lacunae containing the remnants of hypertrophic cells. Many (up to half) of the mineralized longitudinal septa are eroded by chondroclasts which appear identical to osteoclasts (Schenk *et al.*, 1967, 1968). Those that remain (called trabeculae) act as axial calcified cartilaginous scaffolds onto which osteoblasts settle (Figures 2 and 4). This process of vascular invasion carefully balances that involving cell division and assembly of new matrix. In rats the rate of vascular ingrowth is approximately 0.3 mm in 24 h (Hunziker *et al.*, 1987). Any residual uncalcified matrix, which remains in what were pericellular sites, is

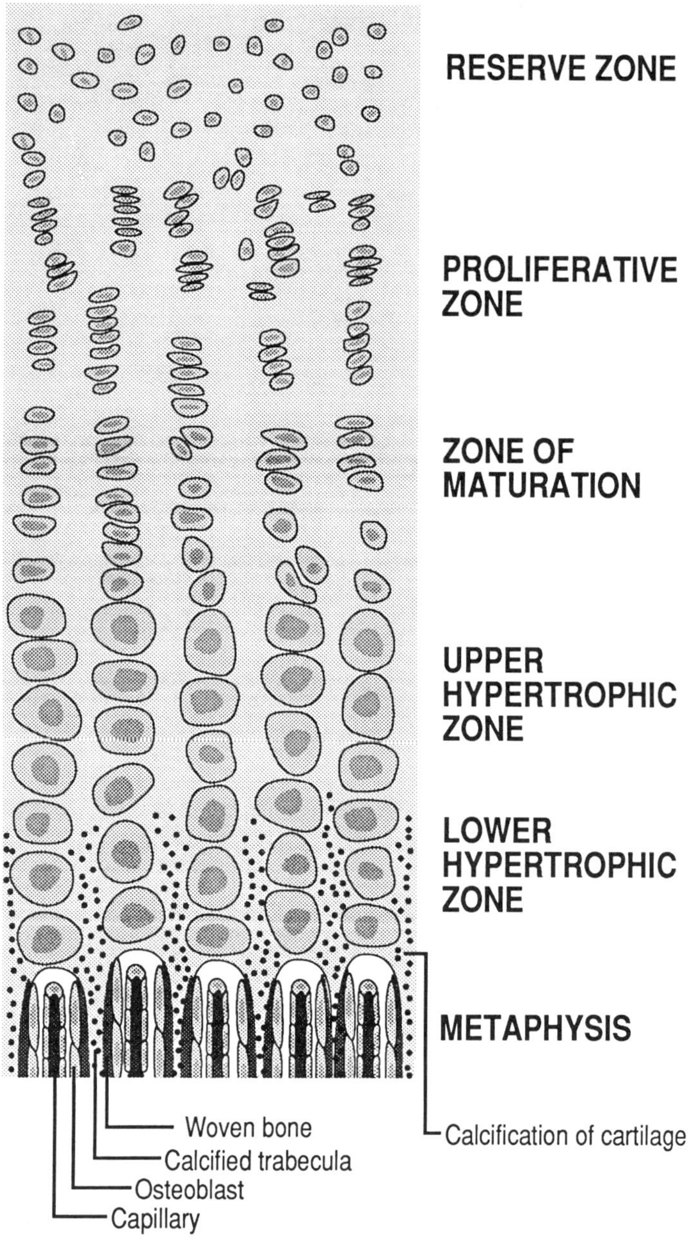

Figure 2. Diagrammatic representation of the organization of a primary mammalian growth plate, based upon that present in the proximal fetal bovine tibia. The region of the metaphysis is more clearly shown in Figure 4.

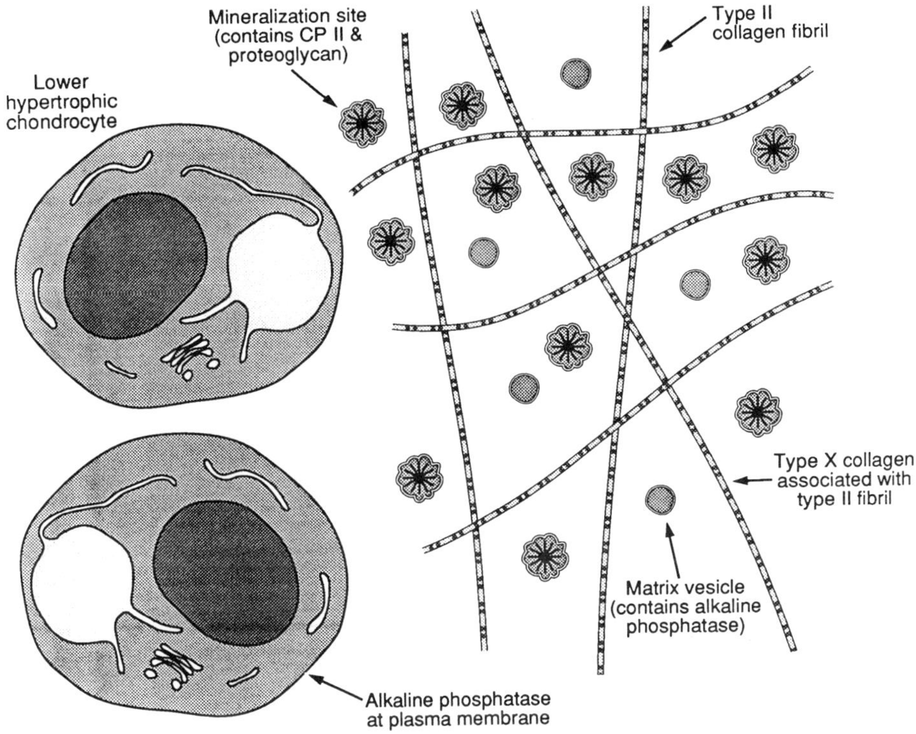

Figure 3. Diagrammatic relationship of collagens, proteoglycan, C-propeptide of type II collagen (CPII), matrix vesicles, and alkaline phosphatase in the calcification of cartilage which occurs in the lower hypertrophic zone. The hypertrophic cells are very enlarged and contain large dilations (usually one per cell) of the rough endoplasmic reticulum. Type II collagen fibrils, matrix vesicles, and focal mineralization sites are drawn to scale.

removed up to the junction with the calcified matrix. This junction is known as the lamina limitans. Osteoblasts secrete osteoid which calcifies, giving rise to woven bone. This covers the calcified cartilaginous trabeculae (Figures 2 and 4). Eventually this woven bone and the calcified trabeculae are resorbed by osteoclasts and replaced by a mature lamellar or cancellous bone.

The time taken for a cell to pass from the upper proliferative zone into the lowermost hypertrophic zone may be up to 3 days. It has been estimated by fluorochrome labelling that one layer of chrondrocytes is eliminated by vascular invasion every 3 h, or 8 layers a day (Hunziker *et al.*, 1987). Kember (1960) found that a layer of approximately five new chondrocytes are produced daily in the proliferating zone of the rat growth plate. Based on these studies, it can be calculated that lower hypertrophic cells probably survive for a total of about 24 h, with the calcification process occupying about half of this time (see also Hunziker *et al.*, 1987). The work of Luder *et al.* (1988) on the

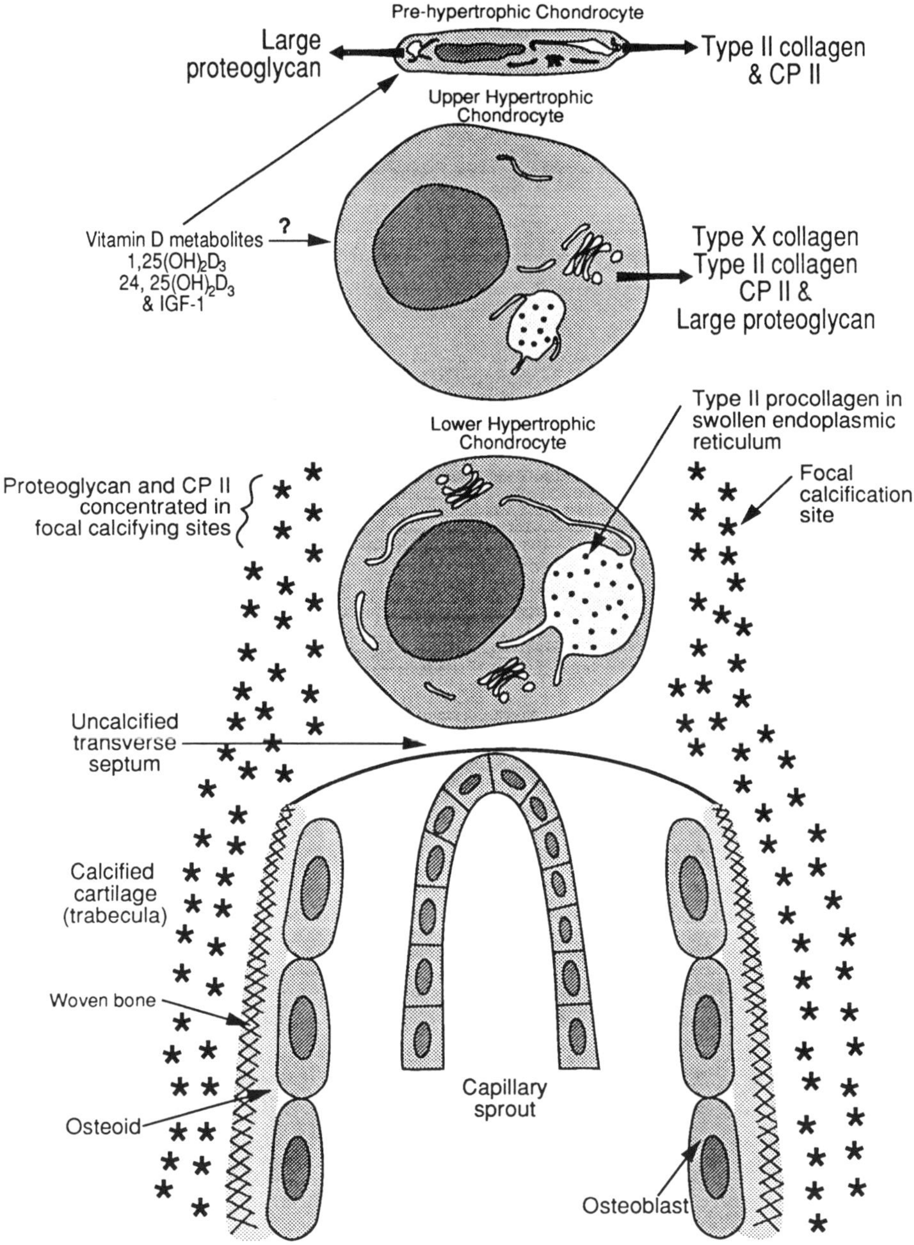

Figure 4. Cellular/matrix changes and interactions in the growth plate. Prehypertrophic cells (cells of the proliferative zone) secrete extracellular matrix under the control of vitamin D metabolites and IGF-I. It is unclear whether upper hypertrophic chondrocytes are also responsive to these molecules. Calcification is initiated in the lower hypertrophic zone in focal sites where the large aggregating proteoglycans and the C-propeptide of type II collagen (CPII) are concentrated. In the metaphysis, osteoblasts secrete osteoid and form woven bone on the calcified trabeculae. Capillary sprouts erode the last transverse septa of the hypertrophic zone which mark the boundary with the metaphysis.

mandibular condyle growth plate in the weanling rat indicates a life span of growth plate chondrocytes of 5.3 days with only 12 h spent as proliferative cells and about 24 h as lower hypertrophic cells.

The Biosynthetic Machinery of the Growth Plate: Matrix Synthesis and Its Regulation

Synthesis

The majority of matrix biosynthesis takes place in the proliferative, maturing, and upper hypertrophic zone as indicated by autoradiographic studies (Greulich and Leblond, 1953; Belanger, 1954). *In situ* hybridization studies of human growth plates have revealed that these zones are also where mRNA for type II collagen is mainly concentrated (Sandberg and Vuorio, 1987). Analyses of mRNA in chick growth plates has revealed that most of the mRNA is present in the proliferative zone, with less present in the upper and lower hypertrophic zones (Leboy *et al.*, 1988). There is little or no evidence for significant matrix synthesis in the lower hypertrophic zones in the chick and in humans (Sandberg and Vuorio, 1987; Leboy *et al.*, 1988). But type II procollagen is detectable in these and upper hypertrophic cells in large dilations of the rough endoplasmic reticulum (Lee and Poole, 1990) suggesting that some is retained after synthesis in the upper hypertrophic zone.

In addition to type II collagen (which is typical of hyaline cartilage, as opposed to fibrocartilage where type I collagen is found), the growth plate contains type IX (Müller-Glauser *et al.*, 1986), type X (Schmid and Conrad, 1982a, b; Gibson *et al.*, 1982), and type XI (Eyre and Wu, 1987) collagens. Type I collagen mRNA is not detectable (Sandberg and Vuorio, 1987; Leboy *et al.*, 1988). Type X is only synthesized by cells of the upper hypertrophic and maturing zones (Schmid and Linsenmayer, 1985; Gibson *et al.*, 1986; Leboy *et al.*, 1988), but is a major biosynthetic product probably accounting for up to 45% of the total collagen synthesized at this time (Reginato *et al.*, 1986).

The large chondroitin sulfate-rich proteoglycans that aggregate with hyaluronic acid are synthesized by growth plate chondrocytes. These contain both chondroitin 4-sulfate and chondroitin 6-sulfate, the latter being more common in the hypertrophic zone of the chick (Kimata *et al.*, 1973; Oohira *et al.*, 1974; Stocum *et al.*, 1979). A small, ubiquitously distributed proteoglycan which contains either a single chain of dermatan sulfate or chondroitin sulfate, and has been called decorin, PG II or DS-PG II when dermatan sulfate is present (Rosenberg *et al.*, 1985), is located in the adjacent epiphyseal cartilage and resting zone. But it is not detectable in the proliferative, maturing, and hypertrophic zones of the growth plate (Poole *et al.*, 1986). Little is known about other matrix molecules which may be synthesized by growth plate chondrocytes, with the exception of the calcium binding protein, osteonectin, the mRNA of which has been detected in growth plate cartilage

(Leboy *et al.*, 1988) although the protein was not detected in the growth plate by immunofluorescence (Termine *et al.*, 1981).

Growth Hormone and Insulin-Like Growth Factor-1

In vivo studies of growth plate physiology have revelaed that local infusion of growth hormone (Isaksson *et al.*, 1982; Russell and Spencer, 1985) and insulin-like growth factor I (Schoenle *et al.*, 1982, 1985) can stimulate growth *in vivo*. Immunohistochemical analyses of rats have revealed that insulin-like growth factor 1 (IGF-1) is mainly concentrated in cells of the proliferative zone (Nilsson *et al.*, 1986). After hypophysectomy, the number of staining cells is markedly reduced. On treatment with growth hormone, either systemically or locally, the number of IGF-1 reactive cells in the proliferative zone is increased. These observations suggest that growth hormone can directly induce synthesis of IGF-1 in chondrocytes. There is general agreement that IGF-1 can stimulate colony formation (Lindahl *et al.*, 1987; Vetter *et al.*, 1986), cell proliferation (Trippel *et al.*, 1989; Makower *et al.*, 1989) and acts primarily on cells of the proliferative zone (Trippel *et al.*, 1989; Makower *et al.*, 1988). Whereas growth hormone has an effect *in vivo*, some studies have failed to confirm that it has a stimulatory effect on cell growth *in vitro* (Burch *et al.*, 1985; Vetter *et al.*, 1985, 1986; Trippel *et al.*, 1989; Makower *et al.*, 1989), although other reports have indicated that it has (Madsen *et al.*, 1983), particularly at lower concentrations (Lindahl *et al.*, 1987). Therefore the effect of growth hormone may be to induce IGF-1 synthesis which then acts in a paracrine or autocrine manner. IGF-1 has potent effects on growth plate proteoglycan synthesis (Makower *et al.*, 1988; Trippel *et al.*, 1989); receptors for this molecule have been identified (Trippel *et al.*, 1988).

Vitamin D

Traditionally, a deficiency of vitamin D is known to cause impairment of cartilage calcification in endochondral ossification, leading to the development of a lengthened uncalcified hypertrophic zone and rickets (see, for example, Atkin *et al.*, 1985). There is now evidence that vitamin D has a direct effect on growth plate chondrocytes. Studies employing autoradiography have indicated that there are probably receptors for 24,25-$(OH)_2$ vitamin D_3 (but not for 1,25-$(OH)_2$ vitamin D_3) in cells of the proliferative zone of vitamin D deficient rats (Fine *et al.*, 1985) and for 1,25-$(OH)_2$ vitamin D_3 on healthy chick embryo growth cartilage chondrocytes (Suda *et al.*, 1985). No receptor was found in resting cartilage. Increases in calcification of growth cartilage are paralleled by an increase in receptor levels (Suda *et al.*, 1985). Rat growth plate chondrocytes contain a vitamin D-dependent calcium-binding protein found in the duodenum (Balmain *et al.*, 1986; Zhou *et al.*, 1986). It is mainly localized in the cytoplasm of maturing chondrocytes but is present in the cytoplasmic processes of hypertrophic cells. Chick embryo growth plate chondrocytes also stain (Zhou *et al.*, 1986). The stimulatory effect of 24,25 $(OH)_2$

vitamin D_3 as well as 1,25 $(OH)_2$ vitamin D_3 on proteoglycan synthesis in growth plate chondrocytes has also been indicated in studies of normal rabbit growth plates (Corvol *et al.*, 1978). In organ culture, maintenance of normal hydration, hypertrophy, columnar arrangement, and preservation of mineral require the presence of both metabolites at physiological concentrations (Plachot *et al.*, 1982). 1,25-$(OH)_2$ vitamin D_3 has been found to stimulate the proliferation of rabbit costal growth plate cells but inhibits biosynthesis of proteoglycans (Takigawa *et al.*, 1988). In contrast, 24,25 $(OH)_2$ vitamin D_3 stimulates proteoglycan synthesis but has no effect on proliferation. More recently, analyses of calcification mediated by rat growth plate chondrocytes *in vitro* revealed that 24,25 $(OH)_2$ vitamin D_3 and 1,25 $(OH)_2$ vitamin D_3 are both required for maximal calcification and type II collagen synthesis (measured as C-propeptide) in healthy rats (Hinek and Poole, 1988). Parallel investigations of chondrocytes from healthy bovine fetuses (Hinek *et al.*, 1987) have also revealed that both 1,25 $(OH)_2$ vitamin D_3 and 24,25 $(OH)_2$ vitamin D_3 are required to support maximal calcification and type II collagen synthesis in vitamin D sufficient animals. These effects are specific for growth plate chondrocytes.

But in the case of vitamin D deficiency, the evidence consistently points to an exclusive requirement for 24,25 $(OH)_2$ vitamin D_3 to support maximal cellular metabolism in rachitic growth plates of the chick (Ornoy *et al.*, 1978; Lidor *et al.*, 1987) and the rat (Atkin *et al.*, 1985; Fine *et al.*, 1985; Hinek and Poole, 1988). Micromass cultures of embryonic calcifying chick chondrocytes have demonstrated that 24,25 $(OH)_2$ vitamin D_3 but not 1,25 $(OH)_2$ vitamin D_3 induces increases in cell proliferation and protein synthesis (Binderman and Sömjen, 1984; Sömjen *et al.*, 1984). In vitamin D deficiency in the chick, smaller aggregating proteoglycans are synthesized (Roughley and Dickson, 1980, 1986). Restoration of calcium results in the production of larger proteoglycans. These effects, which were shown to be specific for growth cartilages, have been confirmed by others (Carrino *et al.*, 1989). In our studies there was no evidence that vitamin D has any effect on the total proteoglycan content of bovine growth cartilages (Hinek and Poole, 1988). In healing rickets in the rat there is a substantial decrease in proteoglycan binding to hyaluronic acid (Reinholt *et al.*, 1985). This suggests that proteoglycan cleavage, which often occurs naturally in cartilage in the part of the molecule adjacent to the hyaluronic acid-binding region and is probably mediated by stromelysin (Nguyen *et al.*, 1989), is increased in the presence of vitamin D and retarded in rickets.

Studies of alkaline phosphatase have revealed that 24,25 $(OH)_2$ vitamin D_3 has a direct stimulatory effect on the activity (and presumably biosynthesis) of this enzyme in cultures of growth plate chondrocytes isolated from vitamin D deficient rachitic chicks (Hale *et al.*, 1986). But alkaline phosphatase in costochondral chondrocytes from healthy rats is stimulated by 1,25-$(OH)_2$ vitamin D_3 and no effect of 24,25-$(OH)_2$ vitamin D_3 was observed (Schwartz *et al.*, 1988a). These effects were specific for growth cartilage as opposed to resting zone chondrocytes. In studies of isolated matrix vesicles and plasma membranes isolated from healthy rats, direct stimulatory effects of both 1,25-$(OH)_2$ vitamin D_3 and 24,25-$(OH)_2$ vitamin D_3 on alkaline phosphatase have been observed (Schwartz *et al.*, 1988b).

Type X collagen is over-produced in vitamin D deficiency in chicks (Reginato *et al.*, 1988). Thus it is regulated differently from type II collagen and alkaline phosphatase.

Cellular Changes Related to Hypertrophy and the Calcification of Cartilage Matrix

Mitochondrial Calcium

Numerous studies have demonstrated that there is an accumulation of calcium in the mitochondria of chondrocytes in the zone of maturation and the upper hypertrophic zone (see, for example, Martin and Matthews, 1970; Arsenis, 1972; Brighton and Hunt, 1974; Hargest *et al.*, 1985). Subsequently there is a loss of mitochondrial calcium in the lower hypertrophic zone when calcification starts in the matrix. Since the mitochondrial calcium is absent or decreased in mitochondria in rachitic rat growth plates that do not mineralize (Matthews *et al.*, 1970), the accumulation of mitochondrial calcium would appear to be related to calcification. Moreover, mitochondrial calcium is fully retained in cultured growth plate chondrocytes only when 1,25-dihydroxy vitamin D_3 is present (Plachot *et al.*, 1982).

Alkaline Phosphatase

The early studies of Robison (1923) and Fell and Robison (1929, 1930) drew attention to the close correlation between the tissue content of alkaline phosphatase and mineralization. Histochemical studies revealed that cellular alkaline phosphatase content increased considerably in hypertrophic cells in association with matrix mineralization (Väänänen, 1980; de Bernard *et al.*, 1986). Biochemical analyses also showed that hypertrophic cartilage contains more alkaline phosphatase than the proliferative zone which contains more than epiphyseal cartilage (Kuhlman and McNamee, 1970; Althoff *et al.*, 1982). Recently, analyses of frozen sections of bovine fetal growth plates have demonstrated a progressive increase in alkaline phosphatase from the proliferative zone through the maturing zone to the hypertrophic zone (M. Alini, Y. Matsui and A.R. Poole, manuscript in preparation).

Alkaline phosphatase is an enzyme with the capacity to liberate inorganic phosphate from organic substrates, or from inorganic substrates such as pyrophosphate. It has been suggested that one role of alkaline phosphatase is to remove pyrophosphate which is inhibitory to the formation of calcium phosphate (see review by Wuthier and Register, 1984). Reaction of inorganic phosphate with calcium leads to the formation of a form of calcium phosphate called hydroxyapatite (Figure 5). Elevated alkaline phosphatase activity is

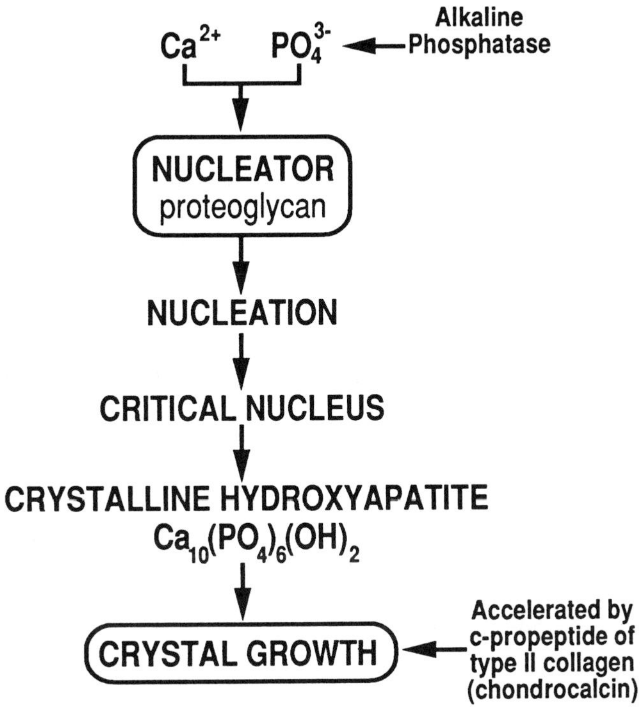

Figure 5. Essential requirements for calcification: a working hypothesis.

seen as an essential requirement in the calcification process ensuring the liberation of sufficient phosphate to form mineral. A deficiency of alkaline phosphatase results in the condition known as hypophosphatasia, characterized by a lack of mineralization of cartilage and bone (Fraser, 1957; Ornoy *et al.*, 1985). The natural substrates of alkaline phosphatase and their locations remain unknown, although phosphoethanolamine is markedly elevated in the urine of patients with hypophosphatasia (Fraser, 1957; Teree and Klein, 1968), suggesting that it is a likely substrate. Moreover, elevated concentrations of acidic phospholipids have been detected in growth plate cartilages in the calcifying lower hypertrophic zone (Boskey *et al.*, 1980). Phosphorus content increases in the hypertrophic matrix in parallel with calcium (Hargest *et al.*, 1985).

Alkaline phosphatase is primarily localized in the plasma membranes of growth plate chondrocytes and in matrix vesicles, as revealed by histochemical and fractionation studies (Ali *et al.*, 1970; Matsuzawa and Anderson, 1971; Väänänen and Korhonen, 1979; Hsu *et al.*, 1985; de Bernard *et al.*, 1986). What actually triggers the activity of alkaline phosphatase is not known, although its synthesis is regulated by vitamin D (see above). Why, for example, isn't it active at an earlier stage in the proliferative zone? Perhaps it is, but there may be insufficient enzyme or substrate available to form the critical levels of inorganic phosphate required for hydroxyapatite synthesis.

Matrix Vesicles

These are small vesicular structures (approximately 80 to 150 nm in diameter) delimited by a unit membrane and found mainly in the proliferative, maturing, and upper hypertrophic zones (Schraer and Gay, 1977; Cecil and Anderson, 1978; Ali *et al.,* 1970; Morris *et al.,* 1983) in the calcifying longitudinal septa (Reinholt and Wernerson, 1988; Anderson, 1989) (Figures 3 and 6). They were first described by Bonucci (1967, 1970) and Anderson (1967, 1969) and Ali *et al.* (1970) using electron microscopy. Analysis of freeze-etched specimens has revealed that they are derived from the plasma membrane and represent vesiculations of membrane which extend out into the longitudinal septa (Cecil and Anderson, 1978). Earlier reports described matrix vesicles as the sites where the earliest mineral is detected and the primary sites of calcification (Anderson, 1969). But others disagree (Thyberg, 1974; Shepard and Mitchell, 1985; Landis and Glimcher, 1982; Poole *et al.,* 1984; Poole and Pidoux, 1989). The highest concentration of vesicles is in the proliferative zone, whilst the lowest is in the lower hypertrophic zone where calcification occurs (Reinholt *et al.,* 1982; Shepard and Mitchell, 1985; Buckwalter *et al.,* 1987a). Clearly, matrix vesicles are present in intimate association with focal mineralizing sites (Poole and Pidoux, 1989), but there is no consistent evidence that they occupy a central location in calcified foci which would be expected if mineral first formed within them and then spread out into the surrounding matrix.

Studies of isolated matrix vesicles have demonstrated that they can actively mediate mineralization (see Wuthier, 1982, for review) but whether they act as true nucleational sites for mineral formation remains controversial. The high content of alkaline phosphatase in matrix vesicles undoubtedly plays a major role in the calcification process. The incorporation of this enzyme within matrix vesicles prevents it from being lost from the extracellular matrix where calcification occurs. It also provides a means whereby it can be concentrated sufficiently for it to provide the high local concentrations of inorganic phosphate that would be required for mineral to form. Therefore, the presence of high local concentrations of this enzyme in matrix vesicles in calcifying sites is probably in itself sufficient to indicate an essential requirement for these structures in the calcification process. The acidic phospholipids, which probably act as substrates for alkaline phosphatase, are present in the plasma membranes of chondrocytes and in the membranes of matrix vesicles. These can also promote or induce hydroxyapatite formation and may play an important role in this process (see Boskey, 1981).

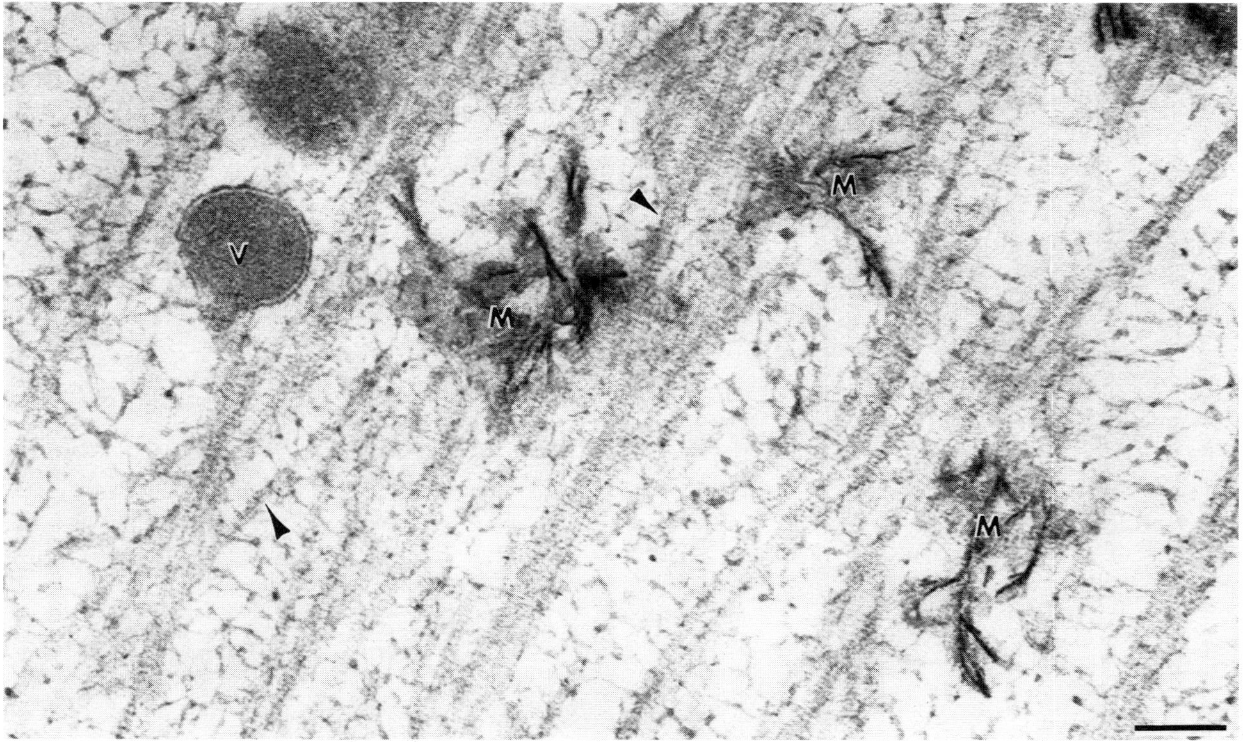

Figure 6. Electron micrograph of the lower hypertrophic zone of the proximal bovine fetal tibial primary growth plate to show a matrix vesicle (V) delimited by a unit membrane. Adjacent focal mineralization sites (M) can be seen. Type II collagen fibrils (arrowheads) with a network of aggregating proteoglycans between them can also be seen. Scale = 100 nm. (From Eunice Lee, unpublished results.)

Type X Collagen

The discovery of type X collagen led to the recognition of its unique association with cellular hypertrophy (Schmid and Linsenmayer, 1985, 1987; Gibson *et al.*, 1986; Leboy *et al.*, 1988). Although it is not synthesized by cells of the proliferative zone, its sudden synthesis by cells in the maturing and hypertrophic zones precedes cartilage calcification (Figure 4). It was surmised that it must in some way be related to cartilage calcification. An earlier claim that it is closely associated with matrix vesicles (Habuchi *et al.*, 1985) is not supported by immunoelectron microscopy (Poole and Pidoux, 1989). The mechanism of regulation of synthesis of type X collagen remains unknown, but it probably results from regulation at the level of the proliferative chondrocyte since proliferative cells can mature in culture into hypertrophic cells synthesizing type X collagen (see below). Since it is not synthesized by chondrocytes, other than those which exhibit cellular hypertrophy, it would appear that the growth plate chondrocytes are specifically regulated as has been indicated by studies demonstrating specific effects of vitamin D metabolites on these cells (see above).

Type II Procollagen

The synthesis of type II procollagen can be monitored by the assay of the C-propeptide of this molecule (Hinek and Poole, 1988) which is released with the N-propeptide on secretion from the cell as fibrils form. There is a direct correlation between the rate of calcification of bovine growth plate chondrocyte cultures and the C-propeptide content (Hinek *et al.*, 1987). Moreover, there is also a direct relationship between stimulation of type II collagen synthesis by vitamin D metabolites and stimulation of calcification in rat growth plate cultures (Hinek and Poole, 1988). Studies currently in progress of chondrocytes isolated from the growth plates of bovine fetuses have revealed that when cells of the lowest density and largest size are cultured, they produce more alkaline phosphatase and C-propeptide of type II collagen than cells of higher density (D. Carey, M. Alini, Y. Matsui and A.R. Poole, manuscript in preparation). This increased synthesis of type II collagen is related to an increased content of the C-propeptide in the extracellular matrix of the hypertrophic zone (M. Alini, Y. Matsui and A.R. Poole, manuscript in preparation).

Cellular Commitment to Hypertrophy: is it Preprogrammed or Under Hormonal Control?

The progressive maturation of chondrocytes leading to cellular hypertrophy inevitably raises such a question. The cartilaginous keel bone of the chick lends itself to studies of this kind since it represents a gradation of stages in chondrocyte development, starting with the proliferative cells of the caudal region and progressing to the hypertrophic cells of the cephalic region. Only the cephalic region undergoes endochondral ossification. When caudal chondrocytes are cultured in type I collagen gels, they mature into hypertrophic chondrocytes and synthesize type X collagen (Gibson and Flint, 1985; Thomas and Grant, 1988). This process is prevented by the presence of aggregating cartilage proteoglycans, chondroitin sulfate or hyaluronic acid. When chick embryo chondrocytes are cultured on agarose-coated dishes they differentiate into hypertrophic cells that synthesize type X collagen (Castagnola *et al.*, 1986). Other studies of organ cultures and cell cultures of prehypertrophic chondrocytes from chick embryos have also revealed that some of the cells become hypertrophic and synthesize type X collagen (Eavey *et al.*, 1988). Thus the maturation of prehypertrophic chondrocytes into hypertrophic cells appears to be preprogrammed but is dependent upon a permissive environment which includes both matrix components and is probably hormonally determined. Vitamin D is an obvious candidate to be investigated in this regard.

Angiogenesis and Growth Plate Chondrocytes

The regular ingrowth of capillaries into the lowermost lacunae of the hypertrophic zone of the growth plate is perhaps the best example of angiogenesis as a physiological event. This highly organized and polarized process proceeds until such time that the growth plate closes. Recent observations have revealed that a molecule or molecules chemotactic for vascular endothelial cells (Howell *et al.*, 1989) and angiogenic in the chick yolk sac membrane (Brown *et al.*, 1987) are produced by growth plate chondrocytes.

The importance of vascularization in the process of cartilage calcification in endochondral ossification is indicated by studies of bone induction *in vitro* and *in vivo*. It is well known that when demineralized bone matrix is injected subcutaneously or intramuscularly in rats, it has the capacity to induce endochondral ossification (see review by Reddi, 1981). Cartilage is formed which then calcifies and is replaced by woven bone and then trabecular bone. Cartilage formation can be induced in culture when embryonic rat skeletal muscle is cultured with demineralized rat bone *in vitro* (Nathanson *et al.*, 1978) or with demineralized rat bone matrix in diffusion chambers *in vivo* (Urist *et al.*, 1978). But the cartilage that forms does not calcify. If, however, the cartilage is transplanted into the muscle of the abdominal wall of a syngeneic recipient it becomes vascularized, calcifies, and forms bone (Urist *et al.*, 1978). These observations suggest that angiogenesis is a requirement for cellular hypertrophy and the calcification of cartilage in endochondral ossification.

Matrix Changes Related to Cellular Hypertrophy and the Calcification of Cartilage Matrix

Degradation of Matrix

Within the zone of maturation there is new evidence for pericellular cleavage of type II collagen (A.R. Poole and I. Pidoux, manuscript in preparation) detected using antibodies to epitopes that are only detectable on exposed type II α-chains and not in fibrillar helical collagen (Dodge and Poole, 1989). Therefore, when triple helical collagen is cleaved, the procollagen unwinds to expose these α-chain epitopes which react with antibodies directed against them. Collagen cleavage later extends throughout the extracellular matrix in the hypertrophic zone, with the exception of the transverse septa. A reduction in collagen fibril density in the territorial matrix has also been reported by Buckwalter *et al.* (1986) who used morphometric analyses. The destruction of type II collagen is associated with the presence of collagenase in the lower hypertrophic zone as well as in the remainder of the growth plate (Brown *et al.*, 1989). But in the lower hypertrophic zone there is a deficiency of the tissue inhibitor of this protease called TIMP. Earlier studies of rachitic growth plates also revealed the presence of active collagenase mainly in the hypertrophic zone (Dean *et al.*, 1985). Chick type X collagen can also be degraded by collagenase (Schmid *et al.*, 1986; Gadher *et al.*, 1989). One of the collagenase cleavage sites in chick type X collagen is adjacent to the globular domain at the C-terminus which is recognizable with electron microscopy. Each pro α chain of chick type X collagen has a molecular weight of 59 kDa. In studies of chick-embryo tibiotarsal cartilages, a time-dependent conversion was observed of type X procollagen to a 49 kDa polypeptide which lacked the globular region (Kielty *et al.*, 1985). These observations suggest that type X collagen may be cleaved by collagenase *in vivo* adjacent to the globular domain.

Neutral metalloproteinases which degrade proteoglycan and/or gelatin and which are resistant to protease inhibitors such as TIMP have been detected in endochondral ossification in embryonic chick bones (Mikuni-Takagaki and Cheng, 1987). Evidence for the presence of a stromelysin-like protease in growth plate has been obtained from biochemical studies (Ehrlich *et al.*, 1982). Stromelysin is known to be synthesized and secreted by chondrocytes in the growth plate (Brown *et al.*, 1989). This metalloprotease can activate procollagenase (Murphy *et al.*, 1987; Fini *et al.*, 1987) as well as degrading cartilage proteoglycans (Galloway *et al.*, 1983). Proteoglycan degradation may occur as a late event when cartilage calcifies (see below). A metalloprotease resembling stromelysin which is associated with isolated matrix vesicles has also been reported (Katsura and Yamada, 1986).

The C-Propeptide of Type II Collagen (Chondrocalcin)

Microimmunochemical analyses of bovine fetal growth plate cartilages have revealed that there is a greater accumulation of the C-propeptide of type II collagen in the hypertrophic zones (M. Alini, Y. Matsui and A.R. Poole, manuscript in preparation) than has been observed for hydroxyproline (type II and type X collagens). This follows a redistribution of the free C-propeptide from a close association with type II collagen fibrils in the proliferative and upper hypertrophic zone to focal calcification sites in the calcifying lower hypertrophic zone (Figure 7) (Poole *et al.,* 1984; E. Lee and A.R. Poole, manuscript in preparation).

The role of the C-propeptide may be to accelerate mineral growth rather than to cause nucleation per se (see Figure 9). Addition of the C-propeptide to artificially mineralizing cultures of nasal chondrocytes, cultured in the presence of β-glycerophosphate and alkaline phosphatase, accelerates mineral growth (A. Hinek and A.R. Poole, unpublished results). The initial rate of $^{45}Ca^{2+}$ incorporation into bovine fetal growth plate cultures is directly correlated with their C-propeptide content (Hinek *et al.,* 1987). The addition of the antibody to the C-propeptide also partly inhibits mineral growth in cultures of growth plate chondrocytes (A. Hinek and A.R. Poole, unpublished results). The C-propeptide is a calcium binding protein (P. Hauschka and L.C. Rosenberg, unpublished results) and binds avidly to hydroxyapatite (Choi *et al.,* 1983). This would favor its role in mineral growth. The concentration of the C-propeptide in calcifying sites may be simply a passive process with the C-propeptide binding to hydroxyapatite. However, the C-propeptide has been detected in demineralized cartilage in the so-called ‘mineralization nodules’ or ‘crystal ghosts’ (Poole *et al.,* 1984), suggesting that it may bind not only to mineral but to other components in the extracellular matrix which are not extractable with EDTA. Binding of purified C-propeptide to aggregating cartilage proteoglycans has also been observed *in vitro* (A.R. Poole, A. Reiner and Y. Matsui, unpublished data). Longitudinal studies of calcification in growth plate cultures have indicated that the concentration of the C-propeptide reaches a maximum when $^{45}Ca^{2+}$ incorporation continues to increase (Hinek and Poole, 1988).

Type X Collagen

This molecule, which is synthesized only in the zones of maturation (maturing zone) and hypertrophy, is not found in cartilages which are not committed to calcification (Schmid and Linsenmayer, 1987). Therefore, its close relationship to the calcification of cartilage was clearly established. Although a direct involvement in nucleation and mineral growth has been proposed, new, more direct studies using immunoelectron microscopy have revealed that it is in fact not present in increased concentrations in sites where focal mineralization is first observed (Poole and Pidoux, 1989). It is not detectable in matrix vesicles, contrary to earlier reports (Habuchi *et al.,* 1985). It is present, however, in close association with type II collagen fibrils. Unlike osteoid, where type I collagen calcifies early in the mineraliation process, type II collagen somehow resists calcification as an early event. Therefore, it has been suggested by Poole and Pidoux (1989) that type X collagen may serve to focally direct calcification to interfibrillar sites, preventing calcification of collagen fibrils. This proposal has no experimental validation, but it does provide an alternative working hypothesis to those already proposed.

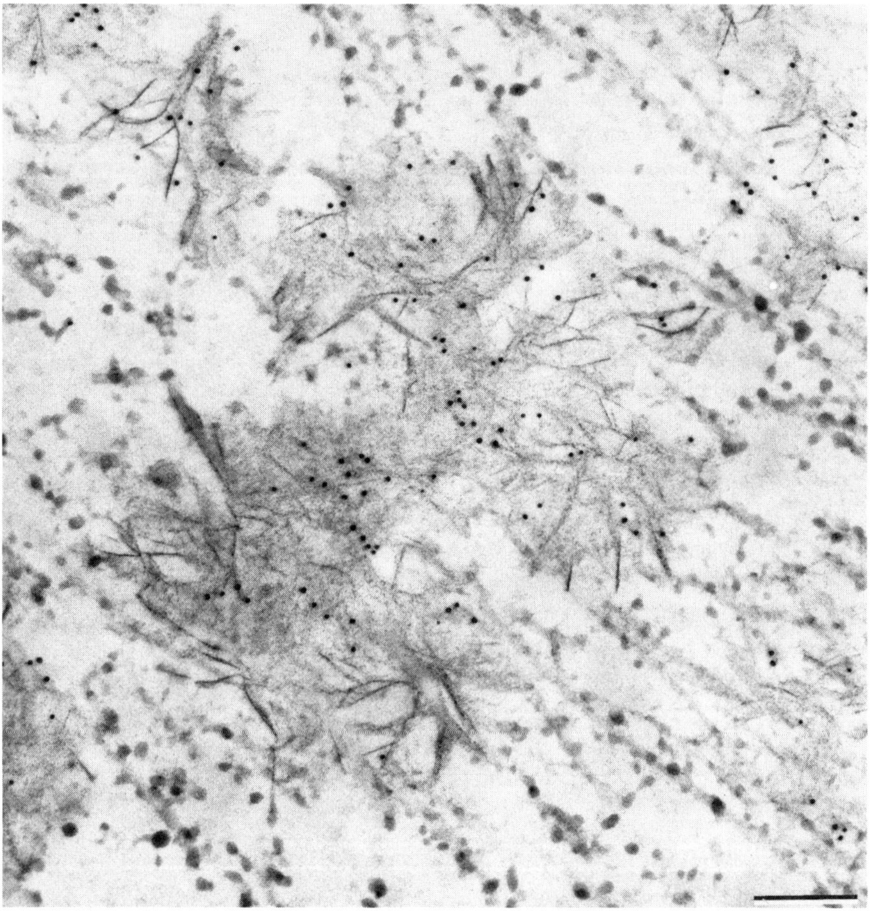

Figure 7. Electron micrograph of the lower hypertrophic zone of the proximal bovine fetal tibial primary growth plate to show focal mineralization sites with which is specifically associated the C-propeptide of type II collagen detected using a colloidal gold labeled antibody. Scale = 100 nm. (From Eunice Lee, unpublished results.)

It also indicates that there is, as yet, no evidence for a direct role of this molecule in nucleation or mineral growth in nucleation sites.

Proteoglycans

The large aggregating proteoglycans of immature cartilage matrix each contain up to 100 chondroitin sulfate chains of molecular weight in the region of 20 kDa and also 30 keratan sulfate chains of approximately 6 kDa (Poole,

1986). They therefore offer a very high negative charge density which can bind calcium (Bowness, 1968; Woodward and Davidson, 1968; Nemeth-Csoka and Sarkozi, 1982; Hunter, 1987a, b, 1988; Dziewiatkowski, 1987). Consequently, the concentration of calcium in cartilage matrix is normally much higher than that in serum (approximately 2.5 m*M*) and is in the range 15 to 60 mmol/kg (Althoff *et al.*, 1982; Hargest *et al.*, 1985). The earliest accumulation of calcium in growth plate matrix is in sites where approximately equimolar concentrations of sulfur (proteoglycan) and calcium are observed (Arsenault and Ottensmeyer, 1983). Therefore, as Hunter (1987a) has proposed, a local increase in phosphate concentration (provided by the action of alkaline phosphatase) would displace calcium bound to chondroitin sulfate (and keratan sulfate) by an ion-exchange mechanism and raise the calcium phosphate product above the threshold required for precipitation of hydroxyapatite. Addadi *et al.* (1987) have provided evidence that a cooperative concentration of calcium by sulfate and carboxyl groups in high negative charge density at a surface leads to nucleation of calcite *in vitro*.

Focal concentrations of proteoglycans in calcification sites in cartilage have been detected by electron spectroscopic imaging and by dye-binding analyses using electron microscopy (Figure 8) (Shepard and Mitchell, 1985; Arsenault and Ottensmeyer, 1983). These focal concentrations of proteoglycans can be artificially produced *in situ* in growth plate cartilages prior to calcification (but not in other noncalcifying cartilages) by the removal of calcium from the extracellular matrix (Figure 9) (A.R. Poole, I. Pidoux, and L.C. Rosenberg, manuscript in preparation; see also reviews by Poole and Rosenberg, 1987; Poole *et al.*, 1989). This observation suggests that calcium plays a role in determining the organization of these molecules. The excessive swelling of growth plate cartilage, compared with other cartilages, caused by the removal of calcium with EGTA (Campo, 1988) also indicates structural and organizational differences in growth plate extracellular matrix which are calcium-dependent. Therefore the high negative charge density of these molecules is increased even further when they become focally concentrated at the time of calcification. This would serve to maximize the local concentration of carboxyl and sulfate groups and therefore of calcium in sites which are about to calcify. The local production of inorganic phosphate (by alkaline phosphatase) would then serve to displace the concentrated calcium leading to precipitation.

The concentration of proteoglycan in these sites caused by the removal of calcium is accompanied by the concentration of link protein (A.R. Poole, I. Pidoux, and L.C. Rosenberg, manuscript in preparation). Therefore, we believe that the proteoglycan is in an aggregated form with hyaluronic acid when it is focally concentrated since link protein is ordinarily tightly bound to the hyaluronic acid-binding region of proteoglycans and to hyaluronic acid,

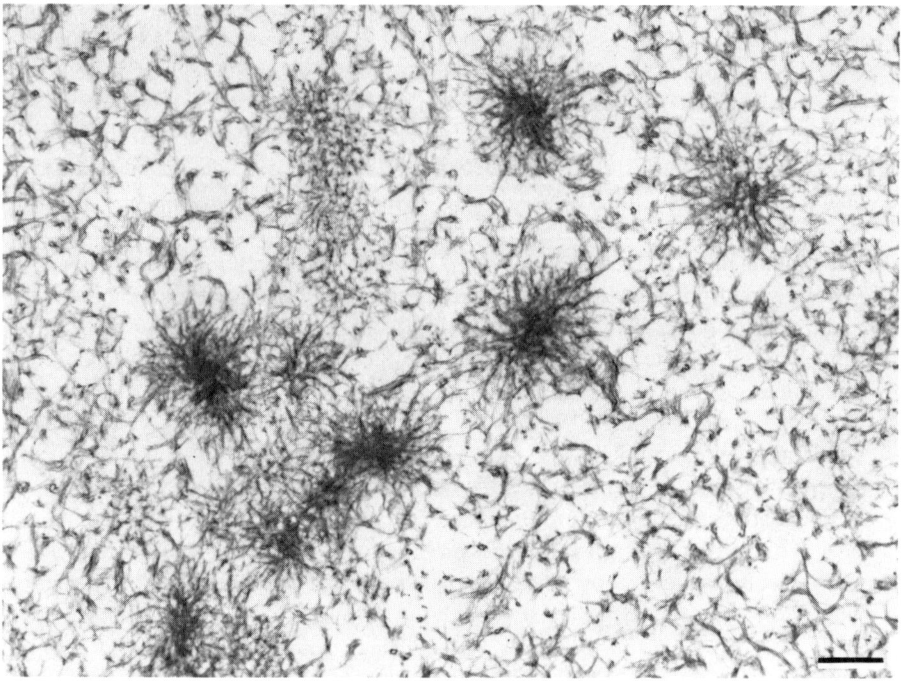

Figure 8. Electron micrograph of the demineralized lower hypertrophic zone of the rat distal femur primary growth plate to show focal concentrations of proteoglycan in calcification sites. The tissue was fixed in acridine orange and glutaraldehyde. Elemental analyses revealed that sulfur was concentrated over the regions of concentrated staining. Scale = 500 nm. (From Shepard, N. and Mitchell, N., Ultrastructural modifications of proteoglycans coincident with mineralization in local regions of rat growth plate. *J. Bone Jt. Surg. Am.* **67**: 455–464 (1985). With permission.)

thereby stabilizing the bond between proteoglycans and hyaluronic acid (Hardingham, 1979; Tang *et al.*, 1979). How this change in aggregated proteoglycan organization occurs is not immediately obvious, but it may involve a direct or indirect interaction of hyaluronic acid with fibrils of type II collagen through the mediation of Ca^{2+} ions. Therefore, when calcium phosphate is formed and mineral growth ensues, a drain on extracellular calcium may be initiated in adjacent precalcification sites leading to a rearrangement and focal collapse of proteoglycan aggregates (Figure 9). Calcium may remain bound to chondroitin sulfate since calcium binds more strongly to chondroitin sulfate than to hyaluronic acid (Hunter *et al.*, 1988). Clearly, once the process has been initiated it should be self-driven; these proposed events are summarized in Figure 9.

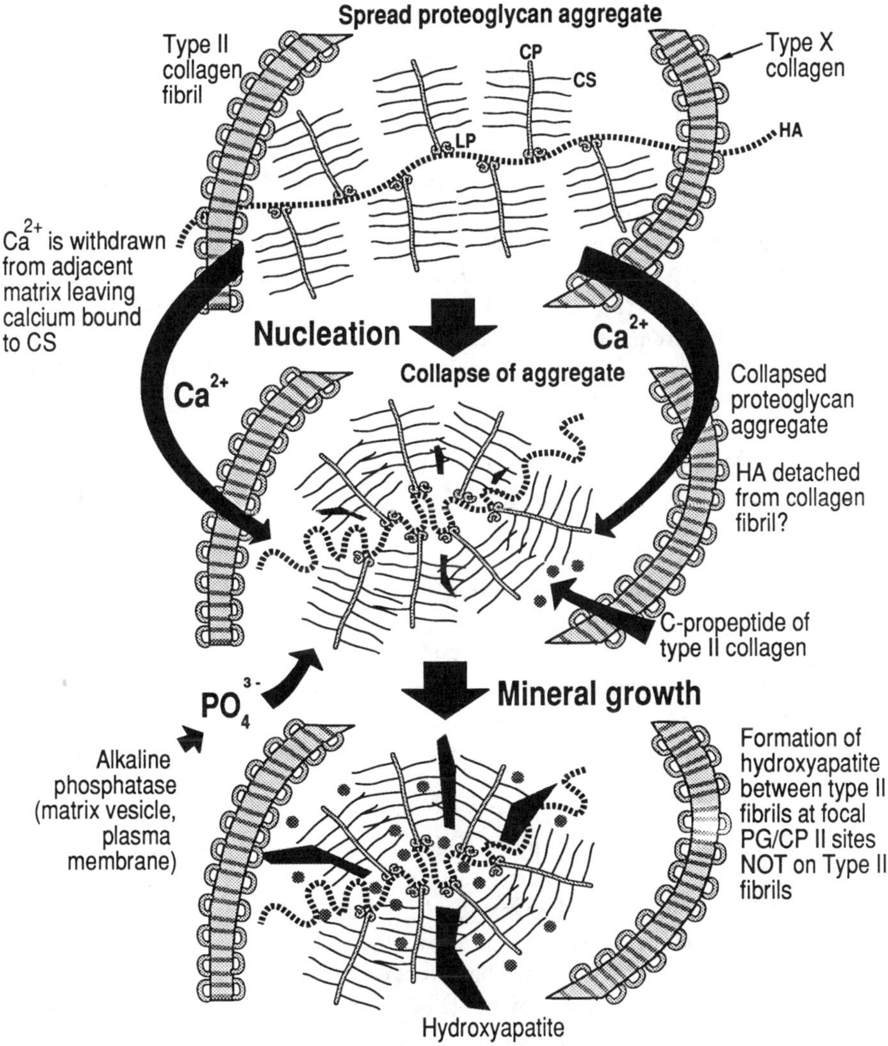

Figure 9. Diagrammatic representation of proposed matrix changes in hypertrophic longitudinal septa during calcification. Calcium may remain bound to chondroitin sulfate (see text) as the proteoglycan aggregates are focally concentrated between type II collagen fibrils where they form nucleation sites. Newly formed crystals are indicated in the middle section. The C-propeptide of type II collagen (CPII) binds to newly forming mineral where it is believed to enhance crystal growth. The core protein (CP), chondroitin sulfate chains (CS) of proteoglycan monomers (PG), and link protein (LP) attached to the hyaluronic acid binding region of proteoglycan monomers and to hyaluronic acid (HA) are indicated at the top of the figure. (Modified from Poole *et al.* (1989) with permission.)

Thus, collapsed proteoglycan aggregates may act as the nucleators of calcification. Molecules such as the C-propeptide of type II collagen may then act to accelerate crystal growth. Type X collagen may serve to direct mineral growth to interfibrillar sites.

Traditionally, as a result of *in vitro* studies of proteoglycans, it has been concluded that these large proteoglycans, particularly proteoglycan aggregates, inhibit the formation of mineral in metastable solutions of calcium and phosphate (Weinstein *et al.*, 1963; DiSalvo and Schubert, 1967; Cuervo *et al.*, 1973; Blumenthal *et al.*, 1979; Chen *et al.*, 1984; and Dziewiatkowski and Majznerski, 1985). Since calcium is an excellent counterion for the sulfate and carboxyl groups of chondroitin sulfate, its capture would be expected to impede calcification. But *in vivo* this calcium would be focally concentrated and capable of displacement by phosphate with resultant precipitation. This may not normally occur in the free solution employed in these earlier studies, although DiSalvo and Schubert (1967) observed the formation of a complex of Ca_3 $(PO_4)_2$ and proteoglycans. Moreover, it is very unlikely that there is a free and uniform solution of these molecules in growth plate cartilage. Thus, these *in vitro* studies should be reevaluated in light of what we now know of the binding of calcium by proteoglycan chondroitin sulfate and keratan sulfate chains, and in particular with respect to the changes that occur in proteoglycan organization.

Studies of proteoglycans in growth plate cartilage have also produced conflicting accounts of what happens to these molecules when cartilage calcifies. As, or just before, cartilage calcifies, a loss of matrix proteoglycan has been observed (Lohmander and Hjerpe, 1975; Boyde and Shapiro, 1980; Franzen *et al.*, 1982; Matukas and Krikos, 1968; Thyberg *et al.*, 1973; Thyberg, 1974). But others have found no evidence for a loss of proteoglycan in the extracellular matrix (Howell and Carlson, 1968; Poole *et al.*, 1982a, b; Scherft and Moskalewski, 1984; Hargest *et al.*, 1985) prior to calcification, although there may be a loss once calcification has been initiated (Shepard and Mitchell, 1982; Hargest *et al.*, 1985). We have recently performed quantitative analyses of proteoglycan content in growth plate cartilage. When the data are expressed per unit volume of extracellular matrix, there is, in fact, a progressive increase in proteoglycan content from the proliferative zone to the lower hypertrophic zone, indicating that these molecules become more concentrated in the extracellular matrix (Matsui *et al.*, 1991). Others using electron probe analysis also observed an increase in sulfur (sulfate) content in the upper hypertrophic zone (Hargest *et al.*, 1985). Earlier studies reported that proteoglycans undergo a reduction in aggregation in calcifying cartilages (Campo and Romano, 1986; Buckwalter *et al.*, 1987b), partial degradation (Campo and Romano, 1986), and a reduction in aggregate size (Buckwalter *et al.*, 1987b). Changes in aggregation were not observed in our recent investigations of the bovine fetal growth plate; neither was there any clear evidence for a reduction in size except for a very small change in the calcifying lower hypertrophic zone (Matsui *et al.*, 1991). There is, however, a net reduction in proteoglycan and collagen content per unit wet weight of tissue as cells enlarge in size and matrix volumes are considerably reduced.

The chemistry of the chondroitin sulfate chains does, however, change, indicating that there are changes in biosynthesis. These probably occur in the upper hypertrophic zone. There is an increased proportion of chondroitin 6-sulfate in hypertrophic cartilages (Kimata *et al.*, 1973; Stocum *et al.*, 1979). Whether chondroitin 6-sulfate is more effective than chondroitin 4-sulfate at binding calcium remains to be established.

Chondrodysplasias

These represent genetic changes which result in skeletal abnormalities. Although we still know very little of the alterations responsible for these changes, some progress has been made. A single mutation affecting the synthesis of a matrix component would appear to be sufficient to produce the characteristic irregularities in growth plate structure—the appearance of a disorganized hypertrophic zone with the loss of a columnar arrangement of chondrocytes in the proliferative, maturing, and hypertrophic zones—or a lack of calcification in the hypertrophic zone.

Kniest dysplasia is characterized by an abnormality of type II collagen fibril formation. The C-propeptide is ordinarily retained within the cell where it can be detected separate from procollagen. Thus, the defect appears to involve a premature cleavage of the C-propeptide from procollagen in intracellular rather than extracellular sites resulting in the release of procollagen lacking the C-propeptide. Fibril formation is thereby impaired (Poole *et al.*, 1988a, b). Yet in the lower hypertrophic zone calcification does occur, and extracellular C-propeptide has been detected in association with mineral. This difference in the distribution of the C-propeptide in the hypertrophic zone points to differences in the processing and/or regulation of processing of procollagen in hypertrophic cells from the mutant.

Achondrogenesis is characterized by nonexpression of type II collagen. Instead, some type I collagen is detected (Eyre *et al.*, 1986). In spondyloepiphysial dysplasia, a single exon deletion has been identified in the gene for type II collagen (Lee *et al.*, 1989).

Abnormalities in aggregating proteoglycans can produce dramatic changes in the structure and organization of the extracellular matrix since proteoglycans, through hydration of the many carboxyl and sulfate groups of their chondroitin sulfate and keratan sulfate chains, are responsible for the hydration and hence volume of the extracellular matrix and its compressive stiffness (Kempson, 1980). In brachymorphic mice, the inability to add sulfate groups to chondroitin during biosynthesis, as a result of a limited synthesis of the sulfate donor 3-phosphoadenosine 5-phosphosulfate, is a consequence of defects in adenosine 5′-phosphosulfate kinase and ATP sulfurylase (Schwartz *et al.*, 1978; Sugahara and Schwartz, 1982). This results in a considerable reduction in matrix volume. Nanomelia in chicks (Argraves *et al.*, 1981) and the mouse mutation, known as cartilage matrix deficiency (*cmd*/*cmd*) (Kimata *et al.*, 1981) are both lethal and result from a lack of synthesis of the core protein of aggregating cartilage proteoglycans.

Hypophosphatasia can be of varying severity. Its cause is related to a deficiency in the alkaline phosphatase of bones and cartilage which leads to impaired mineralization (Fraser, 1957; Ornoy *et al.*, 1985). Studies of growth plates of patients with this defect have revealed that there is a lack of mineral deposition in the lower hypertrophic zone. The hypertrophic zone is extended as in rickets. The lack of alkaline phosphatase, with the attendant defect in mineralization, provides the best evidence in support of the essential role of this enzyme in calcification. Associated with the lack of mineralization is the absence of the C-propeptide of type II collagen from the lower hypertrophic zone (A.R. Poole, I. Pidoux, F. Glorieux, and D. Rimoin, manuscript in preparation). This is, of course, normally found in these sites in association with mineral. Since the C-propeptide ordinarily binds to mineral and since there is no detectable lack of the C-propeptide in nonmineralizing cartilages in these patients, these observations tend to indicate that in the lower hypertrophic zone the C-propeptide is normally retained as a result of its interaction with mineral.

In avian tibial dyschondroplasia the chondrocytes fail to acquire a fully hypertrophic morphology. This results in an uncalcified extracellular matrix and a considerable increase in the thickness of the growth plate. ^{35}S-incorporation (as a measure of proteoglycan synthesis) is reduced (Freedman *et al.*, 1985) as is the content of type X collagen (Bashey *et al.*, 1989).

Conclusions

The primary growth plates of mammalian and avian bones are extremely well organized growth centers in which chondrocytes synthesize a cartilage matrix and then calcify it in a carefully controlled manner. This results in the production of calcified trabeculae on which bone formation is initially centered. What distinguishes chondrocytes of the growth plate from other chondrocytes would appear to be their ability to respond to environmental regulation by vitamin D metabolites in a manner not shared by other chondrocytes. Consequently, they synthesize unique molecules, such as type X collagen, and organize their extracellular matrix in a manner suited to calcification. This is reflected in differences in collagen and proteoglycan organization and composition. Mineralization is initiated and regulated by a specialized chondrocyte called a hypertrophic cell. This has developed by a special process of maturation involving considerable cellular enlargement at the expense of a reduction in matrix volume per unit mass of tissue. Matrix calcification appears to be centered in nucleation sites where the large aggregating proteoglycans are focally concentrated. These would appear to act as a means of ionically focally concentrating calcium. Alkaline phosphatase probably acts to generate inorganic phosphate which can displace calcium resulting in precipitation of hydroxyapatite. Matrix vesicles, which are closely associated with these mineralization sites, can act as a local source of this enzyme. Mineral growth is probably assisted by the C-propeptide of type II collagen. Although many aspects of the regulation, assembly, and calcification of cartilage remain to be established, we now have new insights into the process of endochondral ossification, and these hypotheses can be tested further.

Acknowledgments

This work is funded by the Medical Research Council of Canada and the Shriners of North America. I would like to thank Mark Lepik and Jane Wishart for their artwork and Michele Burman Turner for processing this manuscript. Dr. Eunice Lee, Department of Anatomy, University of Ottawa, generously supplied Figures 6 and 7.

References

L. Addadi, J. Moriadian, E. Shay, N.G. Maroudas, and S. Weiner. A chemical model for the cooperation of sulfates and carbohydrates in calcite crystal nucleation: relevance to biomineralization. *Proc. Natl. Acad. Sci. U.S.A.* **84**: 2732–2736 (1987).

S.Y. Ali, S.W. Sajdera, and H.C. Anderson. Isolation and characterization of calcifying matrix vesicles from epiphyseal cartilage. *Proc. Natl. Acad. Sci. U.S.A.* **67**: 1513–1520 (1970).

J. Althoff, P. Quint, E.R. Krefting, and H.J. Höhling. Morphological studies on the epiphyseal growth plate combined with biochemical and x-ray microprobe analyses. *Histochemistry* **74**: 541–552 (1982).

H.C. Anderson. Electron microscopic studies of induced cartilage development and calcification. *J. Cell Biol.* **35**: 81–92 (1967).

H.C. Anderson. Vesicles associated with calcification in the matrix of epiphyseal cartilage. *J. Cell Biol.* **41**: 59–72 (1969).

H.C. Anderson. Mechanism of mineral formation in bone. *Lab. Invest.* **60**: 320–330 (1989).

W.S. Argraves, P.J. McKeown-Longo, and P.F. Goetinck. Absence of proteoglycan core protein in the cartilage mutant nanomelia. *FEBS Lett.* **131**: 265–268 (1981).

A.L. Arsenault. Microvascular organization at the epiphyseal-metaphyseal junction of growing rats. *J. Bone Miner. Res.* **2**: 143–149 (1987).

A.L. Arsenault and F.P. Ottensmeyer. Quantitative spatial distributions of calcium, phosphorus and sulfur in calcifying epiphysis by high resolution electron spectroscopic imaging. *Proc. Natl. Acad. Sci. U.S.A.* **80**: 1322–1326 (1983).

C. Arsenis. Role of mitochondria in calcification. Mitochondrial activity distribution in the epiphyseal plate and accumulation of calcium and phosphate ions by chondrocyte mitochondria. *Biochem. Biophys. Res. Commun.* **46**: 1928–1935 (1972).

I. Atkin, J.C. Pita, A. Ornoy, A. Agundez, G. Castiglione, and D.S. Howell. Effects of vitamin D metabolites on healing of low phosphate-vitamin D-deficient induced rickets in rats. *Bone* **6**: 113–123 (1985).

N. Balmain, E. Tisserand-Jochem, M. Thomasset, P. Cuisinier-Gleizes, and H. Mathieu. Vitamin D-dependent calcium-binding protein (CaBP-9k) in rat growth cartilage. *Histochemistry* **84**: 161–168 (1986).

R.I. Bashey, R.M. Leach, C.V. Gay, and S.A. Jimenez. Type X collagen in avian tibial dyschondroplasia. *Lab Invest.* **60**: 106–112 (1989).

L.F. Bélanger. Autoradiographic visualization of the entry and transit of ^{35}S in cartilage, bone, and dentine of young rats and the effect of hyaluronidase *in vitro*. *Can. J. Physiol. Biochem.* **32**: 161–169 (1954).

I. Binderman and D. Sömjen. 24,25-dihydroxycholecalciferol induces the growth of chick cartilage *in vitro*. *Endocrinology* **115**: 430–432 (1984).

N.C. Blumenthal, A.S. Posner, L.D. Silverman, and L.C. Rosenberg. Effect of proteoglycans on *in vitro* hydroxyapatite formation. *Calcif. Tissue Int.* **27**: 75–82 (1979).

E. Bonucci. Fine structure of early cartilage calcification. *J. Ultrastruct. Res.* **20**: 33–50 (1967).

E. Bonucci. Fine structure and histochemistry of 'calcifying globules' in epiphyseal cartilage. *A. Zellforsch Mikrosk. Anat.* **103**: 192–217 (1970).

A.L. Boskey. Current concepts of the physiology and biochemistry of calcification. *Clin. Orthop. Rel. Res.* **157**: 225–257 (1981).

A.L. Boskey, A.S. Posner, J.M. Lane, M.R. Goldberg, and D.M. Cordella. Distribution of lipids associated with mineralization in the bovine epiphyseal growth plate. *Arch. Biochem. Biophys.* **199**: 305–311 (1980).

J. Bowness and K. Lee. Effects of chondroitin sulphates on mineralization *in vitro*. *Biochem. J.*, **103**: 382–390 (1967).

A. Boyde and I.M. Shapiro. Energy dispersive X-ray elemental analysis of isolated epiphyseal growth plate chondrocyte fragments. *Histochemistry* **69**: 85–94 (1980).

C.T. Brighton and R.B. Heppenstall. Oxygen tension in zones of the epiphyseal plate, the metaphysis, and diaphysis. An *in vitro* and *in vivo* study in rats and rabbits. *J. Bone Jt. Surg. Am.* **53**: 719–728 (1971).

C.T. Brighton and R.M. Hunt. Mitochondrial calcium and its role in calcification. *Clin. Orthop.* **100**: 406–416 (1974).

C.C. Brown, R.M. Hembry, and J.J. Reynolds. Immunolocalization of metalloproteinases and their inhibitor in the rabbit growth plate. *J. Bone Jt. Surg. Am. Vol.* **71**: 580–593 (1989).

R.A. Brown, C. Taylor, B. McLaughlin, C.D. McFarland, J.B. Weiss, and S.Y. Ali. Epiphyseal growth plate cartilage and chondrocytes in mineralising cultures produce a low molecular mass angiogenic procollagenase activator. *Bone Miner.* **3**: 143–158 (1987).

J.A. Buckwalter, D. Mower, R. Ungar, J. Schaeffer, and B. Ginsberg. Morphometric analysis of chondrocyte hypertrophy. *J. Bone Jt. Surg. Am. Vol.* **68**: 243–255 (1986).

J.A. Buckwalter, D. Mower, and J. Schaeffer. Differences in matrix vesicle concentration among growth plate zones. *J. Orthop. Res.* **5**: 157–163 (1987a).

J.A. Buckwalter, L.C. Rosenberg, and R. Ungar. Changes in proteoglycan aggregates during cartilage mineralization. *Calcif. Tissue Int.* **41**: 228–236 (1987b).

W.M. Burch, G. Corda, J.J. Kopchick, and F.C. Leung. Homologous and heterologous growth hormones fail to stimulate avian cartilage growth *in vitro*. *J. Clin. Endocrinol. Metab.* **60**: 747–750 (1985).

R.D. Campo. Effects of cations on cartilage structure: swelling of growth plate and degradation of proteoglycans induced by chelators of divalent cations. *Calcif. Tissue Int.* **43**: 108–121 (1988).

R.D. Campo and J.E. Romano. Changes in cartilage proteoglycans associated with calcification. *Calcif. Tissue Int.* **39**: 175–184 (1986).

D.A. Carrino, C. Lidor, S. Edelstein, and A.I. Caplan. Proteoglycan synthesis in vitamin D-deficient cartilage: recovery from vitamin D-deficiency. *Connect. Tissue Res.* **19**: 135–147 (1989).

P. Castagnola, G. Moro, F. Descalzi-Cancedda, and R. Cancedda. Type X collagen synthesis during *in vitro* development of chick embryo tibial chondrocytes. *J. Cell Biol.* **102**: 2310–2317 (1986).

R.N.A. Cecil and H.C. Anderson. Freeze-fracture studies of matrix vesicle calcification in epiphyseal growth plate. *Metab. Bone Dis. Relat. Res.* **1**: 89–95 (1978).

C.C. Chen, A.L. Boskey, and L.C. Rosenberg. The inhibitory effect of cartilage proteoglycans on hydroxyapatite growth. *Calcif. Tissue Int.* **36**: 285–290 (1984).

H.O. Choi, L.-H. Tang, T.L. Johnson, S. Pal, L.C. Rosenberg, A. Reiner, and A.R. Poole. Isolation and characterization of a 35,000 molecular weight subunit fetal cartilage matrix protein. *J. Biol. Chem.* **258**: 655–661 (1983).

M.T. Corvol, M.F. Dumontier, M. Garabedian, and R. Rappaport. Vitamin D and cartilage. II. Biological activity of 25-hydroxycholecalciferol and 24,25- and 1,25-dihydroxycholecalciferols on cultured plate chondrocytes. *Endocrinology* **102**: 1269–1274 (1978).

L. Cuervo, J. Pita, and D. Howell. Inhibition of calcium phosphate mineral growth by proteoglycan aggregate fractions in a synthetic lymph. *Calcif. Tissue Res.* **13**: 1–10 (1973).

D.D. Dean, O.E. Muniz, I. Berman, J.C. Pita, M.R. Carreno, J.F. Woessner, Jr., and D.S. Howell. Localization of collagenase in the growth plate of rachitic rats. *J. Clin. Invest.* **76**: 716–722 (1985).

B. de Bernard, P. Bianco, E. Bonucci, M. Constantini, G.C. Lunazzi, P. Martinuzzi, C. Modricky, L. Moro, E. Panfili, P. Polleslo, N. Stagni, and F. Vittur. Biochemical and immunohistochemical evidence that in cartilage an alkaline phosphatase is a Ca^{2+}-binding glycoprotein. *J. Cell Biol.* **103**: 1615–1623 (1986).

J. DiSalvo and M. Schubert. Specific interaction of some cartilage protein polysaccharides with freshly precipitating calcium phosphate. *J. Biol. Chem.* **242**: 705–710 (1967).

G.R. Dodge and A.R. Poole. Immunohistochemical detection and immunochemical analysis of type II collagen degradation in human normal, rheumatoid, and osteoarthritic articular cartilages and in explants of bovine articular cartilage cultured with interleukin 1. *J. Clin. Invest.* **83**: 647–661 (1989).

D.D. Dziewiatkowski. Binding of calcium by proteoglycan *in vitro*. *Calcif. Tissue Int.* **40**: 265–269 (1987).

D.D. Dziewiatkowski and L.L. Majznerski. Role of proteoglycans in endochondral ossification: inhibition of calcification. *Calcif. Tissue Int.* **37**: 560–564 (1985).

R.D. Eavey, T.M. Schmid, and T.F. Linsenmayer. Intrinsic and extrinsic controls of the hypertrophic program of chondrocytes in the avian columella. *Dev. Biol.* **126**: 57–62 (1988).

M.G. Ehrlich, A.L. Armstrong, R.G. Neuman, M.W. Davis, and H.J. Mankin. Patterns of proteoglycan degradation by a neutral protease from human growth-plate epiphyseal cartilage. *J. Bone Jt. Surg. Am. Vol.* **64**: 1350–1354 (1982).

D. Eyre, M.P. Upton, F.D. Shapiro, R.H. Wilkinson, and G.F. Vawter. Non-expression of cartilage type II collagen in a case of Langer-Saldino achondrogenesis. *Am. J. Hum. Genet.* **39**: 52–67 (1986).

D. Eyre and J.-J. Wu. Type XI or 1α 2α 3α collagen. In: *Structure and Function of Collagen Types,* Eds R. Mayne and R.E. Burgeson. Academic Press, New York. pp. 261–281, 1987.

C.E. Farnum and N.J. Wilsman. Morphologic stages of the terminal hypertrophic chondrocytes of growth plate cartilage. *Anat. Rec.* **219**: 221–232 (1987).

H.B. Fell and R. Robison. The growth, development and phosphatase activity of embryonic avian femora and limb-buds cultivated *in vitro*. *Biochem. J.* **23**: 767–784 (1929).

H.B. Fell and R. Robison. The development and phosphatase activity *in vivo* and *in vitro* of the mandibular skeletal tissue of the embryonic fowl. *Biochem. J.* **24**: 1905–1921 (1930).

N. Fine, I. Binderman, D. Sömjen, Y. Earon, S. Edelstein, and Y. Weisman. Autoradiographic localization of 24R,25-dihydroxyvitamin D_3 in epiphyseal cartilage. *Bone* **6**: 99–104 (1985).

M.E. Fini, M.J. Karmilowicz, P.L. Ruby, A.M. Beeman, K.A. Borges, and C.E. Brinckerhoff. Cloning of a complementary DNA for rabbit proactivator. A metalloproteinase that activates synovial cell collagenase, shares homology with stromelysin and transin, and is coordinately regulated with collagenase. *Arthritis Rheum.* **30**: 1254–1264 (1987).

A. Franzen, D. Heinegård, S. Reiland, and S.-E. Olsson. Proteoglycans and calcification of cartilage in the femoral head epiphysis of the immature rat. *J. Bone Jt. Surg. Am. Vol.* **64**: 558–566 (1982).

D. Fraser. Hypophosphatasia. *Am. J. Med.* **22**: 730–746 (1957).

B.D. Freedman, C.V. Gay, and R.M. Leach. Avian tibial dyschondroplasia. II. Biochemical changes. *Am. J. Pathol.* **119**: 191–198 (1985).

S.J. Gadher, T.M. Schmid, L.W. Heck, and D.E. Woolley. Cleavage of collagen type X by human synovial collagenase and neutrophil elastase. *Matrix* **9**: 109–115 (1989).

W.A. Galloway, G. Murphy, J.D. Sandy, J. Gavrilovic, T.E. Cawston, and J.J. Reynolds. Purification and characterization of a rabbit bone metalloproteinase that degrades proteoglycan and other connective tissue components. *Biochem. J.* **209**: 741–752 (1983).

G.J. Gibson, C.H. Bearman, and M.H. Flint. The immunoperoxidase localization of type X collagen in chick cartilage and lung. *Collagen Relat. Res.* **6**: 163–184 (1986).

G.J. Gibson and M.H. Flint. Type X collagen synthesis by chick sternal cartilage and its relationship to endochondral development. *J. Cell Biol.* **101**: 277–284.

G.J. Gibson, S.L. Schor, and M.E. Grant. Effects of matrix molecules on chondrocyte gene expression: synthesis of a low molecular weight collagen species by cells within collagen gels. *J. Cell Biol.* **93**: 767–774 (1982).

R.C. Greulich and C.P. Leblond. Radioautographic visualization of radiocarbon in the organs and tissues of newborn rats following administration of ^{14}C labeled bicarbonate. *Anat. Rec.* **115**: 559–585 (1953).

H. Habuchi, H.E. Conrad, and J.H. Glaser. Coordinate regulation of collagen and alkaline phosphatase levels in chick embryo chondrocytes. *J. Biol. Chem.* **260**: 13029–13034 (1985).

L.C. Hale, M.L.S. Kemick, and R.E. Wuthier. Effect of vitamin D metabolites on the expression of alkaline phosphatase activity by epiphyseal hypertrophic chondrocytes in primary cell culture. *J. Bone Miner. Res.* **1**: 489–495 (1986).

T.E. Hardingham. The role of link-protein in the structure of cartilage proteoglycan aggregates. *Biochem. J.* **177**: 237–247 (1979).

T.E. Hargest, C.V. Gay, H. Schraer, and A.J. Wasserman. Vertical distribution of elements in cells and matrix of epiphyseal growth plate cartilage determined by quantitative electron probe analysis. *J. Histochem. Cytochem.* **33**: 275–286 (1985).

A. Hinek and A.R. Poole. The influence of vitamin D metabolites on the calcification of cartilage matrix and the C-propeptide of type II collagen (chondrocalcin). *J. Bone Miner. Res.* **3**: 421–429 (1988).

A. Hinek, A. Reiner, and A.R. Poole. The calcification of cartilage matrix in chondrocyte culture. Studies of the C-propeptide of type II collagen (chondrocalcin). *J. Cell Biol.* **104**: 1435–1441 (1987).

D.A. Howell and L. Carlson. Alterations in the composition of growth cartilage septa during calcification studied by microscopic X-ray elemental analysis. *Exp. Cell Res.* **51**: 185–196 (1968).

D.S. Howell, D.D. Dean, O.E. Muniz, U.S. Ryan. Hypertrophic cell zone growth cartilage contains heparin-binding growth factors. *Trans. Orthop. Res. Soc.* **14**: 525 (1989).

H.H.T. Hsu, P.A. Munoz, J. Barr, I. Oppliger, D.C. Morris, H.K. Väänänen, N. Tarkenton, and H.C. Anderson. Purification and partial characterization of alkaline phosphatase of matrix vesicles from fetal bovine epiphyseal cartilage. Purification by monoclonal antibody affinity chromatography. *J. Biol. Chem.* **260**: 1826–1831 (1985).

G.K. Hunter. An ion-exchange mechanism of cartilage calcification. *Connect. Tissue Res.* **16**: 111–120 (1987a).

G.K. Hunter. Chondroitin sulfate-derivatized agarose beads: a new system for studying cation binding to glycosaminoglycans. *Anal. Biochem.* **165**: 435–441 (1987b).

G.K. Hunter, K.S. Wong, and J.J. Kim. Binding of calcium to glycosaminoglycans: an equilibrium dialysis study. *Arch. Biochem. Biophys.* **260**: 161–167 (1988).

E.B. Hunziker, R.K. Schenk, and L.-M. Cruz-Orive. Quantitation of chondrocyte performance in growth-plate cartilage during longitudinal bone growth. *J. Bone Jt. Surg. Am. Vol.* **69**: 162–173 (1987).

O.G.P. Isaksson, J.-O. Jansson, and I.A.M. Gause. Growth hormone stimulates longitudinal bone growth directly. *Science* **216**: 1237–1239 (1982).

N. Katsura and K. Yamada. Isolation and characterization of a metalloprotease associated with chicken epiphyseal cartilage matrix vesicles. *Bone* **7**: 137–143 (1986).

N.F. Kember. Cell division in endochondral ossification: a study of cell proliferation in rat bones by the method of tritiated thymidine autoradiography. *J. Bone Jt. Surg. Br. Vol.* **42**: 824–839 (1960).

G.E. Kempson. The mechanical properties of articular cartilage. In: *The Joints and Synovial Fluid,* Vol. 2, Ed L. Sokoloff. Academic Press, New York. pp. 177–238, 1980.

C.M. Kielty, A.P.L. Kwan, D.F. Holmes, S.L. Schor, and M.E. Grant. Type X collagen, a product of hypertrophic chondrocytes. *Biochem. J.* **27**: 545–554 (1985).

K. Kimata, H.-J. Barrach, K.S. Brown, and J.P. Pennypacker. Absence of proteoglycan core protein in cartilage from the *cmd/cmd* (cartilage matrix deficiency) mouse. *J. Biol. Chem.* **256**: 6961–6968 (1981).

K. Kimata, M. Okayama, A. Oohira, and S. Suzuki. Heterogeneity of proteochondroitin sulfates produced by chondrocytes at different stages of cytodifferentiation. *J. Biol. Chem.* **249**: 1646–1653 (1973).

R. Kuhlman and M. McNamee. The biochemical importance of the hypertrophic cartilage cell area to endochondral bone formation. *J. Bone Jt. Surg. Am. Vol.* **52**: 1025–1032 (1970).

W.J. Landis and M.J. Glimcher. Electron optical and analytical observations of rat growth plate cartilage prepared by ultramicrotomy: the failure to detect a mineral phase in matrix vesicles and the identification of heterodispersed particles as the initial solid phase of calcium phosphate deposited in the extracellular matrix. *J. Ultrastruct. Res.* **78**: 227–268 (1982).

P.S. Leboy, I.M. Shapiro, B.D. Uschmann, O. Oshima, and D. Lin. Gene expression in mineralizing chick epiphyseal cartilage. *J. Biol. Chem.* **263**: 8515–8520 (1988).

E. Lee, Y. Matsui, and A.R. Poole. Immunochemical and immunocytochemical detection of the C-propeptide of type II procollagen in hypertrophic chondrocytes of the growth plate. *J. Histochem. Cytochem.* **38**: 659–673 (1990).

B. Lee, H. Vissug, F. Ramirez, D. Rogers, and D. Rimoin. Identification of the molecular defect in a family with spondyloepiphyseal dysplasia. *Science.* **244**: 978–980 (1989).

C. Lidor, I. Atkin, A. Ornoy, S. Dekel, and S. Edelstein. Healing of rachitic lesion in chicks by 24R,25-dihydroxycholecalciferol administered locally into bone. *J. Bone Miner. Res.* **2**: 91–98 (1987).

A. Lindahl, J. Isgaard, L. Carlsson, and O.G.P. Isaksson. Differential effects of growth hormone and insulin-like growth factor 1 on colony formation of epiphyseal chondrocytes in suspension culture in rats of different ages. *Endocrinology* **121**: 1061–1069 (1987).

S. Lohmander and A. Hjerpe. Proteoglycans of mineralizing rib and epiphyseal cartilage. *Biochem. Biophys. Acta* **404**: 93–109 (1975).

H.U. Luder, C.P. Leblond, and K. von der Mark. Cellular stages in cartilage formation as revealed by morphometry, radioautography and type II collagen immunostaining of the mandibular condyle from weanling rats. *Am. J. Anat.* **182**: 197–214 (1988).

K. Madsen, U. Friberg, P. Roas, S. Eden, and O.G.P. Isaksson. Growth hormone stimulates the proliferation of cultured chondrocytes from rabbit ear and rat rib growth cartilage. *Nature (London)* **304**: 545–547 (1983).

A.-M. Makower, J. Wroblewski, and A. Pawlowski. Effects of IGF-1, rGH, FGF, EGF and NCS on DNA—synthesis, cell proliferation and morphology of chondrocytes isolated from rat rib growth cartilage. *Cell Biol. Int. Rep.* **13**: 259–270 (1989).

A.-M. Makower, J. Wroblewski, and A. Pawlowski. Effects of IGF-1, EGF and FGF on proteoglycans synthesized by fractionated chondrocytes of rat rib growth plate. *Exp. Cell Res.* **179**: 498–506 (1988).

J.H. Martin and J.L. Matthews. Mitochondrial granules in chondrocytes, osteoblasts and osteocytes. *Clin. Orthop.* **68**: 273–278 (1970).

Y. Matsui, M. Alini, C. Webber, and A. R. Poole. Cartilage aggregating proteoglycans of the growth plate: extracellular matrix content increases without evidence of degradation and loss of aggregation at the time of calcification. *J. Bone Jt. Surg. Am.* in press (1991).

T. Matsuzawa and H.C. Anderson. Phosphatases of epiphyseal cartilage studied by electron microscopic cytochemical methods. *J. Histochem. Cytochem.* **19**: 801–808 (1971).

J.L. Matthews, J.H. Martin, H.W. Sampson, A.S. Kunim, and J.H. Roan. Mitochondrial granules in the normal and rachitic rat epiphysis. *Calcif. Tissue Res.* **5**: 91–99 (1970).

V.J. Matukas and G.A. Krikos. Evidence for changes in protein polysaccharide associated with the onset of calcification in cartilage. *J. Cell Biol.* **39**: 43–48 (1968).

Y. Mikuni-Takagaki and Y.-S. Cheng. Metalloproteinases in endochondral bone formation: appearance of tissue inhibitor-resistant metalloproteinases. *Arch. Biochem. Biophys.* **259**: 576–588 (1987).

D.C. Morris, H.K. Väänänen, and H.C. Anderson. Matrix vesicle calcification in rat epiphyseal growth plate cartilage prepared anhydrously for electron microscopy. *Metab. Bone Dis. Relat. Res.* **5**: 131–137 (1983).

W. Müller-Glauser, B. Humbel, M. Glatt, P. Strauli, K.H. Winterhalter, and P. Bruckner. On the role of type IX collagen in the extracellular matrix of cartilage: type IX collagen is localized to intersections of collagen fibrils. *J. Cell Biol.* **102**: 1931–1939 (1986).

G. Murphy, M.I. Cockett, P.E. Stephens, B.J. Smith, and A.J.P. Docherty. Stromelysin is an activator of procollagenase. A study with natural and recombinant enzymes. *Biochem. J.* **248**: 265–268 (1987).

M.A. Nathanson, S.R. Hilfer, and R.L. Searls. Formation of cartilage by non-chondrogenic cell types. *Dev. Biol.* **64**: 99–117 (1978).

M. Nemeth-Csoka and A. Sarközi. The effect of proteoglycans of cartilage and oversulphated polysaccharides on the development of calcium-hydroxy-apatite (CHA) crystal formation *in vitro*. *Acta Biol. Acad. Sci. Hung.* **33**: 407–417 (1982).

Q. Nguyen, G. Murphy, P.J. Roughley, and J.S. Mort. Degradation of proteoglycan aggregate by a cartilage metalloproteinase. Evidence for the involvement of stromelysin in the generation of link protein heterogeneity *in situ*. *Biochem. J.* **259**: 61–67 (1989).

A. Nilsson, J. Isgaard, A. Lindahl, A. Dahlström, A. Skottner, and O.G.P. Isaksson. Regulation by growth hormone of number of chondrocytes containing IgGF-1 in rat growth plate. *Science* **233**: 571–574 (1986).

H. Nogami and M.R. Urist. A substratum of bone matrix for differentiation of mesenchymal cells into chondro-osseous tissues *in vitro*. *Exp. Cell Res.* **63**: 404–410 (1970).

A. Oohira, K. Kimata, S. Suzuki, K. Takata, I. Suzuki, and M. Hoshino. A correlation between synthetic activities for matrix molecules and specific stages of cytodifferentiation in developing cartilage. *J. Biol. Chem.* **249**: 1637–1645 (1974).

A. Ornoy, G.E. Adomian, and D.L. Rimoin. Histologic and ultrastructural studies on the mineralization process in hypophosphatasia. *Am. J. Med. Gen.* **22**: 743–758 (1985).

A. Ornoy, D. Goodwin, D. Noff, and S. Edelstein. 24,25-dihydroxyvitamin D is a metabolite of vitamin D essential for bone formation. *Nature (London)* **276**: 517–519 (1978).

J.J. Plachot, M.B. Du Bois, S. Halpern, G. Courtnot-Witmer, M. Garabedian, and S. Balsan. *In vitro* action of 1,25 dihydroxycholecalciferol and 24,25-dihydroxycholecalciferol on matrix organization and mineral distribution in rabbit growth plate. *Metab. Bone Dis. Relat. Res.* **4**: 135–142 (1982).

A.R. Poole. Proteoglycans in health and disease: structures and functions. *Biochem. J.* **236**: 1–14 (1986).

A.R. Poole, Y. Matsui, A. Hinek, and E.E. Lee. Cartilage macromolecules and the calcification of cartilage matrix. *Anat. Rec.* **224**: 167–179 (1989).

A.R. Poole and I. Pidoux. Immunoelectron microscopic studies of type X collagen in endochondral ossification. *J. Cell Biol.* **109**: 2547–2554 (1989).

A.R. Poole, I. Pidoux, A. Reiner, H. Choi, and L.C. Rosenberg. Association of an extracellular protein (chondrocalcin) with the calcification of cartilage in endochondral bone formation. *J. Cell Biol.* **98**: 54–65 (1984).

A.R. Poole, I. Pidoux, and L. Rosenberg. Role of proteoglycans in endochondral ossification: immunofluorescent localization of link protein and proteoglycan monomer in bovine fetal epiphyseal growth plate. *J. Cell Biol.* **92**: 249–260 (1982a).

A.R. Poole, I. Pidoux, A. Reiner, L. Rosenberg, D. Hollister, L. Murray, and D. Rimoin. Kniest dysplasia is characterized by an apparent abnormal processing of the C-propeptide of type II collagen resulting in imperfect fibril assembly. *J. Clin. Invest.* **81**: 579–589 (1988a).

A.R. Poole, A.H. Reddi, and L.C. Rosenberg. Persistence of cartilage proteoglycan and link protein during matrix-induced endochondral bone development: an immunofluorescent study. *Dev. Biol.* **89**: 532–539 (1982b).

A.R. Poole and L.C. Rosenberg. Proteoglycans, chondrocalcin, and the calcification of cartilage matrix in endochondral ossification. In: *Biology of Proteoglycans,* Eds T.N. Wight and R.P. Mecham. Academic Press, New York. pp. 187–210, 1987.

A.R. Poole, L. Rosenberg, L. Murray, and D. Rimoin. Kniest dysplasia: a probable type II collagen defect. *Pathol. Immunopathol. Res.* **7**: 95–98 (1988b).

A.R. Poole, C. Webber, I. Pidoux, H. Choi, and L.C. Rosenberg. Localization of a dermatan sulfate proteoglycan (DS-PGII) in cartilage and the presence of an immunologically related species in other tissues. *J. Histochem. Cytochem.* **34**: 619–625 (1986).

H. Reddi. Cell Biology and biochemistry of endochondral bone development. *Collagen Relat. Res.* **1**: 209–226 (1981).

A.M. Reginato, J.W. Lash, and S.A. Jiminez. Biosynthetic expression of type X collagen in embryonic chick sternum cartilage during development. *J. Biol. Chem.* **261**: 2897–2904 (1986).

A.M. Reginato, I.M. Shapiro, J.W. Lash, and S.A. Jimenez. Type X collagen alterations in rachitic chick epiphyseal growth cartilage. *J. Biol. Chem.* **263**: 9938–9945 (1988).

F.P. Reinholt, B. Engfeldt, D. Heinegård, and A. Hjerpe. Proteoglycans and glycosaminoglycans of epiphyseal cartilage in florid and healing low phosphate, vitamin D deficiency rickets. *Collagen Relat. Res.* **5**: 55–64 (1985).

F.P. Reinholt, B. Engfeldt, A. Hjerpe, and K. Jansson, Stereological studies on the epiphyseal growth plate with special reference to the distribution of matrix vesicles. *J. Ultrastruct. Res.* **80**: 270–279 (1982).

F.P. Reinholt and A. Wernerson. Septal distribution and the relationship of matrix vesicle size to cartilage mineralization. *Bone Miner.* **4**: 63–71 (1988).

R. Robison. The possible significance of hexose phosphoric esters in ossification. *Biochem. J.* **17**: 286–293 (1923).

L.C. Rosenberg, H.U. Choi, L.-H. Tang, T.L. Johnson, S. Pal, C. Webber, A. Reiner, and A.R. Poole. Isolation of dermatan sulfate proteoglycans from mature bovine articular cartilages. *J. Biol. Chem.* **260**: 6304–6313 (1985).

P.J. Roughley and I. Dickson. Factors influencing proteoglycan size in rachitic-chick growth cartilage. *Biochem. J.* **185**: 33–39 (1980).

P.J. Roughley and I.R. Dickson. A comparison of proteoglycan from chick cartilage of different types and a study of the effect of vitamin D on proteoglycan structure. *Connect. Tissue Res.* **14**: 187–197 (1986).

S.M. Russell and E.M. Spencer. Local injections of human or rat growth hormone or of purified human somatomedin-C stimulate unilateral tibial epiphyseal growth in hypophysectomized rats. *Endocrinology* **116**: 2563–2567 (1985).

M. Sandberg and E. Vuorio. Localization of types I, II and III collagen mRNAs in developing human skeletal tissue by *in situ* hybridization. *J. Cell Biol.* **104**: 1077–1084 (1987).

R.K. Schenk, D. Spiro, and J. Wiener. Cartilage resorption in the tibial epiphyseal growth plate of growing rats. *J. Cell Biol.* **34**: 275–291 (1967).

R.K. Schenk, J. Weiner, and D. Spiro. Fine structural aspects of vascular invasion of the tibial epiphyseal plate of growing rats. *Acta Anat.* **68**: 1–17 (1968).

J.P. Scherft and S. Moskalewski. The amount of proteoglycan in cartilage matrix and the onset of mineralization. *Metab. Bone Dis. Relat. Res.* **5**: 195–203 (1984).

T.M. Schmid and H.E. Conrad. A unique low molecular weight collagen secreted by cultured chick embryo chondrocytes. *J. Biol. Chem.* **257**: 12444–12450 (1982a).

T.M. Schmid and H.E. Conrad. Metabolism of low molecular weight collagen by chondrocytes obtained from histologically distinct zones of the chick embryo tibio-tarsus. *J. Biol. Chem.* **257**: 12451–12457 (1982b).

T.M. Schmid and T.F. Linsenmayer. Immunohistochemical localization of short chain cartilage collagen (type X) in avian tissues. *J. Cell Biol.* **100**: 598–605 (1985).

T.M. Schmid and T.F. Linsenmayer. Type X collagen. In: *Structure and Function of Collagen Types,* Eds R. Mayne and R.E. Burgeson. Academic Press, New York. pp. 223–259, 1987.

T.M. Schmid, R. Mayne, J.J. Jeffrey, and T.F. Linsenmayer. Type X collagen contains two cleavage sites for vertebrate collagenase. *J. Biol. Chem.* **261**: 4184–4189 (1986).

E. Schoenle, J. Zapf, C. Hauri, T. Steiner, and E.R. Froesch. Comparison of *in vivo* effects of insulin-like growth factor I and II and of growth hormone in hypophysectomised rats. *Acta Endocrinol.* **108**: 167–174 (1985).

E. Schoenle, J. Zapf, R.E. Humbel, and E.F. Froesch. Insulin-like growth factor 1 stimulates growth in hypophysectomized rats. *Nature (London)* **296**: 252–253 (1982).

H. Schraer and C.V. Gay. Vesicles in newly synthesizing bone observed after ultracryomycrotomy and ultramicroincineration. *Calcif. Tissue Res.* **23**: 185–188 (1977).

Z. Schwartz, G. Knight, L.D. Swain, and B.D. Boyan. Localization of vitamin D_3-responsive alkaline phosphatase in cultured chondrocytes. *J. Biol. Chem.* **263**: 6023–6026 (1988a).

N.B. Schwartz, V. Ostrowski, K.S. Brown, and R.M. Pratt. Defective PAPS-synthesis in epiphyseal cartilage from brachymorphic mice. *Biochem. Biophys. Res. Commun.* **82**: 173–178 (1978).

Z. Schwartz, D.L. Schlader, L.D. Swain, and B.D. Hoyan. Direct effects of 1,25-dihydroxyvitamin D_3 and 24,25-dihydroxyvitamin D_3 on growth zone and resting zone chondrocyte membrane alkaline phosphatase and phospholipase-A_2 specific activities. *Endocrinology* **123**: 2878–2884 (1988b).

N. Shepard and N. Mitchell. Ultrastructural modifications of proteoglycans coincident with mineralization in local regions of rat growth plate. *J. Bone Jt. Surg. Am. Vol.* **67**: 455–464 (1985).

D. Sömjen, A.M. Kaye, and I. Binderman. 24R,25-dihydroxyvitamin D stimulates creatine kinase BB activity in chick cartilage cells in culture. *FEBS Lett.* **167**: 281–284 (1984).

D.L. Stocum, R.M. Davis, M. Leger, and H.E. Conrad. Development of the tibiotarsus in the chick embryo: biosynthetic activities of histologically distinct regions. *J. Embryol. Exp. Morphol.* **54**: 155–170 (1979).

S. Suda, N. Takahashi, T. Shinki, N. Horiuchi, A. Yamaguchi, S. Yoshiki, S. Enomoto, and T. Suda. 1α, 25,-dihydroxyvitamin D_3 receptors and their action in embryonic chick chondrocytes. *Calcif. Tissue Int.* **37**: 82–90 (1985).

K. Sugahara and N.B. Schwartz. Defect in 3′-phosphoadenosine 5′-phosphosulfate synthesis in brachymorphic mice. 1. Characterization of the defect. *Arch. Biochem. Biophys.* **214**: 589–601 (1982).

M. Takigawa, M. Enomoto, E. Shirai, Y. Nishii, and F. Suzuki. Differential effects of 1α-25-dihydroxycholecalciferol and 24R,25-dihydroxycholecalciferol on the proliferation and the differentiated phenotype of rabbit costal chondrocytes in culture. *Endocrinology* **122**: 831–839 (1988).

L.-H. Tang, L. Rosenberg, A. Reiner, and A.R. Poole. Proteoglycans from bovine nasal cartilage. Properties of a soluble form of link protein. *J. Biol. Chem.* **254**: 10523–10531 (1979).

J.D. Termine, H.K. Kleinman, S.W. Whitson, K.M. Conn, M.L. McGarvey, and G.R. Martin. Osteonectin, a bone-specific protein linking mineral to collagen. *Cell* **26**: 99–105 (1981).

T.M. Teree and L. Klein. Hypophosphatasia, clinical and metabolic studies. *J. Pediatr.* **72**: 41–50 (1968).

J.T. Thomas and M.E. Grant. Cartilage proteoglycan aggregate and fibronectin can modulate the expression of type X collagen by embryonic chick chondrocytes cultured in collagen gels. *Biosci. Rep.* **8**: 163–171 (1988).

J. Thyberg. Electron microscopic studies on the initial phases of calcification in guinea pig epiphyseal cartilage. *J. Ultrastruct. Res.* **46**: 206–218 (1974).

J. Thyberg, S. Lohmander, and U. Friberg. Electron microscopic demonstration of proteoglycan in guinea pig epiphyseal cartilage. *J. Ultrastruct. Res.* **45**: 407–427 (1973).

S.B. Trippel, S.D. Chernausek, J.J. van Wyk, A.C. Moses, and H.J. Mankin. Demonstration of type I and type II somatomedin receptors on bovine growth plate chondrocytes. *J. Orthop. Res.* **6**: 817–826 (1988).

S.B. Trippel, M.T. Corvol, M.F. Dumontier, R. Rappaport, H.H. Hung, and H.J. Mankin. Effect of somatomedin-c/insulin-like growth factor I and growth hormone on cultured growth plate and articular chondrocytes. *Fed. Res.* **25**: 76–82 (1989).

M.R. Urist, M. Nakagawa, N. Nakata, and H. Nogami. Experimental myositis ossificans. Cartilage and bone formation in muscle in response to a diffusible bone matrix-derived morphogen. *Arch. Pathol. Lab. Med.* **102**: 312–316 (1978).

H.K. Väänänen. Immunohistochemical localization of alkaline phosphatase in the chicken epiphyseal growth cartilage. *Histochemistry* **65**: 143–148 (1980).

H.K. Väänänen and L.K. Korhonen. Matrix vesicles in chicken epiphyseal cartilage. Separation from lysosomes and the distribution of inorganic pyrophosphatase activity. *Calcif. Tissue Int.* **28**: 65–72 (1979).

U. Vetter, G. Helbing, W. Heit, W. Pirsig, K. Sterzig, and E. Heinze. Clonal proliferation and cell density of chondrocytes isolated from human fetal epiphyseal, human adult articular and nasal septal cartilages. Influence of hormones and growth factors. *Growth* **49**: 239–245 (1985).

U. Vetter, J. Zapf, W. Heit, G. Helbing, E. Heinze, E.R. Froesch, and W.M. Teller. Human fetal and adult chondrocytes. Effect of insulin-like growth factors I and II, insulin and growth hormone on clonal growth. *J. Clin. Invest.* **77**: 1903–1908 (1986).

H. Weinstein, C. Sach, and M. Schubert. Protein polysaccharide in connective tissue: inhibition of phase separation. *Science* **142**: 1073–1075 (1963).

D. Woodward and E. Davidson. Structure-function relationships of protein polysaccharide complexes: specific ion-binding properties. *Proc. Natl. Acad. Sci. U.S.A.* **60**: 201–205 (1968).

R.E. Wuthier. A review of the primary mechanism of endochondral calcification with special emphasis on the role of cells, mitochondria and matrix vesicles. *Clin. Orthop. Relat. Res.* **169**: 219–242 (1982).

R.E. Wuthier and T.C. Register. Role of alkaline phosphatase, a polyfunctional enzyme in mineralizing tissues. In: *The Chemistry and Biology of Mineralized Tissues,* Ed W.T. Butler. Ebso Media, Birmingham, AL. pp. 113–124, 1984.

X.Y. Zhou, D.W. Dempster, S.L. Marion, J.W. Pike, M.R. Haussler, and T.L. Clemens. Bone vitamin D-dependent calcium-binding protein is localized in chondrocytes of growth plate cartilage. *Calcif. Tissue Int.* **38**: 244–247 (1986).

7
Cartilage Degradation

JENNY A. TYLER
Strangeways Research Laboratory
Worts' Causeway
Cambridge, England

Introduction

Articular chondrocytes are remarkable in their capacity to synthesize, organize, and regulate the deposition of their complex surrounding matrix in a highly ordered and efficient manner. At each stage of growth and development the relative rates of synthesis and degradation are adjusted to achieve net growth, remodeling, or a balanced equilibrium. This chapter outlines what is known about the process of degradation in articular cartilage and concentrates on the two major structural components of the matrix, the aggregating proteoglycans and type II collagen.

These molecules have very different structures and rates of turnover which relate to their unique function within the matrix. The relationship between structure and susceptibility to degradation during maturation is presented. The likely contribution of different enzyme/inhibitor systems is examined together with the influence of various mediators on the process of both normal and pathological degradation of cartilage.

Relationship Between Structure and Susceptibility to Degradation

Type II Collagen

The synthesis, secretion, and maturation of collagen into insoluble fibers proceeds via a series of complex processing steps (see Kivirikko and Myllylä, 1984). The different intermediate forms vary considerably in their rate of turnover and susceptibility to degradation. The probable pathways of collagen degradation are illustrated in Figure 1. Much of the work described in this chapter originally relates to the pro $\alpha 1(I)$ chains of type I collagen but can also be considered pertinent to type II collagen.

The Triple Helix

An important early step is the formation of 96% of the molecule into a triple helix just prior to secretion. Three identical $\alpha 1(II)$ procollagen chains (M_r 95,000) become associated by interchain disulfide bonds between the

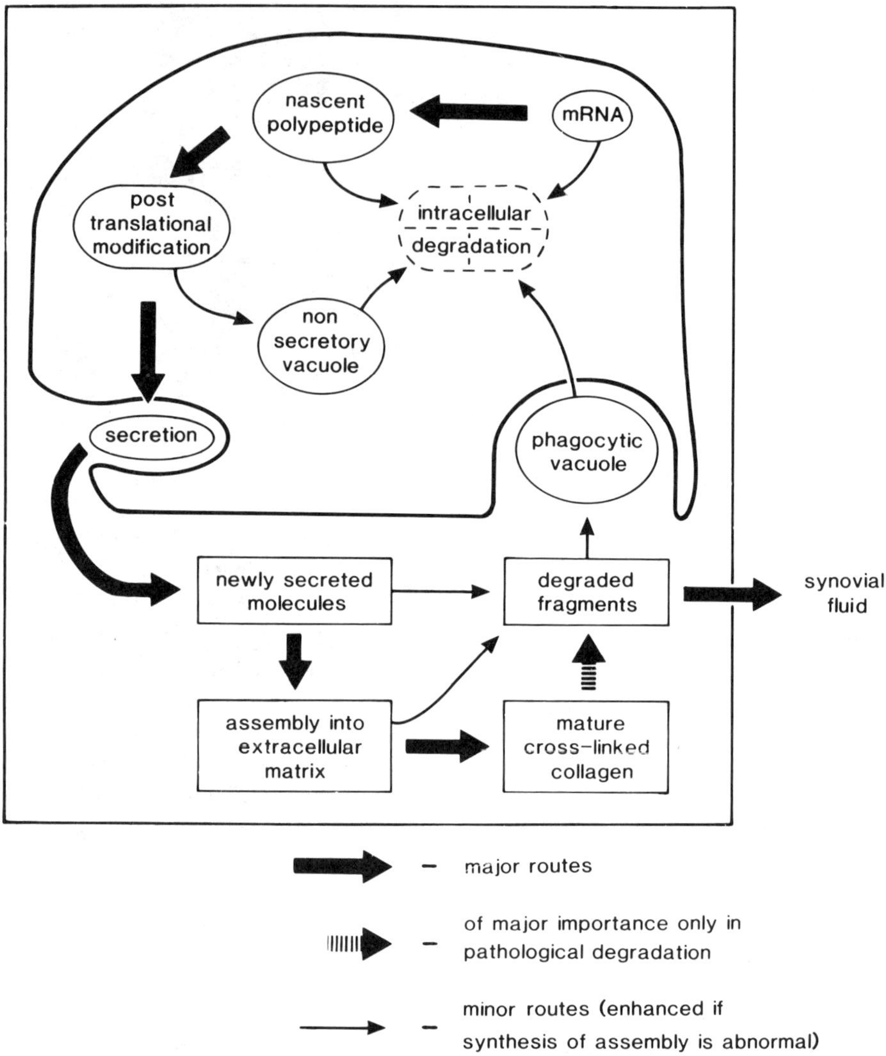

Figure 1. Potential pathways for the degradation of cartilage collagen.

amino-terminal propeptides probably in the cisternae of the rough endoplasmic reticulum (Bruckner *et al.*, 1978). The correctly aligned chains are then thought to spontaneously wind into the helical form (Figure 2a), with inter-chain hydrogen bonds (one per triplet) evenly distributed along the length of the molecule. Correct helix formation is essential for further secretion and processing of the procollagen into fibrils and confers resistance of this portion of the molecule to degradation by all neutral proteinases except the specific collagenase (see "Helical Collagen" *infra*). Several features combine to sta-

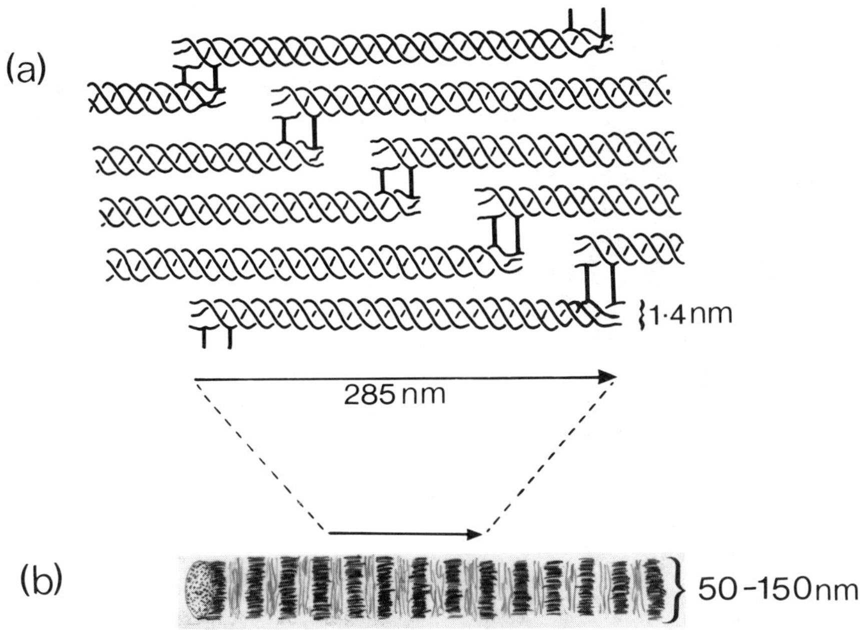

Figure 2. Maturation of extracellular collagen. (a) Diagram (not to scale) indicates the way in which triple helical monomeric collagen molecules become aligned and cross-linked within microfibrillar bundles. (b) Diagram shows how the microfibrils pack together to form a mature collagen fibril that is relatively resistant to degradation. The final diameter ranges from 50 to 80 nm in midzone to 100 to 150 nm in the deeper layers of cartilage. The average length of a fibril is unknown but could range to 1 μm.

bilize and maintain the structure of the long (~285 × 1.4 nm diameter) rigid α helix (see Piez, 1984). Glycines are present as every third residue in a repeating Gly-X-Y sequence, with a high content of proline, frequently in the X position. Previous 4-hydroxylation of any proline residues in the Y position is also essential; there is a direct correlation between stability of the helix and the order and total content of imino (proline and hydroxyproline) residues (Traub and Piez, 1971; Rao and Adams, 1979). The melting temperature of underhydroxylated collagen can be as low as 20°C compared to 40°C in a normal molecule (Berg and Prockop, 1973; Jiminez *et al.*, 1973, Rosenbloom *et al.*, 1973). The considerable contribution of hydroxyproline to helix stability is not understood but may involve additional interchain water-bridged hydrogen bonds (Ramachandran and Chandrasekaran, 1968). Collagen can bind about 20% water, so this could in theory provide one extra bond per triplet. An alteration in any one of these features will lead to local denaturation or melting of the helix, which will slow down or prevent fibril formation and markedly increase the rate of degradation by nonspecific proteinases.

Intracellular Degradation

A significant proportion of procollagen molecules can be completely degraded to small peptides prior to secretion (Bienkowski *et al.*, 1978; Baum *et al.*, 1980). There appears to be a small (1 to 2%) basal level of intracellular degradation which is unaffected by agents that inhibit lysosome function and therefore that probably occurs in a nonlysosomal compartment. This level can be enhanced up to 30% by any condition which leads to the production of defective pro-α chains that cannot form a stable triple helix, as in underhydroxylated collagen (Berg *et al.*, 1980). The enhanced level of intracellular degradation is influenced by the intracellular concentration of cyclic AMP and can be decreased back to the basal level of degradation by inhibitors of lysosomal enzymes, indicating that this part of the process occurs in lysosomes. The mechanism used by the cell to recognize and channel abnormal molecules toward this degradative pathway is not known.

Insoluble Cross-Linked Collagen

Procollagen is secreted from the cell and converted to collagen by the action of N & C terminal propeptidases (Uitto *et al.*, 1979). Subsequent formation of fibers occurs spontaneously by self-assembly of the collagen molecules into a quarter-stagger array (Figure 2a) starting probably with a five stranded microfibril (Smith, 1968). Many of these microfibrillar units then pack together to form larger fibrils and fibers. It is not known where fibril formation is initiated; it is possible that even the procollagen is secreted as aggregated bundles (Trelstad and Birk, 1985).

Within hours after fibril formation intermolecular reducible cross-links form between modified lysine and hydroxylysine residues in the N & C nonhelical telopeptide regions and reactive unmodified residues of adjacent molecules. The initial modification is catalyzed by lysyl oxidase, but subsequent formation of the covalent cross-links is spontaneous. It is possible that these reducible bonds represent an intermediate linkage which stabilizes the quarter-stagger arrangement to promote growth of the fibrils. They are thought to be gradually converted into more stable and complex nonreducible cross-links during maturation (see Miller, 1984; Bailey *et al.*, 1974). Cartilage collagen also contains one nonreducible 3-hydroxypyridium cross-link per molecule which is derived from modified hydroxylysyl residues (Eyre and Oguchi, 1980). The resistance of collagen to digestion with specific collagenase increases in proportion to the number of cross-links (Vater *et al.*, 1979). Even the vulnerable nonhelical regions of molecules become protected as the fibril bundles increase in diameter (Figure 2b) and limit the penetration of proteinases and other molecules. Previous hydroxylation of some lysines, mostly in the Y position, by lysyl hydroxylase is required for the modification of these residues prior to the formation of intermolecular cross-links (Kivirikko and Myllylä, 1980). Up to 6% of α1(II) chains also contain carbohydrate O-linked oligosaccharides in the nonhelical ends. The exact role of this carbohydrate is not clear, but it has been suggested that it may further protect the collagen from proteolysis. The additional hydrophilic groups certainly bind water, leading to an increased lateral spacing of molecules within type II fibrils. They probably also promote interactions with other matrix components (Grynpas *et al.*, 1980). Insoluble highly cross-linked mature collagen is therefore characterized by a very slow rate of degradation which is essential for normal function in providing the tensile strength of the tissue (Bailey *et al.*, 1974; Tanzer, 1973; Kempson, 1976; Bader *et al.*, 1981; Broom, 1988).

Interaction with Other Matrix Components

Type XI collagen accounts for 8 to 10% of cartilage collagen and is thought to be the functional equivalent of type V collagen which is present in other tissues (Upholt and Olsen, this volume). Immunolocalization studies (Vaughan *et al.*, 1988) suggest that type XI forms cofibrils with type II and together with the small dermatan sulfate proteoglycan (DS-PG II) may regulate the size of the collagen fibrils (Vogel *et al.*, 1984). The α1, α2, α3 chains of type XI collagen make up a continuous helix which is more glycosylated than that of type II and lacks the specific cleavage site for collagenase (Eyre *et al.*, 1984). This molecule is therefore probably fairly resistant to proteolysis and may stabilize the fibril network.

Type IX accounts for 5% of cartilage collagen and has four nonhelical regions interspersed between the collagenous domains. It is therefore very sensitive to nonspecific proteolysis (Upholt and Olsen, this volume; Gadher *et al.*, 1988; Grant *et al.*, 1988). Type IX collagen has one chain of chondroitin sulfate (CS) per molecule and is covalently attached to the telopeptide region of type II collagen by covalent pyridinium cross-links (van der Rest and Mayne, 1988, Vaughan *et al.*, 1988). It is possible that degradation of type IX may promote swelling of the fibers by allowing the type II molecules to move apart more easily permitting access of enzymes to collagen within the bundles.

Processing of Extracellular Proteoglycans

In cartilaginous tissues much of the proteoglycans exist as large macromolecular aggregates noncovalently bound to a single unbranched chain of hyaluronic acid. The core protein extends to about 300 nm in length and has several structurally and functionally distinct regions (see Hardingham *et al.*, 1986; Figure 4). There are two globular regions at the N terminal end. G1 forms a specific site for interaction with hyaluronic acid. The function of G2 is not known. The adjacent region is rich in keratan sulfate with a CS attachment region and a third lectin-like globular region (G3) at the C-terminal end. These structures undergo extensive but gradual age-related changes during normal maturation, which vary significantly in different species but which are consistent with a limited proteolysis of the core protein. A diagram representing electron micrographs of proteoglycan aggregates extracted from fetal epiphyseal, calf nasal, and mature nasal bovine cartilage is shown in Figure 3 (Buckwalter and Rosenburg, 1983). This demonstrates how the size of the aggregates decrease markedly with age. The length of the hyaluronate chains were found to be on average 2842 nm with 113 attached monomers in the fetal tissue and 1135 nm with 57 monomers in calf articular cartilage. The length of individual monomers also decreased consistently throughout fetal development and in the older tissue were very heterodisperse. Electron microscopy of rotary-shadowed proteoglycans confirmed that a large proportion of molecules did not have an intact C-terminal globular domain, which is otherwise present at the "free" end of the core protein which does not bind to hyaluronic acid (Paulsson *et al.*, 1987) (Figure 4b). Biochemical analysis revealed that there was a consistent but gradual change in the aggregating proteoglycan extracted from canine (Inerot and Heinegard, 1978) and human (Bayliss and Ali, 1978; Roughley and White, 1980) cartilage of different ages. The molecules from neonatal cartilage were fairly homogeneous. They were large in size, had a high buoyant density, and were rich in chondroitin sulfate relative to protein (Figure 4b). Between birth and maturity the proportion of molecules with a lower buoyant density, smaller size, and decreased content of chondroitin sulfate (Figure 4c) increased and were predominant in the heterogeneous population extracted from mature tissues. A high proportion of aggregating subunits isolated from aged human cartilage had only the binding region left, with no detectable glycosaminoglycans attached (Roughley *et al.*, 1985) (Figure 4d). The idea that the progressive decrease in the length of the CS attachment region results from degradation was supported by experiments on human cartilage of different ages. It was shown using agarose/acrylamide electrophoresis that the major synthetic product stays constant with age and is similar to the CS-rich larger proteoglycan found in the younger tissue (Bayliss *et al.*, 1986). Pulse-chase experiments confirmed a precursor-product relationship between the high and low buoyant density proteoglycans in cultured human cartilage (Oegema and Thompson, 1986) and in rats *in vivo* (Heinegard *et al.*, 1986). In contrast, monolayer chondrocytes derived from calf or steer bovine articular cartilage synthesize different proteoglycans typical of the unlabeled extracted monomers. The relative proportions of the two types did not change with increasing times of labeling (Thonar *et al.*, 1986), indicating that they represent two distinct biosynthetic products. Recent data strongly suggest that there is only one gene for the core protein of the aggregating proteoglycan (Dodge *et al.*, 1987). However, other work also supports the view that a new proteoglycan species with a modified core protein may be produced during aging in addition to the changes induced by degradation (Heingard *et al.*, 1985; Glant *et al.*, 1986).

Relationship of Monomer Size to Function

The glycosaminoglycans in aggregating proteoglycans become hydrated (within the cartilage matrix) to create an immense swelling pressure (Urban

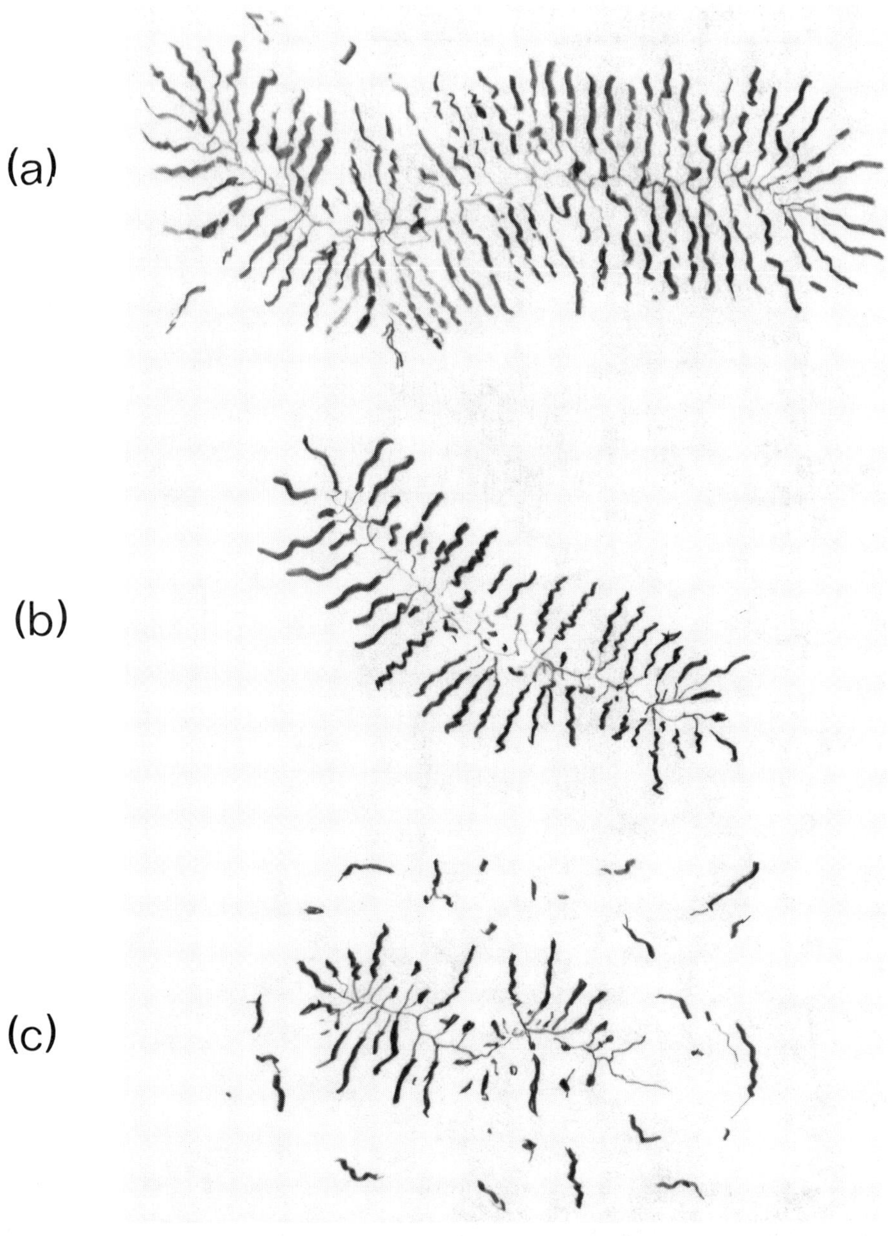

Figure 3. Maturation of proteoglycan aggregates. Electron micrographs show the decrease in size with age of proteoglycan aggregates extracted from (a) fetal bovine epiphyseal cartilage, (b) calf nasal cartilage, and (c) mature bovine nasal cartilage. (From Buckwalter and Rosenburg, 1983. With permission.)

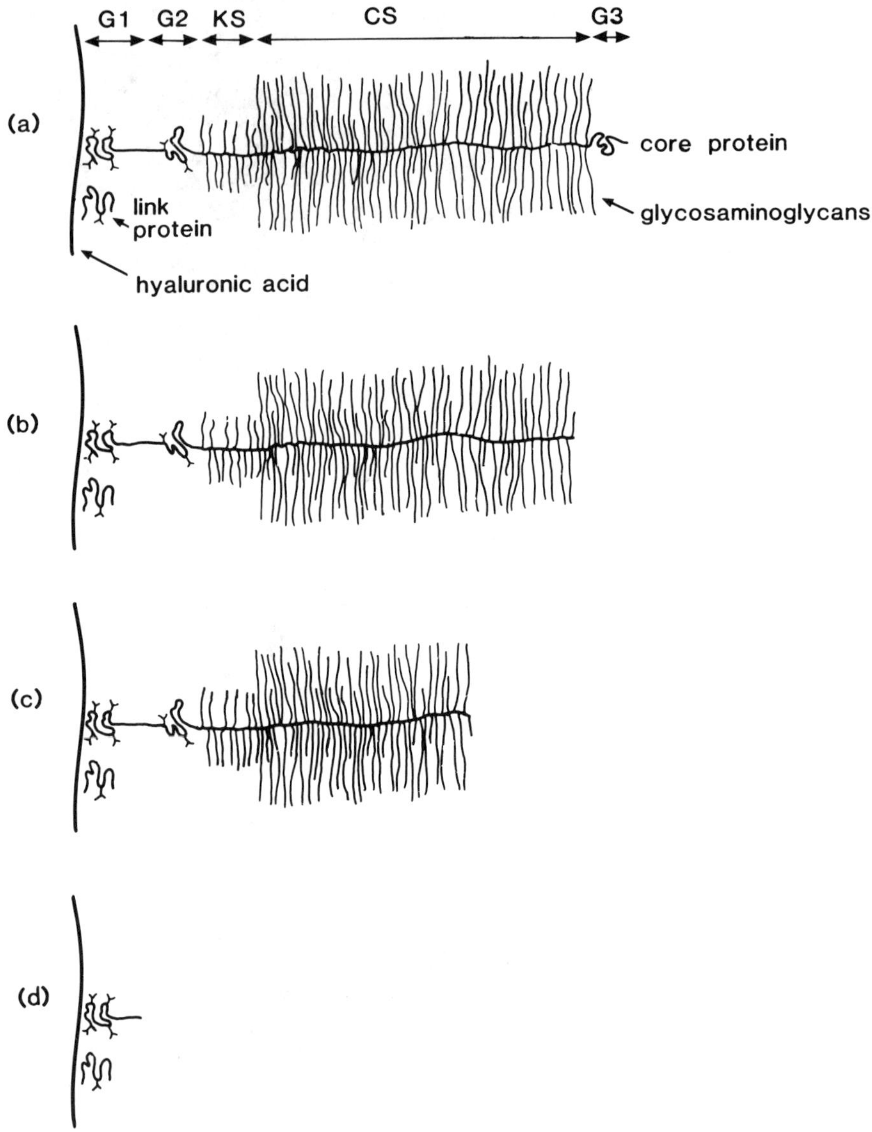

Figure 4. Processing of extracellular proteoglycans. Possible structure of proteoglycan monomers extracted from human cartilage of different ages. (a) The complete, newly synthesized molecule as shown by Hascall (1986). (b) The large CS-rich species which predominates in the extracellular matrix of neonatal cartilage. From 2 to 10 years the proportion of smaller, KS-rich molecules with a low buoyant density gradually increases (c) and becomes the major species by 20 to 58 years of age. (d) Core protein fragments with no attached glycosaminoglycan that have been extracted from adult but not neonatal cartilage. These changes are consistent with a gradual proteolysis of the core protein (see "Processing of Extracellular Proteoglycans").

et al., 1979; Maroudas and Bannon, 1981). It is this high osmotic pressure together with the low hydraulic permeability of the proteoglycan gel that enables cartilage to resist compressive loads (Kempson, 1976; see also Comper, this volume). Aggregate formation has a major influence on the dynamic behavior and fluid flow in cartilage during compression (Hardingham *et al.*, 1987). Physiochemical measurement have shown, however, that in a static system the size and degree of aggregation of the proteoglycans did not contribute to the osmotic pressure, which depended only on the concentration of the negatively charged glycosaminoglycans (Urban *et al.*, 1979). Therefore, even the smallest proteolytic degradation products found in older cartilage are fully functional provided they are retained within the matrix, and the older tissue which contains more monomers albeit of a smaller size is equally able to resist compressive loading. The age-related increase in the amount of hyaluronic acid (Roughley and White, 1980; Holmes *et al.*, 1988) is consistent with this. In contrast to the type II collagen, aggregating proteoglycans are designated to be rapidly and efficiently degraded and replaced in response to environmental influences. It is not clear whether the gradual change to increased numbers of smaller monomers with age is the result of random proteolysis of molecules which remain in the matrix for a long time or a programmed response to coordinate with the decreased space available in the more densely packed matrix of older cartilage.

Link Protein

Although some association has been recognized between sulfated proteoglycans and the cationic collagens, it is the formation of aggregates which is thought to prevent the large proteoglycans diffusing out of the tissue. The noncovalent interaction between the core protein and the hyaluronic acid occurs spontaneously extracellularly and is stabilized by a link protein. The link proteins may have a regulatory function in promoting the formation of larger aggregates which increase resistance of the complex to enzymic degradation, and thermal and pH denaturation (Tang *et al.*, 1979; Frazén, 1981; Buckwalter *et al.*, 1984; Hardingham, 1979).

Three different sizes of link protein with molecular weights ranging from 40 to 51 kDa have been identified in aggregates extracted from bovine nasal cartilage (Baker and Caterson, 1977) and human cartilage (Mort *et al.*, 1983). The two larger forms are versions of the same protein which differ in the degree of glycosylation, and the smallest is thought to be a proteolytic degradation product of the other two (Bonnet *et al.*, 1978; Mort *et al.*, 1985). The three types of link protein can be identified in all ages of human cartilage, but the amount of the largest form decreases in older cartilage with a proportionate increase in the smallest one. Characteristic smaller fragments (25 to 30 kDa) of the link proteins were also identified by immunoblotting techniques in many of the samples, and these fragments were more abundant in older tissue. The fragments remained covalently bound to the undegraded link proteins by disulfide bonds, but could be demonstrated by SDS electrophoresis under reducing conditions (Mort *et al.*, 1983). It is not known to what extent these changes alter the function of the link protein or compromise the stability of proteoglycan/HA aggregates (see Rosenburg and Buckwalter, 1986).

Enzymes Which Can Degrade Cartilage Matrix

Many proteinases that can degrade cartilage matrix components have been identified in connective tissues and inflammatory cells. Four main classes are recognized at present, as shown in Table 1. The likely substrates of these enzymes and their mode of action on proteoglycan and collagen is described below.

Core Protein

When aggregated proteoglycan has been used as a substrate for proteinases, the globular (G1) hyaluronate binding region/link protein complex remains very resistant to proteolysis (Heinegard and Hascall, 1974; Caputo *et al.*, 1980; Campbell *et al.*, 1986). In contrast, analysis using antibody to the G2 globular region which lies between the binding region and the keratan sulfate (KS) attachment region (Figure 5a; Hardingham *et al.*, 1986) indicated that it is very vulnerable to complete proteolysis with stromelysin (Murphy *et al.*, 1990). The extended region of the core protein is also extremely susceptible to degradation. The majority of proteinases cleave within the CS-rich region (Figure 5c) (Roughley, 1978; Stevens *et al.*, 1984; Campbell *et al.*, 1986; Nguyen, 1989). This action, in one step, releases the functional portion of the molecule bearing the glycosaminoglycan chains from that part which anchors the proteoglycan within the matrix. This initial attack is rapidly followed by further cleavages (Figure 5d) leading to different types of limit digestion product depending on the specificity of the enzyme (Table 1). In one, a peptide bearing one or two KS or CS chains results from the action of enzymes with a broad specificity such as papain, elastase, and cathepsin B. In another, 5 to 12 CS chains remain attached to the peptide. This second mode is characteristic of digestion with trypsin, stromelysin, cathepsin D, and most other proteinases (Keiser *et al.*, 1976; Roughley, 1978; Morrison *et al.*, 1973; Caputo *et al.*, 1980; Campbell *et al.*, 1986). Larger fragments containing more than 15 glycosaminoglycan chains were left after digestion of aggregates with plasmin. The initial sites of attack, common to all proteinases, presumably reflect particularly vulnerable areas of the substrate. This may represent stretches of the core protein which bear no glycosaminoglycans, as is the case adjacent to the hyaluronate binding region, or may be between clusters of glycosaminoglycan chains in the CS attachment region. The variation in subsequent cleavages indicates the specificity of the different enzymes for particular amino-acid sequences.

Table 1
Proteinases Which Can Degrade Cartilage Proteoglycan and Collagen

				Substrates				
Proteinase	Class	pH Range	Inhibitors	Core protein[a]	Collagen telopeptides	Insoluble helical collagen	Solubilized helical collagen	Gelatin
Cathepsin B	Cysteine (EC 3.4.22)	3–5.5	α2M cystatins	1–2	√	√	√	√
Cathepsin L								
Cathepsin H								
Cathepsin D	Aspartic (EC 3.4.23)	4–6	α2M	5–12	Limited	x	x	x
Plasmin	Serine	6–9	α2M	>15	?	x	x	√
Elastase	(EC 3.4.21)		α1 PI	1–2	√	x	x	√
Cathepsin G			α1 Anti-chymo-trypsin	5–12	√	x	x	
Stromelysin	Metallo	6–9	α2M	5–12	x	x	x	Limited
Gelatinase	(EC 3.4.24)	7–8	TIMP	?	Limited	x	x	√
Collagenase		7–8		x	x	Limited	√	Limited/?

[a] Numbers represent the average number of glycosaminoglycan chains left attached to the core protein fragments.

(a)

(b)

(c)

(d)

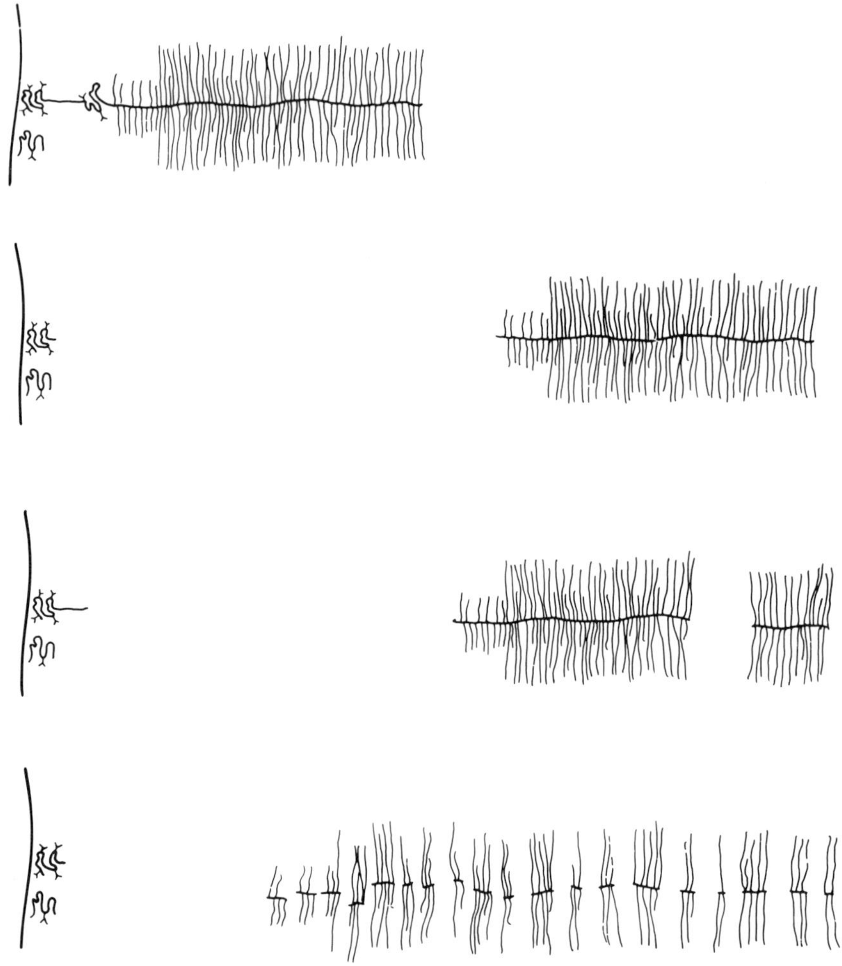

Figure 5. Proteolysis of isolated aggregating proteoglycan. The effect of proteinase action on proteoglycan. (a) Proteoglycan extracted from cartilage usually has already lost the C terminal globular region (G3). (b) The extended core protein is susceptible to cleavage by most proteinases adjacent to the binding region and (c) at a site within the chondroitin sulfate attachment region. (d) Subsequent action to produce peptides bearing from 1 to 15 glycosaminoglycan chains depends on the specificity of individual enzymes (see "Core Protein" and Table 1).

Link Protein

Treatment of link protein with either trypsin (Hascall, 1977; Baker and Caterson, 1977) or a neutral metallo-proteinase isolated from cartilage conditioned culture medium (Campbell *et al.*, 1986; Nguyen, 1989) converted the larger form to the smaller one. In the latter case it was noted that even in limit digests, no further fragmentation occurred to the smaller sizes (25,000 to 30,000) that have been extracted from older human cartilage, suggesting that other factors, perhaps free radical attack, may be the cause. Cathepsin G and elastase appeared to degrade the larger link protein with no apparent change to the smaller form (Keiser *et al.*, 1976).

Glycosaminoglycans

There is no direct evidence for extracellular glycosidase activity in cartilage *in vivo* (Wasteson *et al.*, 1972), although extracellular cleavage of heparan sulfate has been demonstrated by activated T lymphocytes (Naparstek *et al.*, 1984) and platelets (Oosta *et al.*, 1982). Cartilage proteoglycan fragments with intact glycosaminoglycan chains are thought mostly to diffuse into the synovial fluid and via the circulation to the liver where they are endocytosed. Cultured cartilage explants containing prelabeled ^{35}S-proteoglycans have been shown to release a small but constant proportion of this material as very low M_r components (Morrison, 1970; Tyler, 1985a; Wasteson *et al.*, 1972), which in the latter case was identified as free organic sulfate. This suggests that in some instances chondrocytes may be able to phagocytose proteoglycans. It was not clear from these experiments which type of proteoglycan gave rise to sulfate. Receptor-mediated endocytosis of the small dermatan sulfate proteoglycan has been demonstrated in chondrocyte cultures (von Figura *et al.*, 1980). Heparan sulfate proteoglycan in similarly endocytosed in other tissues (David *et al.*, 1979). Whole proteoglycan molecules are a very poor substrate for the endo- and exoglycosidases known to be present in the lysosomal system (Barrett, 1969), so prior digestion of the protein component would be required. A range of lysosomal enzymes then degrade the glycosaminoglycans to monosaccharides and inorganic sulfate (see Poole, 1986).

The measurement of KS epitope in serum has been used as a marker to monitor the degradation of cartilage proteoglycans (Thonar *et al.*, 1985).

Depolymerization of Collagen

Collagen degradation is accomplished in several stages by a range of collagenolytic enzymes (Dingle, 1973; Burleigh, 1977; Perez-Tamayo, 1978; Baici, 1980; Murphy and Reynolds, 1985). Cleavage of the nonhelical regions adjacent to the residues involved in cross-links drastically decreases the resistance of collagen to mechanical loading (Bader *et al.*, 1981) and can effectively solubilize the molecule, leaving the α helix intact (Figure 6a, b). Chymotrypsin, pepsin, pancreatic elastase thermolysin, and to a limited extent trypsin have all been shown to attack these regions on soluble collagen. Many of these proteinases will also act on mature insoluble collagen, the most efficient being pepsin (Burleigh, 1977). In physiological terms, two neutrophil proteinases, cathepsin G and elastase, and stromelysin at high enzyme concentrations ($>$20 μg/ml) can cleave type II procollagen in this way at neutral pH (Starkey *et al.*, 1977; Wu *et al.*, 1991). These enzymes are not thought

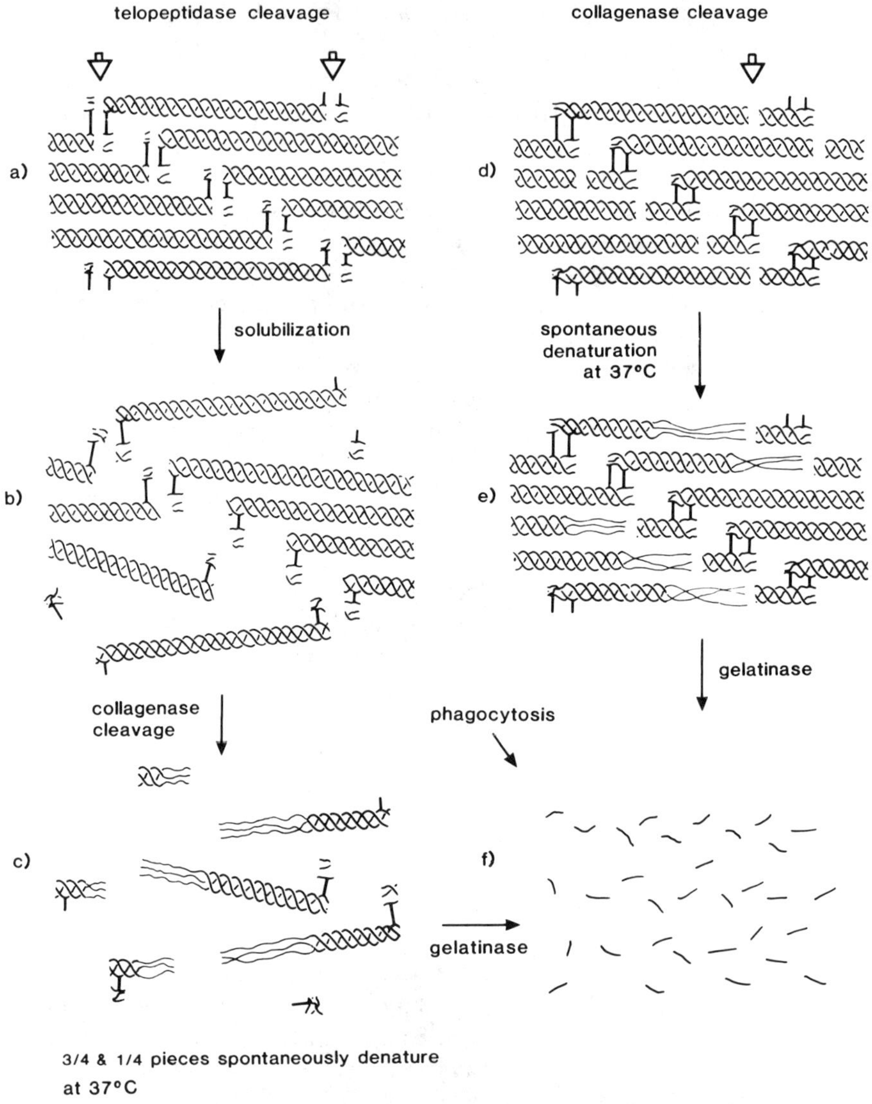

Figure 6. Proteolysis of collagen. The effect of different proteinases on collagen. (a) A telopeptidase cleaves the nonhelical regions adjacent to the cross-links. (b) Soluble monomeric collagen is released from the fibril. While in a helical form it is still resistant to nonspecific proteinases but provides an ideal substrate for collagenase. Alternatively, the molecules could be phagocytosed and degraded to small peptides intracellularly. (c) the $^3/_4$ and $^1/_4$ fragments resulting from collagenase action can be phagocytosed or spontaneously denature at 37°C and become susceptible to enzymes with gelatinase activity. (d) Collagenase can act very slowly on insoluble type II collagen, and the helix denatures spontaneously at 37°C at the points of cleavage. (e) The molecules initially remain attached to the fibril by the cross-links but are readily degraded to small peptides by nonspecific proteinases. There is evidence that some collagenases also have gelatinase activity, and can on their own solubilize cross-linked collagen (see "Helical Collagen").

to be present in chondrocytes, but could make a major contribution to collagen resorption during inflammation (Janoff *et al.,* 1976; Meninger *et al.,* 1980). Cathepsin B, L, and H can rapidly solubilize fibrillar collagen and quickly degrade the complete molecule to small peptides (Figure 6a, b, f). The pH optimum was found to be 4.5 to 5.0 with collagen in solution and 3.6 with insoluble collagen. The rate of degradation of cartilage collagen was ten times slower than that of tendon collagen. These proteinases can act at higher pH values but are very unstable and denature above pH 5.5; they are therefore thought of as acting mainly within lysosomes on phagocytosed material (Burleigh *et al.,* 1974; Evans and Etherington, 1978; Mason *et al.,* 1985). However precursor forms of these enzymes which are stable at neutral pH have now been identified (see "Activation of Proenzyme Precursors" *infra*).

Helical Collagen

The classical specific collagenase is the only enzyme that can act directly on the intact collagen helix at neutral pH (see Harris and Cartwright, 1977; Woolley, 1984). Solubilized α chains are the ideal substrate (Figure 6b, c). Collagenase cleaves all three chains at one specific locus to produce characteristic degradation products. The N-terminal three quarter fragment (TC^A) and the C-terminal one quarter fragment (TC^B) remain helical below 25°C but will spontaneously denature locally above 33°C (Gross and Nagai, 1965; Miller *et al.,* 1976). Active collagenase can bind to insoluble fibrillar collagen (Welgus *et al.,* 1980), and electron micrographs show a complete disruption of fibrils in skin and cartilage incubated with collagenase at 37°C (Woolley et al., 1978; Harris *et al.,* 1970). The degradation products of these samples were not analyzed, but it is logical to assume that some collagenases may be able to make additional cleavages of the denatured random coil molecules (gelatin), but at a very slow rate (Woolley *et al.,* 1975). Leucocyte collagenase will also degrade fibronectin and proteoglycan (Ohlsson and Ohlsson, 1973). Type II collagen, even when solubilized, is 5 to 6 times more resistant to collagenase than soluble type I collagen, and insoluble cross-linked collagen is 100 to 1000 times less susceptible to collagenase than non-cross-linked fibrils (Robertson and Miller, 1972; Woolley *et al.,* 1973; Vater *et al.,* 1979; Tyler and Cawston, 1979). The action of collagenase on mature collagen is therefore very inefficient (Leibovich and Weiss, 1971; Steven, 1976). It can initiate local denaturation of the helix which then becomes susceptible to other proteinases, but does not remove the cross-links or depolymerize the collagen (Gross *et al.,* 1980) (Figure 6d, e). In contrast, bacterial collagenase acts on Gly-X-Y triplets along the length of the helix and digests it to small peptides. In cartilage, the only other enzymes which can attack the native triple-helical molecule are cathepsins B, L, and H at acidic pH as described before. It is not known whether the chondrocyte can create areas of low pH near the cell surface suitable for the extracellular action of these enzymes (Dingle and Knight, 1985). Cathepsin D is ineffective against either soluble collagen or gelatin but might act on the carboxyterminal nonhelical region (Burleigh, 1977). It is often said that digestion by collagenase represents a rate limiting step in the digestion of extracellular collagen, but there are no specific data in support of this idea.

Gelatin

Denaturated regions of the helix can be degraded to small peptides (Figure 6f), either extracellularly by gelatinase or stromelysin or intracellularly following phagocytosis by cathepsins B, L, and H and various exopeptidases in secondary lysosomes (Burleigh *et al.*, 1974, McDonald and Schwabe, 1977). Hydroxyproline cannot be reutilized for protein synthesis. The presence of soluble hydroxyproline in body fluids is therefore often used as an indicator of collagen degradation (Gross, 1974).

Regulation of Enzyme Activity

Genes specifying all of the enzymes listed in Table 1 have now been cloned and sequenced. The synthesis of many of these enzymes has been shown to be altered by a variety of cytokines, growth factors, and hormones. All are produced initially as inactive proforms which are later converted to the active proteinases, which are in turn subject to inhibition by a range of protein inhibitors. The availability of proteolytic activity is therefore very carefully regulated at several stages. (For recent reviews see Mason, 1988; Murphy and Docherty, 1988; Cawston, 1986; Heidtmann and Travis, 1986; Barrett *et al.*, 1986.)

Modulation of Proteinase Synthesis

It has been shown by immunolocalization of cultured monolayers that the same cell can make stromelysin and collagenase at the same time (Murphy *et al.*, 1986). However, a study of growth plate cartilage revealed a consistent but different pattern of production of each enzyme and the tissue inhibitor of metalloproteinases (TIMP) in distinct zones during different stages of development *in vivo* (Brown *et al.*, 1989). An endopeptidase similar to the enzyme (EC 3.4.24.11) isolated from kidney membranes (Fulcher *et al.*, 1982) has also been identified by immunolocalization, mainly in the mid-zone of *ex vivo* cartilage, with none in the deep zone and very little in the superficial layers. (Hembry, unpublished results). The heterogeneity of chondrocytes and their surrounding matrix throughout the depth of articular cartilage is well recognized (Poole *et al.*, 1982; Schenk *et al.*, 1986; Ayedelotte *et al.*, 1986). The different zones probably also have very different rates of degradation, although no detailed data on this topic are available.

Transcription of the metalloproteinases has been shown to be increased, often coordinately, in response to cytokines such as tumor necrosis factor (TNF) and interleukin 1 (IL1) (Murphy *et al.*, 1986) and following exposure to basic fibroblast growth factor (bFGF) or epidermal growth factor (EGF)

(Edwards *et al.*, 1987). Increased levels of neutral-metallo-proteinases and plasminogen activator have also been identified in the culture medium of cartilage or chondrocytes stimulated with IL1 *in vitro* (Tyler, 1985a; Saklatvala *et al.*, 1984; Bunning *et al.*, 1987; Deshmukh-Phadke *et al.*, 1980; McGuire-Goldring *et al.*, 1984; Schnyder *et al.*, 1987; Pasternak *et al.*, 1986; Di Pasquale *et al.*, 1988; Caputo *et al.*, 1987b; Gowen *et al.*, 1984; Shinmei *et al.*, 1988; Nguyen *et al.*, 1989). These enzymes were all in the precursor form and unable to degrade proteoglycan which had been released from the cartilage without pretreatment of the medium to initiate activation (Tyler, 1985a). Measurement of the amounts of latent enzyme secreted into the medium provides a very sensitive marker to demonstrate the activation of chondrocytes by cytokines, but it is not a good indicator of cartilage degradation. Increased synthesis of collagenase can occur at low doses of IL1 when there is no detectable loss of collagen (Tyler, 1988) and extensive degradation of proteoglycan and collagen occurs in cartilage culture with retinol or retinoic acid, which inhibit both basal and IL1-stimulated synthesis of metalloproteinases (Dingle, 1973; Sporn, 1976; Brinkerhoff *et al.*, 1981; Murphy *et al.*, 1986; di Pasquale *et al.*, 1985; Tyler, 1988). It is not known whether large amounts of inactive proforms of these enzymes are also secreted in response to mediators *in vivo* (see "Cytokines" infra).

Increased production of lysosomal enzymes has been demonstrated in many culture systems of resorbing cartilage (see Dingle, 1969; Poole, 1975; Barrett, 1981; Mort *et al.*, 1984) and in cartilage from arthritic joints (see "Rheumatoid Arthritis" and "Osteoarthritis" *infra*). It is not known how this is regulated.

Activation of Proenzyme Precursors

The conversion of proenzymes to the active form probably represents the key step in the degradative process. Knowledge of where and how this activation is achieved is essential in order to understand how the chondrocyte controls the turnover of its extracellular matrix. The lysosomal enzymes are synthesized as high molecular weight precursors on the rough endoplasmic reticulum. An N-terminal signal peptide is removed in the same way as many secreted proteins. In the Golgi, cathepsin D is known to be glycosylated on asparagine residues with mannose-6-phosphate some of which are then phosphorylated. This targets the proenzyme to lysosomal rather than secretory vesicles. Specific proteolytic cleavage to the smaller M_r active form probably occurs within the lysosomal compartment. (Sahagian and Gottesman, 1982; Von Figura and Hasilik, 1986). It is likely that cathepsins B, L, H, and N are transported and activated in the same way, possibly by a common processing enzyme, although this has not yet been identified. The mature smaller M_r forms of these enzymes have a half life of several days at acid pH but rapidly denature irreversibly within minutes above pH 5.5. A secreted form of the higher molecular weight protein has been reported for cathepsin B (Mort and Recklies, 1986). These precursors were active in the pH range 3 to 5.5 and were not denatured in the extracellular fluid by neutral pH. It is not known if similar stable precursors of the cysteine proteinases play a role in the breakdown of cartilage matrix. If they are present, their potent collagenolytic activity would make a major contribution to cartilage degradation.

The metalloproteinases collagenase, gelatinase, and stromelysin are also secreted as inactive precursors. They are processed extracellularly by specific proteolysis to smaller active forms of the enzymes. A conformational change is thought to precede autoactivation which results in the cleavage of about 80 amino acids from the N terminal end of the molecule (Okada *et al.*, 1986; Grant *et al.*, 1987). The process can be efficiently initiated *in vitro* by organomercurials such as 4-aminophenylmercuric acetate (Sachez-Lopez *et al.*, 1988) or by exposure to trypsin, plasmin, kallikrein, and thermolysin (Murphy and Docherty, 1988; Okada *et al.*, 1986), but still requires an autolytic step for complete activation. The smaller M_r active enzymes cannot initiate activation alone but may enhance the effects of other proteolytic activators. It is not known how or where activation of metalloproteinases occurs in cartilage or any other tissue *in vivo*, but plasmin (Werb *et al.*, 1977), an endogenous proteolytic activator (Treadwell *et al.*, 1986), and more recently mast cell tryptase, which is not inactivated by the serum inhibitors (Gruber *et al.*, 1988), have been suggested as possible activating agents.

Plasmin is not normally present in cartilage, but is derived from the action of plasminogen activator on plasminogen. The major action of plasmin is thought to be in fibrinolysis where tissue plasminogen activator requires fibrin for activation. However, tissue-type plasminogen activator as well as a specific inhibitor (PAI 1) can be synthesized by stimulated chondrocytes (Bunning *et al.*, 1987; Yamada *et al.*, 1988), and plasminogen has been shown to be present in some synovial fluids (Clemmenson, 1977). Experiments using cartilage cultured with IL1 show that degradation of both proteoglycan and collagen occurs in the absence of added plasminogen, but the amount of matrix lost can be increased further if it is present due to the activation of latent metalloproteinases in the medium (Tyler, 1988; Collier and Ghosh, 1988; Bunning, unpublished results). It is not known whether metalloproteinases mediate either the basal or IL1-induced resorption of cartilage, but if they are freely secreted *in vivo* in response to cytokines, the presence of plasminogen or tryptase could indirectly lead to enhanced cartilage degradation via this mechanism.

Physiological Inhibitors of Proteinase Activity

α2 Macroglobulin (α2M) is a very important universal inhibitor of all classes of active proteinase. There are, in addition, a range of inhibitors specific for each class except the aspartic proteinases (Barrett and Starkey, 1973). α2M is present in human plasma where it accounts for over 95% of the inhibitory activity toward metalloproteinases (Woolley *et al.*, 1984). It has also been identified in synovial fluid but is too large (M_r 725,000) to penetrate the cartilage matrix. There is recent evidence that α2M may also act as a serum carrier for cytokines (Teodorescu *et al.*, 1988) and growth factors (Huang *et al.*, 1984; 1988).

The cystatins, a large related group of low molecular weight proteins, inhibit the lysosomal cysteine proteinases (Barrett, 1987). Class 1 cystatins (sometimes called stefins) are thought to be mainly cytoplasmic. Class 2 cystatins and the kininogens (class 3) are present mainly in extracellular fluids and as plasma proteins. Some of the inhibitory activity against cysteine proteinases which has been extracted from cartilage matrix is almost certainly due to the cystatins (Killackey, 1983; Roughley, 1978).

α1 Proteinase inhibitor (α1 PI) has a broad specificity toward all mammalian serine proteinases including pancreatic and neutrophil elastase, plasmin, and cathepsin G, and is a potent inhibitor. Other serum inhibitors such as α1-antichymotrypsin and aprotinin are present at a much lower concentration (Travis and Salvesen, 1983; Gebhard *et al.*, 1986), but both have been identified in synovial fluid and may be the source of inhibitors extracted from cartilage (Kuettner *et al.*, 1976; Roughley, 1978; Rifkin and Crowe, 1977). An inhibitor specific for plasminogen activator has also been identified in human articular cartilage (Yamada *et al.*, 1988).

The production of serum inhibitors such as α2M, kininogen, α1 PI, α1 antichymotrypsin, and cystatin is induced by the action of a specific cytokine interleukin 6 (IL6) on hepatocytes as part of the acute phase response to trauma. Their concentration in serum can double within 8 h. In contrast, interleukin 1 (IL1) induces the production of many of the other acute phase reactants but not the proteinase inhibitors (Gauldie, 1989). All connective tissue cells, including chondrocytes, synthesize IL6 constitutively at a very low level which is enhanced several-fold in response to IL1, TNF, and platelet derived growth factor (PDGF) (Sehgal *et al.*, 1987; Tyler; Niefeld; unpublished results). Local production of IL6 by cartilage and other tissues in the joint in response to injury would therefore lead to increased amounts of proteinase inhibitors in the synovial fluid. This could provide a very valuable protection mechanism against rapid cartilage degradation in inflammatory diseases such as rheumatoid arthritis.

There are many reports in the literature of inhibitors of collagenase and other mammalian neutral metalloproteinases. They are probably all forms of TIMP (Cawston, 1986). TIMP can be synthesized locally by chondrocytes and other connective tissue cells and has been identified in extracts of cartilage as well as in serum and synovial fluid (Dean and Woessner, 1984; Bunning *et al.*, 1984; Murphy *et al.*, 1986). TIMP binds very tightly only to the active enzyme to form a 1:1 complex which cannot be reactivated by APMA or proteolysis. IL1, TNF, and growth factors such as EGF and bFGF also induce local production of TIMP (Edwards *et al.*, 1987). In the same study it was interesting to note that transforming growth factor β (TGF-β), when added together with either EGF or FGF, further enhanced TIMP mRNA while decreasing collagenase mRNA.

Mediators of Cartilage Degradation

The aggregating proteoglycans in normal healthy articular cartilage are constantly being degraded and resynthesized at a defined rate which is carefully controlled by the chondrocyte. This balanced equilibrium can be influenced experimentally by many different factors. One of the most important but least understood is mechanical loading; another is the induction of matrix resorption by products of synovial and mononuclear cells, most of which have been shown to be due to the action of cytokines. Both these effects are outlined in detail below. Retinol, retinoic acid, and bacterial lipopolysaccharides will also decrease proteoglycan synthesis, increase degradation, and induce rapid net loss of matrix by a similar mechanism to that described in "Cytokines", *infra*, for the cytokines (Morrison, 1970; Brinkerhoff *et al.*, 1981; Morales, 1984; Jasin, 1983; 1984; Di Pasquale *et al.*, 1985). It is known that chondrocytes can synthesize IL1 and IL6 (Ollivierre *et al.*, 1986; Towle *et al.*, 1987; Rath *et al.*, 1988; Niefled, unpublished results), and it will be interesting to determine whether the effects of altered loading and other mediators are due to the local stimulation of cytokine production. Any drugs or agents that inhibit proteoglycan synthesis will gradually induce matrix depletion (Kalbhen, 1984). This will decrease the resistance of cartilage under load and indirectly lead to mechanical disruption and increased degradation of the matrix.

Mechanical Loading

During normal maturation, articular cartilage tends to become thickest in areas subjected to the greatest load, and it has been observed for many years that altering the loading on a joint has a profound effect on the ability of chondrocytes to maintain their surrounding extracellular matrix (see Helminen *et al.*, 1987). Histological surveys, however, have shown a marked topographical variation between joints, and in different regions of the same joint, and indicate that nonmechanical features also contribute to the thickness of cartilage (Kiviranta *et al.*, 1987a). It is extremely difficult to measure directly either the intensity or frequency of loading locally *in vivo*. Nevertheless, fairly consistent, albeit qualitative, results have emerged from studies of altered loading on cartilage.

Decreased loading. Complete immobilization of one limb, or just unloading with some motion allowed, leads to a thinning of the cartilage and a decreased content of proteoglycan (KS and CS) in sheep (Caterson and Lowther, 1978) and dogs (Olah and Kostensky, 1972; Palmoski *et al.*, 1980). This is accompanied by a decreased synthesis and increased extractability of the proteoglycans and increased water content of the cartilage. Depletion is most noticeable in the superficial zone at the periphery of the femoral condyles, and the degree of softening here corresponds closely to the loss of proteoglycan. Almost no depletion or loss of stiffness is evident in areas which retain some contact force (Kiviranta *et al.,* 1987b). Similar changes leading to matrix loss occurred very rapidly in the rabbit models (Troyer, 1975; Videman *et al.,* 1981a), indicating that degradation is also enhanced. The extensive changes in the rabbit included clear evidence of inflammation, (Michelsson and Hunneyball, 1984), clonal expansion of hypertrophic chondrocytes, and some increased synthesis of CS, all of which are more typical of osteoarthritic lesions than atrophic cartilage. The reason for this species difference is not clear, but younger rabbits and rats are much less affected by immobilization than older animals (Ohta *et al.,* 1981). The total collagen content of cartilage appears unchanged under decreased loading, but there is some evidence of increased biosynthesis, indicating that degradation is also enhanced (Videman *et al.,* 1981).

Increased loading. In general, increased weight-bearing enhances cartilage thickness and the proteoglycan content of the intermediate, deep, and calcified zone, but not the superficial zone. *In vivo,* the effect is highest in areas subject to the greatest load and occurs in all species in response to nonstrenuous exercise, after the application of weighted backpacks or following amputation of a forelimb (Kostensky and Olah, 1972). *In vitro* cyclic compressive loading, but not static loading, will increase cartilage proteoglycan synthesis and encourage aggregate formation (Palmoski and Brandt, 1984; van Kampen *et al.,* 1985). Although chondrocytes rely on diffusion of nutrients from the synovial fluid (McKibbon and Maroudas, 1979), the response to increased loading is not thought to be regulated by an enhanced flow of nutrients.

The resumption of normal weight-bearing and gentle exercise following immobilization, or chemical-induced depletion, can lead to a gradual but complete restoration of proteoglycan content and cartilage function (Sood, 1971; Williams and Brandt, 1984). However, excessive loading or strenuous exercise of such cartilage leads to accelerated matrix degradation. The average size and ability to aggregate of proteoglycan monomers is decreased; the proteoglycan is more easily extracted and there is evidence of irreversible disruption of collagen fibers and surface fibrillation. Again, older animals, rabbits in particular, seem to be especially vulnerable (Palmoski and Brandt, 1981; Vasan, 1983).

Cytokines

Proteoglycans. It is now widely accepted that cytokines induce proteoglycan depletion in cartilage *in vitro* and *in vivo* by increasing the rate of degradation (Dingle *et al.,* 1979; Gowen *et al.,* 1984; Krakauer *et al.,* 1985; Saklatvala *et al.,* 1984; Tyler, 1985a; Pettipher *et al.,* 1986; Hubbard *et al.,* 1988; van den Berg, 1988) and decreasing synthesis (Tyler, 1985b; Saklatvala, 1986; Morales and Hascall, 1988; van den Berg *et al.,* 1988; Benton and

Tyler, 1988). A representative dose response curve for three different commercial preparations of human recombinant IL1α cultured for 72 h with pig articular cartilage is shown in Figure 7. When used alone, IL1 appears to be more potent than TNF or IL6, but some synergy has been demonstrated between them. Their effect can be decreased in the presence of growth factors such as IGF1 (Tyler, 1989), so their final activity will depend on the relative concentration of each present. There is some species specificity; human IL1β acts very poorly on pig cartilage, although pig IL1α and β are both effective (Bird and Saklatvala, 1986). There is a marked change with age which has been noted especially by those working with human explants; often, very little increase in proteoglycan degradation is seen with older human cartilage at doses of IL1 which significantly inhibit synthesis (van den Berg and Bayliss, unpublished results). It is not clear whether this is due to decreased cleavage of the core protein or inability of the degradation products to freely diffuse out of the more densely packed matrix.

Analysis of the degradation products of proteoglycans released by pig cartilage cultured with pig IL1 showed that the average hydrodynamic size of the molecules was only slightly smaller than that of normal monomers. However, degraded material in the medium was not able to aggregate when mixed with hyaluronic acid. The size and charge of the product was consistent with a limited cleavage of the core protein close to the binding region, as shown in Figure 5b. Proteoglycan released from severely depleted cartilage tends to be of smaller average size (Figure 5c), indicating a further cleavage also in the CS region. About 4% of prelabeled proteoglycan was released as free sulfate, indicating that there is also degradation of some glycosaminoglycan chains. The length and degree of sulfation of the CS from the released aggregating proteoglycan appeared to be unchanged, so this sulfate may be derived from one of the smaller proteoglycans (Tyler, 1985a).

Proteoglycan released from bovine nasal cartilage cultured with IL1 purified from human monocytes was shown to still be capable of aggregating with hyaluronic acid (Hubbard *et al.*, 1988). It was postulated that a weakening of collagen-proteoglycan interactions might be responsible for its diffusion out of the matrix. Further characterization, using specific antibodies, of proteoglycan released from pig cartilage cultured with IL1 confirmed that the link protein and cleaved portion of the binding region (M_r ~5 to 80 kDa) with no attached glycosaminoglycans initially remained within the matrix, presumably still attached to hyaluronic acid. With time in culture these epitopes were gradually released into the medium, but at a much slower rate than the nonaggregating glycosaminoglycan-bearing portion of the molecule. Both components were still able to bind to hyaluronic acid; however, they could be displaced in the presence of oligosaccharides, indicating that the interaction is not as stable as in native aggregates. The complex is similar to the one isolated from aged human cartilage as described in "Processing of Extracellular Proteoglycans" *supra*, and shown in Figure 4d. In some ex-

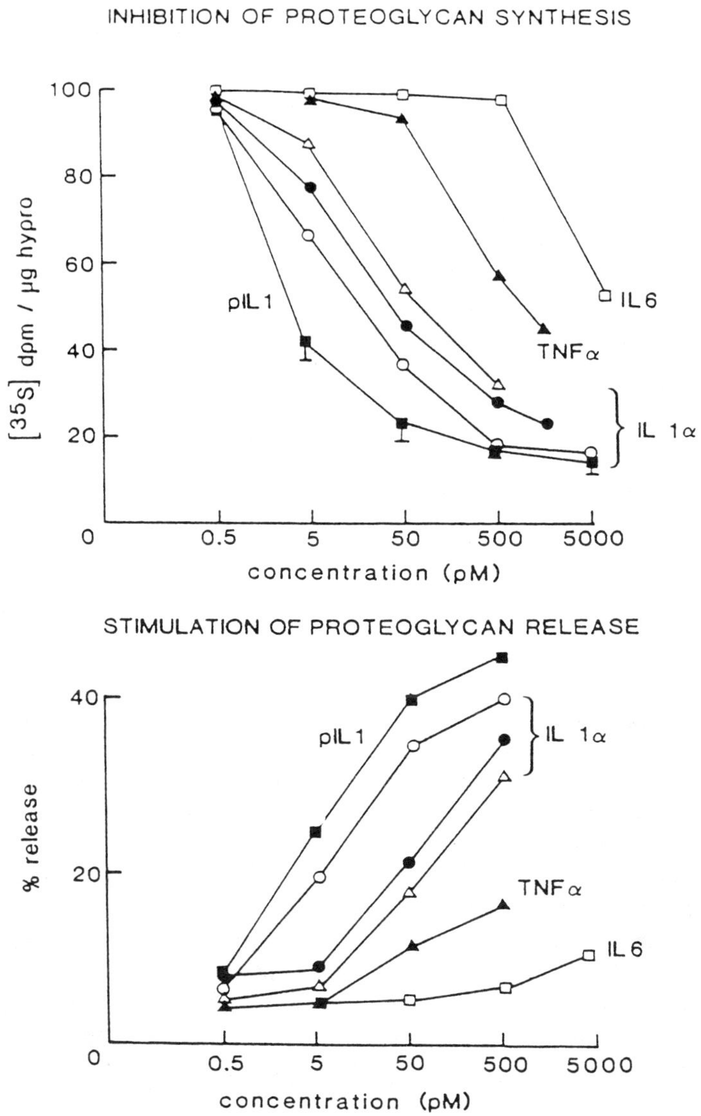

Figure 7. The effect of cytokines on cartilage proteoglycan *in vitro*. Pig articular cartilage was cultured with increasing doses of human recombinant IL1 (Roche, Genzyme, Biogen), TNF (Genzyme), or IL6 (Immunex) for 72 h in Iscove's medium containing supplements of albumin, transferrin, lipoprotein, and ascorbate but no serum. Proteoglycan degradation and synthesis was estimated as described previously (Tyler, 1985a,b).

periments, up to 20% of binding region epitope was unaccounted for and presumed to be completely degraded (Ratcliffe *et al.*, 1986).

A similar study using antibody to the G2 globular region (Figure 4a; Hardingham *et al.*, 1986) showed that none of this epitope is retained in the matrix unattached to glycosaminoglycan. It is either released into the medium as a 70 to 120 K fragment or completely degraded (A. Fosang *et al.*, 1991). A limited degradation of link protein to the smaller form has been demonstrated in human articular cartilage cultured with IL1 (Campbell *et al.*, 1986), which may give rise to the altered stability of the aggregating G1 complex.

Fewer proteoglycan fragments of a similar nature were released from unstimulated cultured cartilage, and extracted from *ex vivo* tissue, indicating that this is a general mechanism for proteoglycan turnover, the rate of which is enhanced by exposure to cytokines.

Collagen. Increased release of proteoglycan can be detected in the medium 4 to 6 h after exposure of cultured cartilage to IL1 (50 pM). In contrast, there is no significant degradation of collagen under these conditions. However, prolonged incubation with higher levels of IL1 leads to extensive resorption of extracellular collagen with a characteristic lag period of between 6 and 8 days before a dose-dependent release of hydroxyproline occurs. Just prior to this there is a slight but significant increase in acid-soluble collagen. Once initiated, degradation rapidly proceeds to completion, and most of the collagen in the medium is in the form of small peptides (Tyler, 1988). It has not been possible to date, even with the use of specific proteinase inhibitors, to isolate and identify intermediate degradation products. A dose-dependent decrease (20 to 80%) in collagen production was observed in both explants and chondrocyte monolayers cultured with IL1 (10 to 100 pM) for 48 h. This was not caused by an increase in the rate of intracellular degradation of newly synthesized collagen, which remained unchanged. Inhibition of collagen synthesis was due to decreased amounts of $\alpha 1$(II) mRNA (Tyler and Benton, 1988).

Increased matrix degradation in cultured cartilage exposed to cytokines is clearly mediated by enhanced proteolytic activity. The characteristic proteoglycan products released remain sensitive to proteolysis and can be readily degraded to smaller pieces by almost any active proteinase (Figure 5d). The fact that this rarely occurs within the matrix demonstrates that even in maximally stimulated cartilage the degradation process mediated by the chondrocyte is not rampant and indiscriminate but precise and discrete. It is not known how this is achieved.

Inhibition of Cytokine-Mediated Degradation

An alternative approach to characterizing degradation products is to try to block the process with inhibitors. This has been tried by many groups in an attempt to understand the mechanism.

Steroids will decrease cartilage breakdown in a co-culture system of cartilage with synovium or with endotoxin, as they are potent inhibitors of cytokine production. However, they do not affect proteoglycan degradation in isolated cartilage cultured with IL1 even though synthesis of stromelysin is inhibited. (Tyler *et al.*, 1982; Sheppeard *et al.*, 1982; Steinberg *et al.*, 1983; Hubbard *et al.*, 1988). Retinoic acid, as well as dexamethasone, have been shown to inhibit both the basal and IL1-enhanced transcription of stromelysin and collagenase mRNA (Murphy *et al.*, 1986; Brinkerhoff *et al.*, 1981). Yet retinoic acid has no effect on the IL1-mediated release of proteoglycan and collagen from cartilage (Tyler, 1988). Other work emphasizes that inhibitors which are very effective on chondrocyte monolayers or in soluble enzyme assays do not act on cartilage explants. Thus, no decrease in degradation was observed using a wide range of specific nontoxic proteinase inhibitors and other nonsteroidal inflammatory drugs on chick limb bones stimulated with retinol (Barrett, 1981), and articular cartilage (Dingle and Tyler, 1986) or bovine nasal cartilage (Saklatvala and Sarsfield, 1988; Hunneyball *et al.*, 1988) cultured with IL1.

More recently, several pharmaceutical companies have prepared synthetic inhibitors of the metalloproteinases. These are mostly substrate analogues bearing carboxylic acid, thiol, or hydroxamic acid ligands and appear to be more effective against stromelysin than collagenase. From the few reports that have been published, only the hydroxamate derivatives significantly decreased loss of proteoglycan from IL1 and retinol stimulated rabbit cartilage. Again, concentrations two to three orders of magnitude higher were required with cartilage compared to inhibition of isolated stromelysin (Di Pasquale *et al.*, 1986; Caputo *et al.*, 1987a). Other agents shown in previous studies to decrease proteoglycan degradation include chloroquine derivatives (Rainsford, 1985), cycloheximide, iodoacetic acid, and cytochalasins (Dingle and Tyler, 1986; Saklatvala and Sarsfield, 1988). This suggests that new protein synthesis, energy, and membrane movement are required for some aspects of the IL1 response. These compounds are of little therapeutic value, however, as they are also potent inhibitors of cartilage matrix synthesis and are likely to induce degenerative changes in cartilage *in vivo* (Kalbhen, 1985).

Specific proteinase inhibitors have been successfully used to demonstrate decreased matrix degradation in cultured bone (Eeckhout *et al.*, 1988; Everts *et al.*, 1988) and synovium (Reynolds *et al.*, 1988). The inability to demonstrate an effect with similar inhibitors in cartilage may reflect the high level of endogenous inhibitors already present in the matrix which have either been synthesized by the chondrocytes or sequestered from the serum (see ''Physiological Inhibitors of Proteinase Activity'', *supra*). Proforms of acid or neutral proteinases may be activated locally, then quickly inhibited so that subsequent addition of exogenous inhibitors makes no difference to the rate of degradation. Specific antibodies which could distinguish between the inactive precursor, activated enzyme, and inhibitor/enzyme complex would be required to test this idea. Alternatively, the degradation process may involve a protected region of the matrix or cell membrane that is not accessible to exogenous inhibitors. It is known that newly synthesized proteoglycans are not preferentially degraded in the presence of IL1 (Tyler, 1985b; van den Berg *et al.*, 1989). If the same cell is involved in synthesis and degradation at the same time, the cell membrane may have to have distinct specialized regions controlling each of these processes.

Constant exposure of cartilage to IL1 is required to enhance the degradation of proteoglycan and collagen, and the effects do not persist once the cytokine is removed. Protein inhibitors of IL1 have been identified in body fluids (Dayer and Seckinger, 1988), but it is not known if they can prevent IL1-induced loss of matrix. In common with its effects on other cells, IL1 acts on cartilage by binding to specific high affinity receptors on the chondrocyte surface (Bird and Saklatvala, 1986). There is clear evidence of heterogeneity of receptors from different cell types (Matsushima *et al.*, 1986; Horuk *et al.*, 1987), and the sequence of one has recently been published (Sims *et al.*, 1988). This indicates that it may be possible to design tissue-specific IL1 antagonists.

Rheumatoid Arthritis

Rheumatoid arthritis is a systemic inflammatory disease characterized by primary synovitis in the joints. There is a marked hyperplasia of the synovial lining due initially to proliferation of intimal cells but also from recruitment of monocyte/macrophages. Initially, lymphocytes and plasma cells are difficult to detect (Henderson and Pettipher, 1985). Many studies show that cartilage degradation is associated with this phase of the disease and occurs typically in marginal areas where the aggressive cell mass comes in contact with the matrix. The activated cells can migrate inward over the cartilage surface and penetrate into the tissue (Harris, 1981; Fassbender, 1986). One of the earliest changes noted in the hypertrophic intimal cells is an increase in the number of lysosomes and in the activity of their enzymes, both intra- and extracellularly (Muirden, 1972; Barrett and Saklatvala, 1981; Henderson, 1981; Poole *et al.*, 1975; Mort *et al.*, 1984). Cells adjacent to cartilage lesions have been shown to contain phagocytosed matrix within lysosomes (Barland *et al.*, 1964; Harris *et al.*, 1977). Increased collagenase and prostaglandin production is also evident and is thought to be mediated by the action of IL1 and TNF on the synovial cells and macrophages (Abe et al., 1973; Cawston *et al.*, 1984; Woolley, 1984; Roberts and Dean, 1986; Krane *et al.*, 1982; Dayer and Seckinger, 1988). The destructive proliferative phase is relatively short-lived and the pannus becomes vascularized, less cellular, and eventually modulates to a fibrous tissue. Following angiogenesis there is an infiltration of inflammatory leucocytes, macrophages, lymphocytes, and plasma cells. It has often been observed that the cartilage surface remains resistant to degradation even in highly inflamed joints showing extensive erosion of subchondral bone. This supports the view that in most instances proteinase inhibitors are present in excess in the synovial fluid (see ''Physiological Inhibitors of

Proteinase Activity'', *supra*). To be effective, therefore, enzymes would have to be released or activated close to the cartilage surface. Collagenase has been localized to the pannus/cartilage junction (Woolley *et al.*, 1977). Phagocytosis of immune complexes (Sumi *et al.*, 1986; Cooke *et al.*, 1984) or other debris by neutrophil leucocytes at sites of cartilage erosion might also lead to local release of, and degradation by, elastase, cathepsin G, and collagenase; see ''Depolymerization of Collagen'', *supra* (Barrett, 1978; Oronsky and Buermann, 1976; Campbell and Campbell, 1988).

Increased amounts of prostaglandin E2 are produced by chondrocytes in response to cytokines, mechanical pressure (Chrisman *et al.*, 1981; Somijen *et al.*, 1980), and inflammatory mediators such as histamine and bradykinin (Taylor *et al.*, 1986; Benton *et al.*, 1989). Prostaglandins are potent stimulators of bone resorption via increase levels of cyclic AMP (Tashjian *et al.*, 1987; Vaes, 1988), but they appear to have little direct effect on the basal or stimulated degradation of cartilage *in vitro* (Benton *et al.*, 1989), and similar results are obtained in the presence of indomethacin which inhibits prostaglandin production (McGuire *et al.*, 1983). However, PGE_2 may promote proteoglycan depletion by decreasing chondrocyte metabolism (Fulkerson *et al.*, 1981). Activated macrophages and other phagocytes produce hydroxyl radicals which have been reported to attack proteoglycan and collagen and may contribute to the widespread destruction of the cartilage matrix (Greenwald, 1976; Schalkwijk, 1985).

Almost any insult to the joint could provide a trigger for this series of events whether due to allergy, infection, or injury. It is not known where or how an inflammatory cycle is initiated, but local production of cytokines is probably involved (Eastgate *et al.*, 1988; Hopkins *et al.*, 1988). There is now overwhelming evidence that IL1 and TNF can regulate most of the local and systemic effects of inflammation, while interferon γ directs many of the immune responses and is the major macrophage activating factor (for reviews see Oppenheim *et al.*, 1986; Larrick and Kunkel, 1988; Unanue and Allen, 1987; Sherry and Cerami, 1988). Once activated, fibroblasts, monocytes, and T cells produce colony stimulating factors which will bring large numbers of primed, often immune-specific, effector cells to the site of injury, leading to a cascade of further cytokine production (Clarke and Kamen, 1987). Many of the dendritic and macrophage-like cells in the rheumatoid synovial tissue express class II MHC antigens, probably induced by interferon γ, and there is an accumulation of activated mainly CD_4 T helper lymphocytes. This indicates that the cycle might be sustained by a T cell mediated autoimmune response against degradation fragments of the cartilage matrix (Todd *et al.*, 1988) or bacterial proteins that cross-react with cartilage components (van der Broek *et al.*, 1988). Increased levels of IL6 which are found in synovial fluid from patients with rheumatoid but not osteoarthritis (Houssiau *et al.*, 1988) may promote local differentiation of B lymphocytes (Kishimoto and Hirano, 1988) and plasma cells secreting antibody against type II collagen have been identified in rheumatoid joints (Mestecky, 1975). Animal models of antigen induced arthritis using homologous or heterologous antigens closely mimic some aspects of the human disease and can be characterized by extensive cartilage degradation (Trentham, 1977; Glant, 1982; Kresina *et al.*, 1988; Poole *et al.*, 1988). However, depletion due to necrosis of the chondrocytes is not uncommon (Howson *et al.*, 1986). In one report of such models (Pettipher and Henderson, 1987) the depletion of cartilage proteoglycan persisted only in animals with active cell-mediated immunity. It is not known whether the depletion was caused by degradation of proteoglycan or inhibition of synthesis. Chondrocytes exposed to interferon γ also express class II MHC antigens, but it is not clear whether such cells can activate T lymphocytes (Goldring *et al.*, 1986).

In addition to acting as an autoantigen, type II collagen and type IX collagen have been shown to induce IL1 and PGE_2 production by cultured monocytes (Dayer and Seckinger, 1988).

Osteoarthritis

Osteoarthritis results from a failure of the chondrocyte to maintain the balance between synthesis and degradation of the extracellular matrix. Inflammation, when present, is thought to be a late and transient event arising probably from products of cartilage degradation inducing secondary synovitis in the adjacent tissue (Boniface *et al.*, 1988).

Treatment with traditional anti-inflammatory drugs fails to halt cartilage degradation and often contributes to its deterioration (Brandt, 1985; Hess and Herman, 1986; Burkhardt and Ghosh, 1987). Changes in the morphology and properties of osteoarthritic cartilage which have been noted as characteristic in both the human disease and in animal models have been extensively documented elsewhere (Collins, 1949; Ghadially, 1983; Gardner, 1983; Pond and Nuki, 1973; Moskowitz *et al.*, 1985; Muir, 1986). The molecular basis for these changes is not understood but aging, genetic, and environmental factors are significant variables.

Progressive but focal depletion of proteoglycan is a common feature, despite increased DNA and matrix synthesis, indicating that degradation of the matrix is markedly accelerated. Consistent with this is the identification of increased proteolytic activity in osteoarthritic cartilage by a number of investigators (Sapolsky *et al.*, 1973; Bayliss and Ali, 1978; Howell, 1984; Ehrlich, 1985; Pelletier *et al.*, 1983, 1985, 1988; Nojima *et al.*, 1986; Vittorio *et al.*, 1986; Dean *et al.*, 1987). A close correlation was found between enzyme activity and severity of the lesions. Increased levels of proteinases were also detected in inflamed synovium when this was present (Martel-Pelletier, 1987). These enzymes can degrade both proteoglycan and collagen as described earlier and, in contrast to rheumatoid arthritis, are probably derived as much if not more from the chondrocytes as from other sources. There are clearly also major changes in matrix synthesis with evidence of extensive remodeling. The identity of mediators involved in the initiation and/or progression of this disease has not been established. However, IL1 and TNF have been measured in the synovial fluid of patients with osteoarthritis even when there are no overt signs of local or systemic inflammation (di Giovine *et al.,* 1986). These cytokines could induce focal depletion of cartilage proteoglycan leading to a greater, possibly unbalanced, loading on the cells. This pressure, together with the increased penetration of mitogens and hormones into the tissue, may lead to the enhanced metabolic activity typical of early OA in animal models. Whether or not these changes progress to an irreversible degenerative condition would depend on whether the mechanical stability of the depleted joint had been significantly altered (see "Mechanical Loading", *supra*). Recent studies have shown that insulin-like growth factor 1 (IGF1) can enhance proteoglycan synthesis and dramatically decrease proteoglycan degradation in cultured cartilage exposed to IL1 and TNF (Tyler, 1989). In the final stages of OA there appears to be an as yet unidentified threshold of injury beyond which the ability of chondrocytes to replace the lost material deteriorates. The relative concentration of cytokines and growth factors and the intensity of loading at each stage may determine the rate of progress and final outcome of the disease.

Clinical Markers

The production of reliable diagnostic markers to detect cartilage degradation at an early stage prior to radiological thinning is a major goal of many laboratories. A number of investigators have utilized biochemical and immunological assays to measure both the protein and glycosaminoglycan component of proteoglycan fragments in serum and synovial fluid (see Lohmander, 1988; Risteli and Risteli, 1987). Recent investigations have also demonstrated immunolocalization of denatured collagen within sections of degraded cartilage (Dodge and Poole, 1989). Specific markers representative of the degree of cartilage erosion may eventually be available for use in long-term prospective studies. This is essential, if patient groups at risk of future cartilage breakdown are to be identified, and to test new drugs designed to preserve cartilage integrity.

Acknowledgments

I am extremely grateful to my parents and friends for their encouragement and support during the writing of this chapter. I would like to thank Eileen Surrey for her patient and expert typing of the manuscript and Chris Green for the illustrations. I am indebted to my colleagues at the Strangeways Research Laboratory for many interesting discussions on cartilage degradation and to the Medical Research Council, Arthritis and Rheumatism Council, and the Wellcome Trust for financial support.

References

S. Abe, M. Shinmei, and Y. Nagai. Synovial collagenase and joint disease: significance of latent collagenase with special reference to rheumatoid arthritis. *J. Biochem. (Tokyo)* **73**: 1007 (1973).

M.B. Aydelotte, R. Schleyerbach, B.J. Zeck, and K.E. Kuettner. Articular chondrocytes cultured in agarose gel for study of chondrocyte chondrolysis. In: *Articular Cartilage Biochemistry,* Eds K.E. Kuettner, R. Schleyerbach, and V.C. Hascall. Raven Press, New York. pp. 235–256, 1986.

D.L. Bader, G.E. Kempson, A.J. Barrett, and W. Webb. The effects of leucocyte elastase on the mechanical properties of adult human articular cartilage in tension. *Biochem. Biophys. Acta* **677**: 103–108 (1981).

A. Baici. Structure and breakdown of proteoglycans collagen and elastin. In: *Proteinases and Tumour Invasion,* Ed P. Strauli, Raven Press, New York, pp. 17–47, 1980.

A.J. Bailey and N.D. Light. Intermolecular cross-linking in fibrotic collagen. *Ciba Found. Symp.* **114** (Fibrosis): 80–96 (1985).

A.J. Bailey, S.P. Robbins, and G. Balian. Biological significance of the intermolecular crosslinks of collagen. *Nature (London)* **251**: 105–109 (1974).

J. Baker and B. Caterson. The purification and cyanogen bromide cleavage of the ''link proteins'' from cartilage proteoglycans. *Biochem. Biophys. Res. Commun.* **77**: 1–10 (1977).

P. Barland, A.B. Novikoff, and D. Hammerman. Lysosomes in the synovial membrane in rheumatoid arthritis: a mechanism for cartilage erosion. *Trans. Assoc. Am. Physicians* **77**: 239–247 (1964).

A.J. Barrett. Properties of lysosomal enzymes. In: *Lysosomes in Biology and Pathology,* Eds J.T. Dingle and H.B. Fell. North Holland, Amsterdam. pp. 245–312, 1969.

A.J. Barrett. The cystatins: a new class of peptidase inhibitors. *TIBS* **12**: 193–196 (1987).

A.J. Barrett. The possible role of neutrophil proteinases in damage to articular cartilage. *Agents Actions* **8**: 11–18 (1978).

A.J. Barrett. Which enzymes degrade cartilage matrix? In: *Cellular Interactions,* Eds J.T. Dingle and J.L. Gordon. Elsevier/North-Holland, Amsterdam. pp. 185–194, 1981.

A.J. Barrett, N.D. Rawlings, M.E. Davies, W. Machleidt, G. Salvesen, and V. Turk. Cysteine proteinase inhibitors of the cystatin superfamily. In: *Proteinase Inhibitors,* Eds A.J. Barrett and G. Salvesen. Elsevier, New York, pp. 515–570, 1986.

A.J. Barrett and P.M. Starkey. The interaction of α2-macroglobulin with proteinases. *Biochem. J.* **133**: 709–724 (1973).

A.J. Barrett and J. Saklatvala. Proteinases in joint disease. In: *Textbook of Rheumatology,* Eds W.N. Kelley, E.D. Harris, and S. Ruddy. W.B. Saunders, Philadelphia. pp. 195–209, 1981.

B.J. Baum, J. Moss, S.D. Breul, R.A. Berg, and R.G. Crystal. Effect of cyclic AMP on the intracellular degradation of newly synthesized collagen. *J. Biol. Chem.* **255**: 2843–2847 (1980).

M.T. Bayliss. Proteoglycan structure in normal and osteoarthritic human cartilage. In: *Articular Cartilage Biochemistry,* Eds K.E. Kuettner, R. Schleyerbach, and V.C. Hascall. Raven Press, New York. pp. 295–308, 1986.

M. Bayliss and S.Y. Ali. Isolation of proteoglycans from human articular cartilage. *Biochem. J.* **169**: 123–132 (1978).

M.T. Bayliss and S.Y. Ali. Studies on cathepsin B in human articular cartilage. *Biochem. J.* **171**: 149–154 (1978).

H.P. Benton, T.R. Jackson, and M.R. Hanley. Identification of a novel inflammatory stimulant of chondrocytes: early events in cell activation by bradykinin receptors on pig articular chondrocytes. *Biochem. J.* **258**: 861–867 (1989).

H.P. Benton and J.A. Tyler. Inhibition of cartilage proteoglycan synthesis by interleukin 1. *Biochem. Biophys. Res. Commun.* **154**: 421–428 (1988).

R.A. Berg and D.J. Prockop. The thermal transition of a nonhydroxylated form of collagen. Evidence for a role for hydroxyproline in stabilizing the triple-helix of collagen. *Biochem. Biophys. Res. Commun.* **52**: 115–120 (1973).

R.A. Berg, M.L. Schwartz, and R.S. Crystal. Regulation of the production of secretory proteins: intracellular degradation of newly synthesized defective collagen. *Proc. Natl. Acad. Sci. U.S.A.* **77**: 4746–4750 (1980).

R.S. Bienkowski, M.J. Cowan, J.A. McDonald, and R.G. Crystal. Degradation of newly synthesized collagen. *J. Biol. Chem.* **253**: 4356–4363 (1978).

T.A. Bird and J. Saklatvala. Identification of a common class of high affinity receptors for both types of porcine interleukin 1 on connective tissues. *Nature (London),* **324**: 263–265 (1986).

R.J. Boniface, P.R. Cain, and C.H. Evans. Articular responses to purified cartilage proteoglycans. *Arthritis Rheum.* **31**: 258–266 (1988).

F. Bonnet, J.-P. Perin, and P. Jollers. Isolation and chemical characterisation of two distinct "link proteins" from bovine nasal cartilage proteoglycan complex. *Biochim. Biophys. Acta* **532**: 242–248 (1978).

K. Brandt. Drug induced changes in cartilage. Do NSAIDS influence the outcome of degenerative joint disease? In: *Degenerative Joints,* Vol. 2, Eds G. Verbruggen and E.M. Veys. Elsevier, Amsterdam. pp. 315–320, 1985.

C.E. Brinkerhoff, C.E. Vater, and E.D. Harris. Effect of retinoids on rabbit synovial fibroblasts and chondrocytes. In: *Cellular Interactions,* Eds J.J. Dingle and J.L. Gordon. Elsevier/North-Holland, Amsterdam. pp. 215–230, 1981.

N.D. Broom. Connective tissue function and malfunction: a biomechanical perspective. *Pathology* **20**: 93–104 (1988).

C.C. Brown, R.M. Hembry, and J.J. Reynolds. Immunolocalization of metalloproteinases and their inhibitor (TIMP) in the rabbit growth plate. *J. Bone Jt. Surg. Am.* **71A** 580–593 (1989).

P. Bruckner, P. Bachinger, R. Timple, and J. Engel. Three conformationally distinct domains in the amino-terminal segment of type II procollagen and its rapid triple helix coil transition. *Eur. J. Biochem.* **90**: 595–603 (1978).

J.A. Buckwalter and L.C. Rosenburg. Structural changes during development in bovine fetal epiphyseal cartilage. *Collagen Relat. Res.* **3**: 489–504 (1983).

J.A. Buckwalter, L. Rosenburg, and L. Tang. The effect of link protein on proteoglycan aggregate structure. *J. Biochem.* **259**: 5361–5363 (1984).

R.A.D. Bunning, A. Crawford, H.J. Richardson, G. Opdenakker, J. van Damme, and R.G.G. Russell. IL1 preferentially stimulates the production of tissue-type plasminogen activator by human articular chondrocytes. *Biochim. Biophys. Acta* **924**: 473–482 (1987).

R.A.D. Bunning, G. Murphy, S. Kumar, P. Phillips, and J.J. Reynolds. Metalloproteinase inhibitors from bovine cartilage and body fluids. *Eur. J. Biochem.* **139**: 75–80 (1984).

D. Burkhardt and P. Ghosh. Laboratory evaluation of antiarthritic drugs as potential chondroprotective agent. *Semin. Arthritis Rheum.* **17**: 3–34 (1987).

M.C. Burleigh. Degradation of collagenase by non-specific proteinase. In: *Proteinases in Mammalian Cells and Tissues*, Ed A.J. Barrett. North-Holland, Amsterdam. pp. 285–310, 1977.

M.C. Burleigh, A.J. Barrett, and G.S. Lazarus. Cathepsin B1 A lysosomal enzyme that degrades native collagen. *Biochem. J.* **137**: 387–398 (1974).

E.J. Campbell and M.A. Campbell. Pericellular proteolysis by neutrophils in the presence of proteinase inhibitors: effects of substrate opsinization. *J. Cell Biol.* **106**: 667–679 (1988).

I.K. Campbell and P.J. Roughley. The action of human articular cartilage metalloproteinase on proteoglycan and link protein. *Biochem. J.* **237**: 117–122 (1986).

G.B. Caputo, D.K. Maccallum, J.H. Kimura, J. Schrode, and V.C. Hascall. Characterisation of fragments produced by clostripain digestion of proteoglycans from the Swarm rat chondrosarcoma. *Arch. Biochem. Biophys.* **204**: 220–233 (1980).

C.B. Caputo, L.A. Sygowski, D.J. Wolanin, S.P. Patton, R.G. Caccese, A. Shaw, R.A. Roberts, and G. Di Pasquale. Effects of synthetic metalloproteinase inhibitors on cartilage autolysis *in vitro*. *J. Pharmacol. Exp. Ther.* **240**: 460–465 (1987a).

C.B. Caputo, D.J. Wolanin, R.A. Roberts, L.A. Sygowski, S.P. Patlin, R.G. Cacese, A. Shaw, and G. Di Pasquale. Proteoglycan degradation by a chondrocyte metalloprotease. *Biochem. Pharm.* **36**: 995–1002 (1987b).

B. Caterson and D.A. Lowther. Changes in the metabolism of the proteoglycans from sheep articular cartilage in response to mechanical stress. *Biochim. Biophys. Acta* **540**: 412–422 (1978).

T.E. Cawston. Metalloproteinase inhibitors. In: *Proteinase Inhibitors,* Eds A.J. Barrett and G. Salvesen. Elsevier, New York, 589–606, 1986.

T.E. Cawston, E. Mercer, M. de Silva, and B. Hazleman. Metalloproteinases and collagenase inhibitors in rheumatoid synovial fluid. *Arthritis Rheum.* **27**: 285–290 (1984).

O.D. Chrisman, I.M. Ladenbauer-Bellis, M. Paryabi, and S. Goeltz. The relationship of mechanical trauma and the early biochemical reactions of osteoarthritic cartilage. *Clin. Orthop.* **161**: 275–284 (1981).

S.C. Clarke and R. Kamen. The human hematopoietic colony-stimulating factors. *Science* **236**: 1229–1237 (1987).

I. Clemmenson, R. Donde, and R.B. Andersen. The primary plasmin inhibitor in rheumatoid synovial fluid. *Arthritis Rheum.* **20**: 1354–1358 (1977).

S. Collier and P. Ghosh. The role of plasminogen in interleukin 1 mediated cartilage degradation. *J. Rheumatol.* **15**: 1129–1137 (1988).

D.H. Collins. *The Pathology of Articular and Spinal Diseases,* Edward Arnold, London. 1949.

T.D.V. Cooke, M. Sumi, and M. Maeda. Deleterious interactions of immune complexes in cartilage of experimental immune arthritis. I. The erosion of pannus-free hyaline cartilage. *Clin. Orthop. Relat. Res.* **193**: 235–245 (1984).

G. David and M.R. Bernfield. Collagen reduces glycosaminoglycan degradation by cultured mammary epithelial cells: possible mechanism for basal lamina formation. *Proc. Natl. Acad. Sci. U.S.A.* **76**: 786–790 (1979).

J.-M. Dayer and P. Seckinger. IL1 and IL1 inhibitors and their relation to connective tissue destruction. In: *The Control of Tissue Damage,* Ed A.M. Glauert. Elsevier, Amsterdam. pp. 151–163, 1988.

D.D. Dean, W. Azzo, J. Martel-Pelletier, J.-P. Pelletier, and J.F. Woessner, Jr. Levels of metalloproteinases and tissue inhibitor of metalloproteinases in human osteoarthritic cartilage. *J. Rheumatol.,* **14**: 43–48 (1987).

D.D. Dean and J.F. Woessner. Extracts of human cartilage contain an inhibitor of tissue metalloproteinases. *Biochem. J.* **218**: 277–280 (1984).

K. Deshmukh-Phadke, S. Nanda, and K. Lee. Macrophage factor that induces neutral protease secretion by normal rabbit chondrocytes. *Eur. J. Biochem.* **104**: 175–180 (1980).

F. di Giovine, J. Manson, G. Nuki, and G. Duff. The presence of tumour necrosis factor activity in synovial exudates fluids from patients with OA and RA. *Br. J. Rheumatol.* **25**: 5108 (1986).

J.T. Dingle. The extracellular secretion of lysosomal enzymes. In: *Lysosomes in Biology & Pathology,* Eds J.T. Dingle and H.B. Fell. North-Holland, Amsterdam. pp. 421–436, 1969.

J.T. Dingle. The role of lysosomal enzymes in skeletal tissues. *J. Bone Jt. Surg.* **55B**: 87–95 (1973).

J.T. Dingle and G. Knight. The role of the chondrocyte microenvironment in the degradation of cartilage matrix. In: *Degenerative Joints, Vol 2,* Eds G. Verbruggen and E.M. Veys. Elsevier, New York, pp. 69–85, 1985.

J.T. Dingle and J.A. Tyler. Role of intercellular messengers in the control of cartilage matrix dynamics. In: *Articular Cartilage Biochemistry,* Eds K.E. Kuettner, R. Schleyerbach, V.C. Hascall. Raven Press, New York. pp. 181–191, 1986.

J.T. Dingle, J. Saklatvala, R.M. Hembry, J.A. Tyler, H.B. Fell, and R. Jubb. A cartilage catabolic factor from synovium. *Biochem. J.* **184**: 177–180 (1979).

G. Di Pasquale, R. Caccese, R. Pasternak, J. Conaty, S. Hubbs, and K. Perry. Proteoglycan- and collagen-degrading enzymes from human interleukin 1-stimulated chondrocytes from several species: proteoglycanase and collagenase inhibitors as potentially new disease-modifying anti-arthritic agents. *Proc. Soc. Exp. Biol. Med.* **183**: 262–267 (1986).

G.D. Di Pasquale, K.W. Perry, D. Dea, and R. Caccese. Comparison of spontaneous and retinoic acid stimulated rabbit articular cartilage degradation *in vitro. Int. J. Tissue React.* **7**: 397–404 (1985).

G.R. Dodge and A.R. Poole. Immunohistochemical detection and immunochemical analysis of type II collagen degradation in human normal, rheumatoid and osteoarthritic articular cartilages and in explants of bovine articular cartilage cultured with interleukin 1. *J. Clin. Invest.* **83**: 647–661 (1989).

K. Doege, M. Sasaki, E. Horigan, J.R. Hassell, and Y. Yamada. Complete primary structure of the rat cartilage proteoglycan core protein deduced from cDNA clones, *J. Biol. Chem.* **262**: 17757–17767 (1987).

J.A. Eastgate, N.C. Wood, F.S. diGiovine, J.A. Symons, F.M. Grinlinton, and G.W. Duff. Correlation of plasma Il1 levels with disease activity, *Lancet*, Sept. 24, 706–709 (1988).

D.R. Edwards, G. Murphy, J.J. Reynolds, S.E. Whitman, A.J.P. Docherty, P. Angel, and J.K. Heath. Transforming growth factor β modulates the expression of collagenase and metalloproteinase inhibitor. *EMBO* **6**: 1899–1904 (1987).

Y. Eeckhout, J.-M. Delaisse, P. Ledent, and G. Vaes. The proteinases of bone resorption. In: *The Control of Tissue Damage,* Ed A. Glauert. Elsevier, pp. 297–313, 1988.

M. Ehrlich. Degradative enzyme systems in osteoarthritic cartilage. *J. Orthop. Res.* **3**: 170–184 (1985).

P. Evans and D.J. Etherington. The substrate specificity of human placental collagenolytic cathepsin. *Connect. Tissue Res.* **6**: 126 (1978).

V. Everts, W. Beertsen, and R. Schrëder. Effects of the proteinase inhibitors leupeptin and E-64 on osteoclastic bone resorption. *Calcif. Tissue Int.* **43**: 172–178 (1988).

D. Eyre and H. Oguchi. The hydroxypyridinium cross-links of skeletal collagen: their measurement, properties and a proposed pathway of formation. *Biochem. Biophys. Res. Commun.* **92**: 403–410 (1980).

D.R. Eyre, J.J. Wu, and D.E. Woolley. All three chains of 1α, 2α, 3α collagen from hyaline cartilage resist human collagenase. *Biochem. Biophys. Res. Commun.* **118**: 724–729 (1984).

H.G. Fassbender. Joint destruction in various arthritic diseases. In: *Articular Cartilage Biochemistry*, Eds K.E. Kuettner, R. Schleyerbach, and V.C. Hascall. Raven Press, New York, pp. 371–391, 1986.

A.J. Fosang, J.A. Tyler, and T.E. Hardingham. Effect of IL1 and IgF1 on the release of proteoglycan components and hyaluronan from pig articular cartilage in explant culture. *Matrix* **11** (1991).

A. Franzen, S. Bjornsson, and D. Heinegard. Cartilage proteoglycan aggregate formation. *Biochem. J.* **197**: 669–674 (1981).

I.S. Fulcher, R. Matsas, A.J. Turner, and A.J. Kenny. Kidney neutral endopeptidase and the hydrolysis of enkephalin by synaptic membranes show similar sensitivity to inhibitors. *Biochem. J.* **203**: 519–522 (1982).

J.P. Fulkerson, D.P. Williams, and O.D. Christman. Studies of adult and young dog articular depletion by prostaglandin E in tissue culture. *Trans. Orthop. Res. Soc.* **6**: 282 (1981).

S. Gadher, D.R. Eyre, V.C. Duance, S.F. Wotton, L.W. Heck, T.M. Schmid, and D.E. Woolley. Susceptibility of cartilage collagens type II, IX, X and XI to human synovial collagenase and neutrophil elastase. *Eur. J. Biochem.* **175**: 1–7 (1988).

D.L. Gardner. The nature and causes of osteoarthroses. *Br. Med. J.* **286**: 418–424 (1983).

J. Gauldie. Interleukin 6 in the inflammatory response. In: *The Therapeutic Control of Inflammatory Diseases,* Eds A.J. Lewis, N.F. Doherty, and N.R. Ackerman. Elsevier, Amsterdam, pp. 38–46 (1989).

W. Gebhard, H. Tschesche, and H. Fritz. Biochemistry of aprotinin and aprotinin-like inhibitors. In: *Proteinase Inhibitors,* Eds A.J. Barrett and G. Salveson. Elsevier, New York, pp. 375–388, 1986.

F.N. Ghadially. Ageing and osteoarthritic changes in articular cartilage. In: *Fine Structure of the Synovial Joints,* Ed F.N. Ghadially. Butterworths, London, 1983.

T. Glant. Induction of cartilage degradation in experimental arthritis produced by allogenic and xenogenic proteoglycan antigens. *Connect. Tissue Res.* **9**: 137–144 (1982).

T.T. Glant, K. Mikecz, P.J. Roughley, E. Buzas, and R. Poole. Age-related changes in protein-related epitopes of human articular-cartilage proteoglycans. *Biochem. J.* **236**: 71–75 (1986).

M.B. Goldring, L.J. Sandell, M.L. Stephenson, and S.M. Krane. Immune interferon suppresses levels of procollagen mRNA and type II collagen synthesis in cultured human articular and costal chondrocytes. *J. Biol. Chem.* **261**: 9049–9056 (1986).

M. Gowen, D.D. Wood, E.J. Ihrie, J.E. Meats, and R.G.G. Russel. Stimulation by human IL1 of cartilage breakdown and production of collagenase and proteoglycanase by human chondrocytes but not by human osteoblasts *in vitro. Biochem. Biophys. Acta* **797**: 186–193 (1984).

G.A. Grant, A.Z. Eisen, B.L. Marmer, W.T. Roswit, and G.I. Goldberg. The activation of human skin fibroblast procollagenase. Sequence identification of the major conversion products. *J. Biol. Chem.* **262**: 5886–5889 (1987).

M.E. Grant, S. Ayad, A.P.L. Kwan, G.P. Bates, J.T. Thomas, and J. McClure. The structure and synthesis of cartilage collagens. In: *The Control of Tissue Damage,* Ed A. Glauert. Elsevier, New York, pp. 13–28, 1988.

R.A. Greenwald, W.W. May, and D. Lazarus. Degradation of cartilage proteoglycans by superoxide radicals. *Arthritis Rheum.* **19**: 799 (1976).

J. Gross, J.H. Highberger, D. Johnson-Wint, and D. Biswas. Mode of action and regulation of tissue collagenase. In: *Collagenase in Normal and Pathological Connective Tissues,* Eds D.E. Woolley and J.M. Evanson. Wiley, Chichester. pp. 11–35, 1980.

J. Gross, J.H. Highberger, and F.O. Schmitt. Collagen structures considered as states of aggregation of a kinetic unit: the tropocollagen particle. *Proc. Natl. Acad. Sci. U.S.A.* **40**: 679 (1974).

J. Gross and Y. Nagai. Specific degradation of the collagen molecular by tadpole collagenolytic enzyme. *Proc. Natl. Acad. Sci. U.S.A.* **54**: 1197–1204 (1965).

B.L. Gruber, L.B. Schwartz, N.S. Ramamurthyj, A.-M. Irani, and M.J. Marchese. Activation of latent rheumatoid synovial collagenase by human mast cell tryptase. *J. Immunol.* **140**: 3936–3942 (1988).

M.D. Grynpas, D.R. Eyre, and D.A. Kirschner. Collagen type II differs from type I in native molecular packing. *Biochem. Biophys. Acta* **626**: 346–355 (1980).

T.E. Hardingham. The role of link-protein in the structure of cartilage proteoglycan aggregates. *Biochem. J.* **177**: 237–247 (1979).

T.E. Hardingham, M. Beardmore-Grey, D.G. Dunham, and A. Ratcliffe. Cartilage proteoglycans. *Ciba Found. Symp.* **124**: 30–46 (1986).

T.E. Hardingham, H. Muir, M.K. Kwan, W.M. Lai, and V.C. Mow. Viscoelastic properties of proteoglycan solutions with varying proportions present as aggregates. *J. Orthop. Res.* **5**: 36–46 (1987).

E.D. Harris, Jr. Pathogenesis of rheumatoid arthritis. In: *Textbook of Rheumatology,* Ed W.N. Kelley, E.D. Harris, and S. Ruddy. W.B. Saunders, Philadelphia. pp. 896–927, 1981.

E.D. Harris and E.C. Cartwright. Mammalian collagenases. In: *Proteinases in Mammalian Cells and Tissues.* Ed A.J. Barrett. North-Holland, Amsterdam. pp. 249–283, 1977.

E.D. Harris, Jr., J.M. Evanson, D.R. Di Bona, and S.M. Krane. Collagenase and rheumatoid arthritis. *Arthritis Rheum.* **13**: 83 (1970).

E.D. Harris, Jr., A.M. Glauert, and A.H.G. Murley. Intracellular collagen fibres at the pannus-cartilage junction in rheumatoid arthritis. *Arthritis Rheum.* **20**: 657–665 (1977).

V.C. Hascall. Introduction. In: *Functions of the Proteoglycans.* Ciba Foundation Symposium 124, 1–8, 1986.

V.C. Hascall. Interaction of cartilage proteoglycans with hyaluronic acid. *J. Supramol. Struct.* **7**: 101–120 (1977).

H. Heidtmann and J. Travis. Human alpha-1-proteinase inhibitor. In: *Proteinase Inhibitors*, Eds A.J. Barrett and G. Salvezen. Elsevier, Amsterdam. pp. 441–456, 1986.

D. Heinegard, A. Franzén, E. Hedborn, and Y. Sommarin. Common structures of the core proteins of interstitial proteoglycans. In: *Functions of the Proteoglycans.* Ciba Foundation Symposium 124, Wiley, Chicester, 69–88, 1986.

D. Heinegard and V.C. Hascall. Aggregation of cartilage proteoglycans. *J. Biol. Chem.* **249**: 4250–4256 (1974).

D. Heinegard, J. Wieslander, J. Sheehan, M. Paulsson, and Y. Summarin. Separation and characterization of two populations of aggregating proteoglycans from cartilage. *Biochem. J.* **225**: 95–106 (1985).

H.J. Helminen, J. Jurvelin, I. Kiviranta, K. Paukkonen, A.-M. Saamanen, and M. Tammi. Joint loading effects on articular cartilage. A historical review. In: *Joint Loading, Biology and Health of Articular Structures,* Ed H.J. Helminen. John Wright & Sons, Bristol, pp. 1–46, 1987.

B. Henderson. Alteration in the activity of lysosomal hydrolases in the synovial lining cell population of the knee joint of the rabbit during the development of chronic arthritis. *Pathol. Res. Pract.* **172**: 363–371 (1981).

B.J. Henderson and E.R. Pettipher. The synovial lining cell: biology and pathobiology. *Semin. Arthritis Rheum.* **15**: 1–32 (1985).

E.V. Hess and J.H. Herman. Cartilage metabolism and anti-inflammatory drugs in osteoarthritis. *Am. J. Med.* **81**: 36–43 (1986).

M.W.A. Holmes, M.T. Bayliss, and H. Muir. Hyaluronic acid in human articular cartilage. Age-related changes in content and size. *Biochem. J.* **250**: 435–441 (1988).

S.J. Hopkins, M. Humphreys, and M.I.V. Jayson. Cytokines in synovial fluid. I. The presence of biologically active and immunoreactive IL1. *Clin. Exp. Immunol.* **72**: 422–427 (1988).

R. Horuk, J.J. Huang, M. Covington, and R.C. Newton. A biochemical and kinetic analysis of the interleukin-1 receptor. *J. Biol. Chem.* **262**: 16275–16278 (1987).

F.A. Houssiau, J.P. Devogelaer, and J. van Damme. Interleukin 6 in synovial fluid and serum of patients with rheumatoid arthritis and other inflammatory arthritides. *Arthritis Rheum.* **31**: 784–788 (1988).

D.S. Howell. In: *Osteoarthritis—Diagnosis and Management,* Eds R.W. Moskowitz, D.S. Howell, V.M. Goldberg, and H.J. Mankin. W.B. Saunders, Philadelphia. pp. 129–146, 1984.

P. Howson, N. Shepard, and N. Mitchel. The antigen induced arthritis model: the relevance of the method of induction to its use as a model of human disease. *J. Rheumatol.* **132**: 379–390 (1986).

J.S. Huang, S.S. Huang, and T.F. Deuel. Specific covalent binding of platelet-derived growth factor to human plasma α2-macroglobulin. *Proc. Natl. Acad. Sci. U.S.A.* **81**: 342–346 (1984).

S.S. Huang, P. O'Grady, and J.S. Huang. Human transforming growth factor β α2-macroglobulin is a latent form of TGFβ. *J. Biol. Chem.* **263**: 1535–1541 (1988).

J.R. Hubbard, D.R. Mattmueller, J.J. Steinberg, D.P. Poppas, and C.B. Sledge. Effect of steroid hormones on endotoxin cartilage degradation. *Mol. Cell. Biochem.* **79**: 31–37 (1988).

J.R. Hubbard, J.J. Steinberg, M.S. Bednar, and C.B. Sledge. Effect of purified human interleukin 1 in cartilage degradation. *J. Orthop. Res.* **6**: 180–187 (1988).

I.M. Hunneyball, M.J. Crossley, and A. Crawford. Pharmacological manipulation of IL1 production and action and its consequences for the control of connective tissue damage. In: *The Control of Tissue Damage,* Ed A.M. Glauert. Elsevier, Amsterdam, pp. 165–178, 1988.

S. Inerot, D. Heinegard, L. Andell, and S.-E. Olsson. Articular cartilage proteoglycans in ageing and osteoarthritis. *Biochem. J.* **169**: 143–156 (1978).

A. Janoff, G. Feinstein, C.J. Malemud, and J.M. Elias. Degradation of cartilage proteoglycan by human leucocyte granule neutral proteases—a model of joint injury 1. Pentration of enzyme into rabbit articular cartilage and release of $^{35}SO_4$-labelled material from the tissue. *J. Clin. Invest.* **57**: 615–624 (1976).

H.E. Jasin. Bacterial lipopolysaccharides induce *in vitro* degradation of cartilage matrix through chondrocyte activation. *J. Clin. Invest.* **72**: 2014–2019 (1984).

H.E. Jasin. Factors controlling articular cartilage degradation by activation of chondrocytes. In: *Advances in Inflammation Research,* 5, Ed G.A. Weissman. Raven Press, New York. pp. 87–105, 1983.

S. Jimenez, M. Harsch, and J. Rosenbloom. Hydroxyproline stabilizes the triple helix of chick tendon collagen. *Biochem. Biophys. Res. Commun.* **52**: 106–114 (1973).

D.A. Kalbhen. Degenerative joint disease following chemical injury—chemically induced osteoarthrosis. In: *Degenerative Joints* Vol 2, Eds G. Verbruggen and E.M. Veys. Elsevier, New York, pp. 299–311, 1985.

H. Keiser, R.A. Greenwald, G. Feinstein, and A. Janoff. Degradation of cartilage proteoglycan by human leukocyte granule neutral proteases—a model of joint injury. *J. Clin. Invest.* **57**: 625–632 (1976).

G.E. Kempson. The effects of proteoglycan and collagen degradation on the mechanical properties of adult human articular cartilage. In: *Dynamics of Connective Tissue Macromolecules,* Eds P.M.C. Burleigh and A.R. Poole. North-Holland/Elsevier, Amsterdam, pp. 277–305, 1975.

G.E. Kempson, D.A. Tuke, J.T. Dingle, A.J. Barrett, and P.H. Horsefield. The effects of proteolytic enzymes on the mechanical properties of adult human articular cartilage. *Biochim. Biophys. Acta* **428**: 741–760 (1976).

J.J. Killackey, P.J. Roughley, and J.S. Mort. Proteinase inhibitors of human articular cartilage. *Collagen Relat. Res.* **3**: 419–430 (1983).

T. Kishimoto and T. Hirano. Molecular regulation of β lymphocyte response. *Annu. Rev. Immunol.* **6**: 485–512 (1988).

I. Kiviranta, J. Jurvelin, M. Tammi, A.-M. Saamanen, and H.J. Helminen. Weight bearing controls glycosaminoglycan concentration and articular cartilage thickness in the knee joints of young beagle dogs. *Arthritis Rheum.* **30**: 801–809 (1987b).

J. Kiviranta, M. Tamimi, J. Jurvelin, and H.J. Helminen. Topographical variation of glycosaminoglycan content and cartilage thickness in canine knee joint cartilage. *J. Anat.* **150**: 265–276 (1987a).

K.I. Kivirikko and R. Myllyla. Biosynthesis of the collagens. In: *Extracellular Matrix Biochemistry,* Eds K.A. Piez and A.H. Reddi. Elsevier, New York, pp. 83–118, 1984.

K.I. Kivirikko and R. Myllyla. The hydroxylation of prolyl and lysyl residues. In: *The Enzymology of Post-Translational Modifications of Protein,* Vol. 1, Eds R.B. Freedman and H.C. Hauskino. Academic, London, pp. 53–104 1980.

J.A. Knight, R.W. Stephens, G.R. Bushell, P. Ghosh, and T.K.F. Taylor. Neutral protease inhibitors from human intervertebral disc and femoral head articular cartilage. *Biochim. Biophys. Acta* **584**: 304–310 (1979).

K.S. Kostensky and E.H. Olah. Effect of increased functional demand on the glycosaminoglycan content of the articular cartilage. *Acta Biol. Acad. Sci. Hung.* **23**: 75–82 (1972).

T. Krakauer, J.J. Oppenheim, and H.E. Jasin. Human interleukin 1 mediates cartilage matrix degradation. *Cell. Immunol.* **91**: 92–99 (1985).

S.M. Krane, S.R. Goldring, and J-M. Dayer. Interactions among lymphocytes, monocytes and other synovial cells in the rheumatoid synovium. In: *Lymphokines,* Vol 7, Eds E. Pick and M. Landy. Academic, Orlando. pp. 75–136, 1982.

T.F. Kresina, J.U. Yoo, and V.M. Goldberg. Evidence that a humoral immune response to autologous cartilage proteoglycan can participate in the induction of cartilage pathology. *Arthritis Rheum.* **31**: 248–257 (1988).

K.E. Kuettner, J. Hiti, R. Eienstein, and E. Harper. Collagenase inhibition of cationic proteins derived from cartilage and aorta. *Biochem. Biophys. Res. Commun.* **72**: 40–46 (1976).

J.W. Larrick and S.L. Kunkel. The role of tumour necrosis factor and interleukin 1 in the immunoinflammatory response. *Pharm. Res.* **5**: 129–139 (1988).

S.J. Leibovich and J.B. Weiss. Failure of human rheumatoid synovial collagenase to degrade either normal or rheumatoid arthritic polymeric collagen. *Biochim. Biophys. Acta* **25**: 109 (1971).

S. Lohmander. Proteoglycans of joint cartilage. Structure, function, turnover and role as markers of joint disease. *Ballière's Clin. Rheum.* **2**: 37–62 (1988).

A. Maroudas and C. Bannon. Measurement of swelling pressure in cartilage and comparison with the osmotic pressure of constituent proteoglycans. *Biorheology* **18**: 619–632 (1981).

J. Martel-Pelletier and J.-P. Pelletier. Neutral proteases in human osteoarthritic synovium: quantification and characterisation. *J. Rheumatol.* **14**: 38–43 (1987).

N.W. Mason, G.D.J. Green, and A.J. Barrett. Human liver cathepsin 1 *Biochem. J.* **226**: 233–241 (1985).

R.W. Mason. The role of cysteine proteinase in elastin degradation. In: *The Control of Tissue Damage,* Vol. 18, Ed A. Glauert. pp. 259–267, 1988.

K. Matsushima, T. Akahoshi, M. Yamada, Y. Furutani, and J.J. Oppenheim. Properties of a specific IL1 receptor on human Epstein Barr virus-transformed B lymphocytes: identity of the receptor for IL1 and IL1 β. *J. Immunol.* **136**: 4496–4501 (1986).

J.K. McDonald and C. Schwabe. Intracellular exopeptidases proteinases. In: *Mammalian Cells and Tissues,* Ed A.J. Barrett. Elsevier/North-Holland, Amsterdam, pp. 311–391, 1977.

M.B. McGuire-Goldring, J.E. Meats, D.D. Wood, E.J. Ihrie, N.M. Ebsworth, and R.G.G. Russell. *In vitro* activation of human chondrocytes and synoviocytes by a human IL1-like factor. *Arthritis Rheum.* **27**: 654–662 (1984).

M.B. McGuire-Goldring, G. Murphy, M. Gowen, J.E. Meats, N.M. Ebsworth, C. Poll, J.J. Reynolds, and R.G.G. Russell. Effects of retinol and dexamethasone on cytokine-mediated control of metalloproteinases and their inhibitors by human articular chondrocytes and synovial cells in culture. *Biochim. Biophys. Acta* **763**: 129–139 (1983).

B. McKibben and A. Maroudas. Nutrition and metabolism. In: *Adult Articular Cartilage,* Ed M.A.R. Freeman. Pitman Medical, Tunbridge Wells, pp. 461–486, 1979.

H. Meninger, R. Putzier, W. Mohr, D. Wessinghage, and K. Tillmann. Granulocyte elastase at the site of cartilage erosion by rheumatoid synovial tissue. *Z. Rheumatol.* **39**: 149–156 (1980).

J. Mestecky and E.J. Miller. Presence of antibodies specific to cartilage-type collagen in rheumatoid synovial tissue. *Clin. Exp. Immunol.* **22**: 453–456 (1975).

J.E. Michelsson and I.M. Hunneyball. Inflammatory involvement in rabbit knee following immobilization and resulting in osteoarthritis. *Scand. J. Rheumatol.* **13**: 273–281 (1984).

E.J. Miller. Chemistry of the collagens and their distribution. In: *Extracellular Matrix Biochemistry,* Eds K.A. Piez and A.H. Reddi. Elsevier, New York, pp. 41–81, 1984.

E.J. Miller, E.D. Harris, E. Chung, J.E. Finch, P.A. McCrosckery, and W.T. Butler. Cleavage of type II and III collagens with mammalian collagenase: site of cleavage and primary structure at the NH_2-terminal portion of the smaller fragment released from both collagens. *Biochemistry* **15**: 787–792 (1976).

T.I. Morales, L.M. Wahl, and V.C. Hascall. The effect of bacterial lipopolysaccharides on the biosynthesis and release of proteoglycans from calf articular cartilage cultures. *J. Biol. Chem.* **259**: 6720–6729 (1984).

R.I.G. Morrison. In: Ed E.A. Balazs. The Chemistry and Molecular Biology of the Intercellular Matrix. Academic Press, New York. pp. 1683–1691, 1970.

R.I.G. Morrison. A.J. Barrett, and D. Prior. Cathepsins B1 and D. Action on human cartilage proteoglycans. *Biochim. Biophys. Acta* **302**: 411–419 (1973).

J.S. Mort, B. Caterson, A.R. Poole, and P.J. Roughley. The origin of human cartilage proteoglycan link-protein heterogeneity and fragmentation during ageing. *Biochem. J.* **232**: 805–812 (1985).

J.S. Mort, A.R. Poole, and P.J. Roughley. Age-related changes in the structure of proteoglycan link proteins present in normal human articular cartilage. *Biochem. J.* **214**: 269–272 (1983).

J.S. Mort and A.D. Recklies. Inter-relationship of active and latent secreted human cathepsin β precursors. *Biochem. J.* **233**: 57–63 (1986).

J.S. Mort, A.D. Recklies, and A.R. Poole. Extracellular presence of the lysosomal proteinase cathepsin B in rheumatoid synovium and its activity at neutral pH. *Arthritis Rheum.* **27**: 509–515 (1984).

R.W. Moskowitz, D.S. Howell, V.M. Goldberg, and H.J. Mankin, Jr., Eds. *Osteoarthritis: Diagnosis & Management.* W.B. Saunders, Philadelphia, pp. 1–585, 1985.

H. Muir. Current and future trends in articular cartilage research and osteoarthritis. In: *Articular Cartilage Biochemistry,* Eds K.E. Kuettner, R. Schleyerbach, and V.C. Hascall. Raven Press, New York, pp. 423–441, 1986.

K.D. Muirden. Lysosomal enzymes in synovial membrane in rheumatoid arthritis. Relationship to joint damage. *Ann. Rheum. Dis.* **31**: 265–271 (1972).

G. Murphy and A.J.P. Docherty. Molecular studies in the connective tissue metalloproteinases and their inhibitors TIMP. In: *The Control of Tissue Damage,* Ed A. Glauert. **16**: 223–241 (1988).

G. Murphy, R.M. Hembry, and J.J. Reynolds. Characterization of a specific antiserum to rabbit stromelysin and demonstration of the synthesis of collagenase and stromelysin by stimulated rabbit articular chondrocytes. *Collagen Relat. Res.* **6**: 351–364 (1986).

G. Murphy and J.J. Reynolds. Current views of collagen degradation. Progress towards understanding the resorption of connective tissues. *Bio. Essays* **2**: 55–60 (1985).

G. Murphy, R.M. Hembry, C.E. Hughes, A.J. Fosang, and T.E. Hardingham. *Biochem. Soc. Trans* **18**: 812–815 (1990).

J.B. Murphy, K. Allison, J. Subhalter, and R. Lanzer. Purification and partial amino acid sequence of a bovine cartilage-derived collagenase inhibitor. *J. Biol. Chem.* **261**: 4154–4159 (1986).

Y. Naparstek, I.R. Cohen, Z. Fuks, and I. Vlodavsky. Activated T lymphocytes produce a matrix degrading heparan sulphate endoglycosidase. *Nature (London),* **310**: 241–244 (1984).

Q. Nguyen, G. Murphy, P.J. Roughley, and J.S. Mort. Proteoglycan aggregate degradation by a cartilage metalloproteinase. Evidence for the involvement of stromelysin in the generation of link protein heterogenecity *in situ. Biochem. J.* in press (1989).

T. Nojima, C.A. Towle, H.J. Mankin, and B.V. Treadwell. Secretion of higher levels of active proteoglycanases from human osteoarthritic chondrocytes. *Arthritis Rheum.* **29**: 292–295 (1986).

T.R. Oegema, Jr. and R.C. Thompson, Jr. Metabolism of chondrocytes derived from normal and osteoarthritic human cartilage. In: *Articular Cartilage Biochemistry,* Eds K.E. Kuettner, R. Schleyerbach, and V.C. Hascall. Raven Press, New York, pp. 257–270, 1986.

K. Ohlsson and I. Olsson. The neutral proteases of human granulocytes: isolation and partial characterization of two granulocyte collagenase. *Eur. J. Biochem.* **36**: 473 (1973).

N. Ohta, N. Kawai, W. Kawaji, and H. Hirano. Morphological changes in rabbit articular cartilages experimentally induced by joint contracture—in association with ageing. *Okajimas. Folia. Anat. Jpn.* **58**: 205–220 (1981).

Y. Okada, H. Nagase, and E.D. Harris. A metalloproteinase from human rheumatoid synovial fibroblasts that digests connective tissue matrix components. Purification and characterization. *J. Biol. Chem.* **261**: 14245–14255 (1986).

E.H. Olah and K.S. Kostensky. Effect of altered functional demand on the glycosaminoglycan content of the articular cartilage of dogs. *Acta Biol. Acad. Sci. Hung.* **23**: 195–200 (1972).

F. Olliviere, U. Gubler, C.A. Towle, C. Laurencin, and B.V. Treadwell. Expression of IL1 gene in human and bovine chondrocytes. A mechanism for autocrine control of cartilage matrix degradation. *Biochim. Biophys. Res. Commun.* **141**: 904–911 (1986).

G.M. Oosta, L.V. Favreau, D.L. Beeler, and R.D. Rosenberg. Purification and properties of human platelet heparitinase. *J. Biochem.* **257**: 11249–11255 (1982).

J.J. Oppenheim, E.J. Koracs, K. Matsushima, and S.K. Durum. There is more than one interleukin 1. *Immunol. Today* **7** (No. 2): 45–56 (1986).

A.L. Oronsky and C.W. Buermann. Phagocytic release of human leukocyte neutral proteases. Analysis of cartilage degradation products. *Arch. Biochem. Biophys.* **176**: 539–546 (1976).

M.J. Palmoski and K.D. Brandt. Running inhibits the reversal of atrophic changes in canine knee cartilage after removal of a leg cast. *Arthritis Rheum.* **24**: 1329–1337 (1981).

M.J. Palmoski and K.D. Brandt. Effects of static and cyclic compressive loading on articular cartilage plugs *in vitro*. *Arthritis Rheum.* **27**: 675–681 (1984).

M.J. Palmoski, R.A. Colyer, and K.D. Brandt. Joint motion in the absence of normal loading does not maintain normal articular cartilage. *Arthritis Rheum.* **23**: 325–334 (1980).

R.D. Pasternak, J. Hubbs, R.C. Caccese, R.L. Marks, J.M. Conaty, and G. Di Pasquale. Interleukin-1 stimulates the secretion of proteoglycan- and collagen-degrading proteases by rabbit articular chondrocytes. *Clin. Immunol. Immunopathol.* **41**: 351–367 (1986).

M. Paulssen, M. Morgelin, H. Wiedemann, M. Beardmore-Gray, D. Dunham, T. Hardingham, D. Heinegard, R. Timpl, and J. Engel. Extended and globular protein domains in cartilage proteoglycans. *J. Biochem.* **245**: 763–772 (1987).

J.-P. Pelletier and J. Martel-Pelletier. Cartilage degradation by neutral proteoglycanases in experimental osteoarthritis. *Arthritis Rheum.* **28**: 1393–1401 (1985).

J.P. Pelletier, J. Martel-Pelletier, and C.J. Malemud. Canine osteoarthritis: effects of endogenous neutral metalloproteoglycanases on articular cartilage proteoglycans. *J. Orthop. Res.* **6**: 379–388 (1988).

J.-P. Pelletier, J. Martel-Pelletier, D.S. Howell, L. Ghandur-Mhaymneh, J.E. Enis, and J.F. Woessner, Jr. Collagenase and collagenolytic activity in human osteoarthritic cartilage. *Arthritis Rheum.* **26**: 63–68 (1983).

R. Pérez-Tamayo. Pathology of collagen degradation. *Am. J. Pathol.* **92**: 508–566 (1978).

E.R. Pettipher and B. Henderson. The relationship between cell-mediated immunity and cartilage degradation in antigen-induced arthritis in the rabbit. *Br. J. Exp. Pathol.* **68**: 364–378 (1987).

E.R. Pettipher, G.A. Higgs, and B. Henderson. Interleukin 1 induces leukocyte infiltration and cartilage proteoglycan degradation in the synovial joint. *Proc. Natl. Acad. Sci. U.S.A.* **83**: 8749–8753 (1986).

K.A. Piez. Molecular and aggregate structures of the collagens. In: *Extracellular Matrix Biochemistry,* Eds K.A. Piez and A.H. Reddi. Elsevier, New York. pp. 1–39, 1984.

M.J. Pond and G. Nuki. Experimentally induced osteoarthritis in the dog. *Ann. Rheum. Dis.* **32**: 387–388 (1973).

A.R. Poole. Immunocytochemical studies of the secretion of a proteolytic enzyme cathepsin D in relation to cartilage breakdown. In: *Dynamics of Connective Tissue Macromolecules,* Eds P.M.C. Burleigh and A.R. Poole. North-Holland/Elsevier, Amsterdam. pp. 357–379, 1975.

A.R. Poole. Proteoglycans in health and disease: structures and functions. *Biochem. J.* **236**: 1–14 (1986).

A.R. Poole, T.T. Glant, and K. Mihecz. Autoimmunity to cartilage collagen and proteoglycan and the development of chronic inflammatory arthritis. In: *The Control of Tissue Damage,* Ed A. Glauert. pp. 55–65, 1988.

A.R. Poole, I. Pidoux, A. Reiner, and L. Rosenburg. An immunoelectron microscope study of the organisation of proteoglycan monomer, link protein, and collagen in the matrix of articular cartilage. *J. Cell Biol.* **93**: 921–937 (1982).

K.D. Rainsford. Preliminary investigations on the pharmacological control of catabolin-induced cartilage destruction *in vitro*. *Agents Actions*. **16**: 55–57 (1985).

G.N. Ramachandran and R. Chandrasekharan. Interchain hydrogen bonds via bound water molecules in the collagen triple helix. *Biopolymers* **6**: 1649–1658 (1968).

N.V. Rao and E. Adams. Collagen helix stabilization by hydroxyproline in $(Ala\text{-}Hyp\text{-}Gly)_n$. *Biochem. Biophys. Res. Commun.* **86**: 654–660 (1979).

A. Ratcliffe, J.A. Tyler, and T.E. Hardingham. Articular cartilage cultured with interleukin 1. Increased release of link protein hyaluronate-binding region and other proteoglycan fragments. *Biochem. J.* **228**: 571–580 (1986).

N.C. Rath, A.L. Oronsky, and S.S. Kerwar. Synthesis of IL1-like activity by normal rat chondrocytes in culture. *Clin. Immunol. Immunopathol.* **47**: 39–46 (1988).

J.J. Reynolds, C.E. Lawrence, and J. Gavrilovic. Model systems for studying the destruction of the extracellular matrix and their usefulness in testing inhibitors. In: *The Control of Tissue Damage,* Ed A. Glauert. pp. 281–296, 1988.

D.B. Rifkin and R.M. Crowe. Isolation of a protease inhibitor from tissues resistant to tumor invasion. *Hoppe-Seyler's Z. Physiol. Chem.* **358**: 1525–1531 (1977).

L. Risteli and J. Risteli. Analysis of extracellular matrix proteins in biological fluids. In: *Methods in Enzymology,* Vol. 145, Ed L.W. Cunningham. Academic Press, New York. pp. 391–411, 1987.

C.R. Roberts and R.T. Dean. Degradation of cartilage by macrophages in culture: evidence for the involvement of an enzyme which is associated with the cell surface. *Connect. Tissue Res.* **14**: 199–212 (1986).

D.B. Robertson and E.J. Miller. Cartilage collagen: inability to serve as a substrate for collagenase active against skin and bone collagen. *Biochim. Biophys. Acta* **289**: 247–250 (1972).

L. Rosenburg and J.A. Buckwalter. Cartilage proteoglycans. In: *Articular Cartilage Biochemistry,* Ed K.E. Kuettner, R. Schleyerbach, and V.C. Hascall. Raven Press, New York. pp. 39–54, 1986.

J. Rosenbloom, M. Harsch, and S. Jimenez. Hydroxyproline content determines the denaturation temperature of chick tendon collagen. *Arch. Biochem. Biophys.* **158**: 478–484 (1973).

P.J. Roughley. A comparative study of the glycosaminoglycan peptides obtained after degradation of cartilage proteoglycan by different proteinases and their use in the characterization of different proteoglycans. *Connect. Tissue Res.* **6**: 145–153 (1978).

P.J. Roughley and R.J. White. Age-related changes in the structure of the proteoglycan subunits from human articular cartilage. *J. Biochem.* **255**: 217–224 (1980).

P.J. Roughley, R.J. White, and A.R. Poole. Identification of a hyaluronic acid-binding protein that interferes with the preparation of high buoyant density proteoglycan aggregates from adult human cartilage. *Biochem. J.* **231**: 129–138 (1985).

G.C. Sahagain and M.M. Gottesman. The predominant secreted protein of transformed murine fibroblasts carries the lysomal mannose-6-phosphate recognition marker. *J. Biol. Chem.* **259**: 11145–11150 (1982).

J. Saklatvala. Tumour necrosis factor stimulates resorption and inhibits synthesis of proteoglycan in cartilage. *Nature (London).* **322**: 547–549 (1986).

J. Saklatvala, L.M.C. Pilsworth, S.J. Sarsfield, J. Gavrilovic, and J.K. Heath. Pig catabolin is a form of interleukin 1. *Biochem. J.* **224**: 461–466 (1984).

J. Saklatvala and S.J. Sarsfield. How do IL1 and TNF induce degradation of proteoglycan in cartilage? In: *The Control of Tissue Damage,* Ed A.M. Glauert. Elsevier, Amsterdam. pp. 97–108, 1988.

R. Sanchez-Lopez, R. Nicholson, M.-C. Gesnel, L.M. Matrisian, and R. Breathnach. Structure-function relationships in the collagenase family member transin. *J. Biol. Chem.* **263**: 11892–11899 (1988).

J. Schalkwijk, W.B. vanden Berg, L.B.A. van de Putte, and L.B.A. Joosten. Hydrogen peroxide suppresses the proteoglycan synthesis of intact articular cartilage. *J. Rheumatol.* **12**: 205–210 (1985).

R.K. Schenk. P.S. Eggli, and E.B. Hunziker. Articular cartilage morphology. In: *Articular Cartilage Biochemistry,* Ed K. Kuettner. Raven Press, New York, pp. 3–20, 1986.

J. Schnyder, T. Payne, and C.A. Dinarello. Human monocyte on recombinant IL1's are specific for the secretion of a metalloproteinase from chondrocytes. *J. Immunol.* **138**: 496–503 (1987).

P.B. Sehgal, L.T. May, I. Tamm, and J. Vilcek. Human β2 interferon and differentiation factor BSF-Z are identical. *Science* **235**: 731–732 (1987).

H. Sheppeard, L.M.C. Pilsworth, B. Hazleman, and J.T. Dingle. Effects of antirheumatic drugs on the production and action of porcine catabolin. *Ann. Rheum. Dis.* **41**: 463–468 (1982).

B. Sherry and A. Cerami. Cachectin/tumour necrosis factor exerts endocrine paracrine and autocrine control of inflammatory responses. *J. Cell Biol.* **107**: 1269–1277 (1988).

M. Shinmei, T. Kikuchi, K. Masuda, and Y. Shimomura. Effects of interleukin-1 and anti-inflammatory drugs on the degradation of human articular cartilage. *Drugs* **35**: 33–41 (1988).

J.E. Sims, C.J. March, D. Cosman, M.B. Widmer, H. Robson MacDonald, C.J. McMahan, C.E. Grubin, J.M. Wignall, J.L. Jackson, S.M. Call, D. Friend, A.R. Alpert, S. Gillis, D.L. Urdah, and S.K. Dower. cDNA expression cloning of the IL-1 receptor, a member of the immunoglobulin superfamily. *Science* **241**: 585–589 (1988).

J.W. Smith. Molecular pattern in native collagen. *Nature (London)* **219**: 157–158 (1968).

D. Somjen, I. Hinderman. A. Berger, and A. Harell. Bone remodelling induced by physical stress is prostaglandin E2 mediated. *Biochem. Biophys. Acta* **627**: 91–100 (1980).

S.C. Sood. A study of the effects of experimental immobilization on rabbit articular cartilage. *J. Anat.* **108**: 497–507 (1971).

M.B. Sporn, N.W. Dunlop, D.L. Newton, and W.R. Henderson. Relationships between structure and activity of retinoids. *Nature (London)* **263**: 110–113 (1976).

P.M. Starkey, A.J. Barrett, and M.C. Burleigh. The degradation of articular collagen by neutrophil proteinase. *Biochim. Biophys. Acta* **483**: 386–397 (1977).

J.J. Steinberg and C.B. Sledge. Synovial factors and chondrocyte-mediated breakdown of cartilage. Inhibition by hydrocortisone. *J. Orthop. Res.* **1**: 13–21 (1983).

F.S. Steven. Polymer collagen fibrils. An example of substrate mediated steric obstruction of enzymic digestion. *Biochim. Biophys. Acta* **452**: 151 (1976).

J.W. Stevens, Y. Oike, C. Handley, V.C. Hascall, A. Hampton, and B. Caterson. Characteristics of the core protein of the aggregating proteoglycan from the Swarm rat chondrosarcoma. *J. Cell Biochem.* **26**: 247–259 (1984).

M. Sumi, M. Maeda, and T.D.V. Cooke. Deleterious interactions of immune complexes with tibial cartilage of antigen-induced arthritic rabbits. *Clin. Orthop.* **212**: 260–274 (1986).

L. Tang, L. Rosenberg, A. Reiner, and A.R. Poole. Proteoglycans from bovine nasal cartilage. *J. Biol. Chem.* **254**: 10523–10531 (1979).

M.L. Tanzer. Cross-linking of collagen. *Science* **180**: 561–566 (1973).

A.H. Tashjian, Jr., E.F. Voelkel, M. Lazzaro, D. Grad, T. Bostia, and I. Levine, Tumour necrosis factor alpha (cachetin) stimulates bone resorption in mouse calvariae via a prostaglandin-mediated mechanism. *Endocrinology* **120**: 2029–2036 (1987).

D.J. Taylor, J.R. Yoffe, D.M. Brown, and D.E. Woolley. Histamine stimulates prostaglandin E production by rheumatoid synovial cells and human articular chondrocytes in culture. *Arthritis Rheum.* **29**: 160–165 (1986).

M. Teodorescu, J.L. Skosey, and C. Schlesinger. Covalent disulphide binding of IL1 to α2-macroglobulin. In: *Monokines and Other Nonlymphocytic Cytokines,* Ed M.C. Powanda. Alan R. Liss, New York. pp. 209–212, 1988.

E.J.-M.A. Thonar, J.A. Buckwalter, and K.E. Kuettner. Maturation-related differences in the structure and composition of proteoglycans synthesized by chondrocytes from bovine articular cartilage. *J. Biol. Chem.* **261**: 2467–2474 (1986).

E.J.-M.A. Thonar, M.E. Lenz, G.K. Klintworth, B. Caterson, L.M. Packman, P. Glickman, R. Katz, J. Huff, and K.E. Kuettner. Quantification of keratan sulfate in blood as a marker of cartilage catabolism. *Arthritis Rheum.* **28**: 1367–1376 (1985).

J.A. Todd, H. Acha-Orbea, J.I. Bell, N. Chao, Z. Fronek, C.O. Jacob, M. McDermott, A.A. Sinhaj, L. Timmerman, L. Steinman, and H.O. McDevitt. A molecular basis for MHC Class II associated autoimmunity. *Science* **240**: 1003–1009 (1988).

C.A. Towle, M.E. Trice, F. Ollivierre, B.J. Awbrey, and B.V. Treadwell. Regulation of cartilage remodeling by IL1: evidence for autocrine synthesis of IL1 by chondrocytes. *J. Rheumatol.* **14** (Suppl. 14): 11–13 (1987).

W. Traub and K.A. Piez. The chemistry and structure of collagen. *Adv. Protein Chem.* **25**: 243–252 (1971).

J. Travis and G.S. Salvesen. Human plasma proteinase inhibitors. *Am. J. Biochem.* **52**: 655–709 (1983).

B.V. Treadwell, J. Neidel, M. Pavia, C.A. Towle, M.E. Trices, and H.J. Mankin. Purification and characterisation of collagenase activator protein synthesized by articular cartilage. *Arch. Biochem. Biophys.* **251**: 715–723 (1986).

R.L. Trelstad and D.E. Birk. The fibroblast in morphogenesis and fibrosis: cell topography and surface-related functions. *Ciba Found. Symp.* **114**: 4–19 (1985).

D.E. Trentham, A.S. Townes, and A.H. Kang. Autoimmunity to type II collagen: an experimental model of arthritis. *J. Exp. Med.* **146**: 857–868 (1977).

H. Troyer. The effect of short-term immobilization on the rabbit knee joint cartilage. A histochemical study. *Clin. Orthop.* **107**: 249–257 (1975).

J.A. Tyler. Articular cartilage cultured with catabolin (pig interleukin 1) synthesizes a decreased number of normal proteoglycan molecules. *Biochem. J.* **227**: 869–878 (1985b).

J.A. Tyler. Chondrocyte-mediated depletion of articular cartilage proteoglycans *in vitro*. *Biochem. J.* **225**: 493–507 (1985a).

J.A. Tyler. Insulin-like growth factor 1 can decrease degradation and promote synthesis of proteoglycan in cartilage exposed to cytokines. *Biochem. J.*, **260**: 543–548 (1989).

J.A. Tyler. The influence of interleukin 1 and insulin like growth factor 1 on the integrity of cartilage matrix. In: *The Control of Tissue Damage*, Ed A.M. Glauert. Elsevier, Amsterdam. pp. 129–219, 1988.

J.A. Tyler and H.P. Benton. Synthesis of type II collagen is decreased in cartilage cultured with interleukin 1 while the rate of intracellular degradation remains unchanged. *Collagen Relat. Res.* **8**: 393–405 (1988).

J.A. Tyler and T.E. Cawston. Properties of pig synovial collagenase. *Biochem. J.* **189**: 349–357 (1979).

J.A. Tyler, H.B. Fell, and C.E. Lawrence. The effects of cortisol on porcine articular tissues. *J. Pathol.* **137**: 335–351 (1982).

J. Uitto, R.E. Allan, and K.L. Polak. Conversion of type II procollagen to collagen. Extracellular removal of the amino-terminal and carboxy-terminal extensions. *Eur. J. Biochem.* **99**: 97–103 (1979).

E.R. Unanue and P.M. Allen. The basis for the immunoregulatory role of macrophages and other accessory cells. *Science* **236**: 551–557 (1987).

J. Urban, A. Maroudas, M.T. Bayliss, and J. Dillon. Swelling pressure of proteoglycans at the concentrations found in cartilaginous tissues. *Biorheology* **16**: 447–464 (1979).

G. Vaes. Cellular biology and biochemical mechanisms of bone resorption. *Clin. Orthop.* **231**: 239–271 (1988).

M.F. van den Broek, W.B. van den Berg, O.J. Arntz, and L.B. van der Putte. Reaction of bacterium-primed murine T cells to cartilage components: a clue for the pathogenesis of arthritis? *Clin. Exp. Immunol.* **72**: 9–14 (1988).

W.B. van den Berg, F.A.J. van der Loo, W.A. Zwarts, and I.G. Otterness. Effects of murine recombinant IL1 on intact homologous articular cartilage. *Ann. Rheum Dis.* **47**: 855–863 (1988).

M. van der Rest and R. Mayne. Type IX collagen proteoglycan from cartilage is covalently cross-linked to Type II collagen. *J. Biol. Chem.* **263**: 1615–1618 (1988).

G.P.J. van Kampen, J.P. Veldhuijzen, R. Kinjer, R.J. van de Stadt, and C.A. Schipper. Cartilage response to mechanical force in high-density chondrocyte cultures. *Arthritis Rheum.* **28**: 421–424 (1985).

N. Vasan. Effects of physical stress on the synthesis and degradation of cartilage matrix. *Connect. Tissue Res.* **12**: 49–58 (1983).

C.A. Vater, E.D. Harris, Jr., and R.C. Siegel. Native cross-links in collagen fibrils induced resistance to human synovial collagenase. *Biochem. J.* **181**: 639–645 (1979).

L. Vaughan, M. Mendler, S. Huber, P. Bruckner, K.H. Winterhalter, M.I. Irwin, and R. Mayne. D-Periodic distribution of collagen type IX along cartilage fibrils. *J. Cell Biol.* **106**: 991–997 (1988).

T. Videman, I. Eronen, and T. Candolin. ^{3}H proline incorporation and hydroxyproline concentration in articular cartilage during the development of osteoarthritis caused by immobilization. A study *in vivo* with rabbits. *Biochem. J.* **200**: 435–440 (1981).

T. Videman. I. Eronen, and C. Friman. Glycosaminoglycan metabolism in experimental osteoarthritis caused by immobilization. The effects of different periods immobilization and follow-up. *Acta Orthop. Scand.* **52**: 11–21 (1981a).

N. Vittorio, J.D. Crissman, C.N. Hopson, and J.H. Herman. Histologic assessment of cathepsin D in osteoarthritic cartilage. *Clin. Exp. Rheumatol.* **4**: 221–230 (1986).

K.G. Vogel, M. Paulsson, and D. Heinegard. Specific inhibition of type I and type II collagen fibrillogenesis by the small proteoglycan of tendon. *J. Biochem.* **223**: 587–597 (1984).

K. von Figura and A. Hasilik. Lysosomal enzymes and their receptors. *Annu. Rev. Biochem.* **55**: 167–193 (1986).

K. von Figura, H. Mittelviejhaus, R. Prinz, M. Ducheme, and T. Krieg. Endocytosis of sulfated proteoglycans by chicken chondrocytes. In: *Biology of the Articular Cartilage in Health and Disease,* Ed H. Gaster. pp. 189–195, 1980.

A. Wasteson, U. Lindahl, and A. Allen. Mode of degradation of the chondroitin sulphate proteoglycan in rat costal cartilage. *Biochem. J.* **130**: 729–738 (1972).

H.G. Welgus, J.J. Jeffrey, G.P. Stricklin, W.T. Roswit, and A.Z. Wisen. Characteristics of the action of human skin fibroblast collagenase on fibrillar collagen. *J. Biol. Chem.* **255**: 6806–6813 (1980).

Z. Werb, C.L. Mainardi, C.A. Vater, and E.D. Harris. Endogenous activation of latent collagenase by rheumatoid synovial cells. Evidence for a role of plasminogen activator. *N. Engl. J. Med.* **296**: 1017–1023 (1977).

J.M. Williams and J.D. Brandt. Exercise increases osteophyte formation and diminishes fibrillation following chemically induced articular injury. *J. Anat.* **139**: 599–611 (1984).

D.E. Woolley. Mammalian collagenases. In: *Extracellular Matrix Biochemistry,* Eds K.A. Piez and A.H. Reddi. Elsevier, New York. pp. 119–157, 1984.

D.E. Woolley, R.W. Glanville, M.J. Crossley, and J.M. Evanson. Purification of rheumatoid synovial collagenase and its action on soluble and insoluble collagen. *Eur. J. Biochem.* **54**: 611–622 (1975).

D.E. Woolley, R.W. Glanville, K.A. Lindberg, A.J. Bailey, and J.M. Evanson. Action of human skin collagenase on cartilage collagen *FEBS Lett.* **34**: 267–269 (1973).

D.E. Woolley, M.J. Crossley, and J.M. Evanson. Collagenase at sites of cartilage erosion in the rheumatoid joint. *Arthritis Rheum.* **20**: 1231–1239 (1977).

D.E. Woolley, R.W. Glanville, D.R. Roberts, and J.M. Evanson. Purification, characterisation and inhibition of human skin collagenase. *Biochem. J.* **169**: 265–276 (1978a).

J.J. Wu, M.W. Lark, L.E. Chun, and D.R. Eyre. Molecular sites of stromelysis cleavage in collagen Types II, IX, X and XI of cartilage *J. Biol. Chem.* in press (1991).

H. Yamada, R.W. Stephens, T. Nakagawa, and D. McNicol. Human articular cartilage contains an inhibitor of plasminogen activator. *J. Rheumatol.* **15(7)**: 1138–1143 (1988).

INDEX

A

Achondrogenesis, 201
Achondrogenesis-hypochondrogenesis, type II, 14
Adenine nucleotides
  balance of and initiation of protein synthesis, 121—122
  effect of ADP ribosylation on, 109
  measurements of, 121
Adenosine diphosphate (ADP), 97
  in energy metabolism, 98
  oxygen metabolism and, 116—117
  phosphorylation of, 98—99
  ribosylation, 108—109
    and calcium transport, 110—111
    NAD-dependent, 110—111, 122—123
Adenosine monophosphate (AMP), 97
Adenosine triphosphate (ATP)
  ATP/ADP ratio, 99
    and initiation of mineralization, 121
  in energy metabolism, 97—98
  generation of, 98—99
Albumin
  permeability of proteoglycans to, 86
  solution space for, 84—85
Alkaline phosphatase
  and cartilage matrix, 188—190
  deficiency of, 189
  functions of, 188—189
  lack of in hyperphosphatasia, 202
  in phosphate generation, 203
  in plasma membranes, 189—190
  vitamin D_3 effect on, 187—188
alpha1(II) Collagen gene
  exon organization of, 4—7
  molecular cloning of, 2—4
  regulation of, 7—12
alpha1(II) Procollagen chains, triple-helix structure of, 213—214
alpha1(IX) Collagen gene
  amino-terminal amino acid sequence of, 23—24
  exon structure of, 23
  molecular cloning of, 21—24
  multiple domain structure of, 24—27
alpha1(X) Collagen gene
  molecular cloning of, 42—43
  regulation of, 43—44
alpha1(XI) Collagen gene, 14—15
  domain structure of, 16
  molecular cloning of, 15—16
alpha1(XII) Collagen gene
  monoclonal antibody against, 28
  mRNA of, 28
  multiple domain structure of, 27—28
alpha2(IX) Collagen gene
  exon structure of, 22
  molecular cloning of, 21—24
  multiple domain structure of, 24—27
alpha2(XI) Collagen gene, 14—15
  domain structure of, 16
  molecular cloning of, 15—16
alpha1 Proteinase inhibitor, 231
Amino propeptide regions, similarity of type II and type V, 7
Amino trypsin cleavage site, 32
Anchorin CII, type VI, 2
Angiogenesis, 193
  in inflammatory disease, 238
Arthritis, cartilage breakdown in, 87, see also Rheumatoid arthritis
Articular cartilage, oxygen tension in, 105—106
Asparagine glycosylation sites, 34
Avian tibial dyschondroplasia, 202

B

Binding proteins, 133—134
Biphasic matrix, 61
B lymphocytes, in inflammation, 239—240
Bone
  factors inducing growth of, 143—145
  factors in formation and remodeling of, 141
  IGFs isolated from, 134
  oxygen supply to cartilage in, 107
  as source of TGF-beta, 142—143
  type V collagen in, 7
Bone inducing protein, 144
Bone-matrix derived factors, chondrogenesis and, 163

Bone morphogenetic protein (BMP), 142, 144
  activity of, 143—144

C

Cadherins, 158
Calcification, see also Mineralization
  essential requirements for, 189
  in focal sites, 180—181
  initiation of, 180—181
  matrix changes in hypertrophic longitudinal septa during, 199
  matrix vesicles in, 190
  process of, 183—185
  type II procollagen in, 192
  type X collagen in, 192
Calcification sites, proteolgycan concentration at, 197—198
Calcium
  bound to chondroitin sulfate, 198—200
  in chondrogenesis, 158—160, 171
  concentration of, 203
  efflux of from mitochondria, 119
  entry of into cells, 112—114
  of extracellular matrix, 158—160
  intracellular concentration of, 112—113
  mitochondrial, 188
  transport of
    and cartilage energy metabolism, 112—119
    by chondrocytes, 110—111, 116—119
  uptake of by chondrocytes, 118
Calcium channel blockers, 157
Calcium dam, 112
Calcium extrusion phenomenon, 118
Calcium ions
  and chondrogenesis *in vitro*, 156—158
  intracellular level of, 157
Calcium phosphate granules, and cartilage mineralization, 114—115
Calmodulin mRNA levels, 113
Calvarium
  aberrant chondrogenesis in, 160
  calcium-deficient, 159
  embryonic, 159—160
Carboxyl propeptide region, 17
Carboxyl telopeptides, 17
Carboxyl terminal globular domain, of core protein, 33, 35
Cartilage
  active genes of, 1—45
  articular, 61
  breakdown of in arthritis, 87
  components of, 2
  compositional analysis of, 61—63
  degradation of, 213—241
  differentiation of, 9
  endochondral ossification of, 43
  energy metabolism in, 97—123
  factors in formation of, 141
  growth and differentiation factors in, 131—147
  growth and oxygen supply to, 104—105
  growth factors affecting, 133
  histogenesis of, 153
  hypertrophic, type X collagen gene function in, 44—45
  IGFs effects on, 136
  IGFs isolated from, 134
  impaired calcification of, 186
  in vitro studies of, 81
  metabolism of glutamine and glutamate in, 101—102
  nutrient requirements and energy generation by, 99—102
  osmotic flow in, 77—79
  osmotic pressure of, 62
  reviews of physicochemical properties of, 60—61
  type IX collagen in, 19—21
  type XI collagen in, 15
  vascularization in formation of, 193
Cartilage-derived factors (CDFs), 136
  interaction of with IGFs, 134
Cartilage-derived growth factor (CDGF), 136
  effects of, 137
Cartilage extracellular matrix
  components of, 28
  physicochemical aspects of, 59—92
Cartilage/fluid interfaces, partitioning of macromolecules at, 84—88
Cartilage-inducing factors (CIFs), 138
  function of, 142—143
Cartilage matrix protein, 40
  cloning of, 40—41
  genes encoding, 1
  regulation of, 41
Cartilage proteoglycan core protein
  molecular cloning of, 29—36
  regulation of, 36—37
  structure of, 28—29
Cartilage proteolgycans, structure-function of, 90—92

Cartilage stimulating activity (CSA), 143
CAT gene, 9
Cathepsin D
  and collagen digestion, 227—228
  glycosylation of, 229
Cathepsin G, 225
Cathepsins, in collagen degradation, 227
Cationic proteins, in macromolecular partitioning, 85—88
Cell-cell interactions
  in chondrogenesis, 155—156
  classification of, 155—156
  polyionic molecules in, 161
Cell-extracellular matrix interactions, 164
Cell-to-matrix interactions, 155
Cellular condensation
  in chondrogenic differentiation, 160
  and *in vivo* chondrogenesis, 153—154
Charge-charge interactions, 87
Chondroblasts
  cartilage matrix secreted by, 169—170
  dedifferentiated state in, 138
  growth factors affecting, 133
  mitogenic response of, 140
Chondrocalcin, 194—195
Chondrocytes
  5′ promoter in, 21
  angiogenesis and, 193
  articular, 213
  assembly of, 153
  calcium transport by mitochondria of, 110—111, 116—119
  cartilage matrix synthesis by, 202
  CIFs in, 142
  decreased metabolism of, 239
  dedifferentiated, 9
  enhanced differentiation of, 110
  failure of in osteoarthritis, 240
  GAG synthesis by, 137
  glutamine in energy economy of, 122
  human costal, 9
  hypertrophy of, 42—44
  hypoxia of, 123
  IGF-I and IGF-II competition for binding to, 135
  interleukin-1 effects on, 141
  measurement of oxygen consumption by, 106
  mechanisms of energy generation and use in, 97—123
  metabolism of, 60, 103
  NAD in riboxylation reactions of, 122—123
  oxygen supply to, 105
  oxygen tension and growth of, 106—107
  pericellular matrix of, 63, 87
  prehypertrophic, 193
Chondrodysplasias, 201—202
Chondrogenesis, 153
  calcium and polyions in regulation of, 171
  calcium ions and *in vitro,* 156—158
  cell-cell interactions in, 155—156
  in vitro, 154—155
  in vivo and cellular condensation, 153—154
  ionic regulation of, 153—171
  matrix calcium and, 158—160
  and mineralization, 123
  model for polyionic regulation of, 169—170
  morphogenetic phase of, 154
  and oxygen tension, 106—107
  polyionic molecules in, 161—169, 171
  pyridine nucleotides and, 107—112
  studies of using chick tissues, 8—9
Chondrogenic differentiation, 160
Chondrogenic phenotype, expression of, 142
Chondroitin, hydraulic conductivity of, 74—75
Chondroitin 6-sulfate, 201
  interaction of with poly-L-lysine, 162
Chondroitin sulfate
  calcium bound to, 198—200
  in chondrogenesis, 164
  decreased synthesis of, 106
  high sulfation density of, 165
  linkages of, 74—75
  osmotic pressure of, 65
  sulfate and carboxyl groups of, 200
Chondroitin sulfate chains
  anionic charges of, 91
  change in chemistry of, 201
  cylindrical domain concentration of, 83—84
  immobilization of, 82
  interactions between, 82—83
  kinetic segments of, 82
  segmental mobility of, 66—69
  sulfation of, 88
Chondroitin sulfate-rich proteoglycans, 65, 218
  in matrix synthesis, 185

Chondron, heterogeneity of, 63
Chondroprogenitor cells, 164—165
Chondrosarcoma proteoglycan, composition of, 33
Chromosomal location
  of type II collagen genes, 13—14
  of type IX collagen genes, 28
  of type XI collagen genes, 17—18
Chromosome 12, human type II collagen on, 13—14
Chymotrypsin, 225
Clinical markers, in cartilage degradation diagnosis, 241
Collagen, see also specific types
  in calcification of cartilage, 183
  in cartilage degradation, 213, 236
  chromosomal location of genes, 13—14, 17—18, 28
  depolymerization of in cartilage degradation, 225—227
  genes encoding, 1
  helical in cartilage degradation, 227—228
  insoluble cross-linked, 216—217
  in matrix calcification, 193—197, 202—203
  in matrix synthesis, 185—186
  pore sizes in, 85
  potential pathways for degradation of, 214
  proteolysis of, 226
  type II, 1—2
    gene of, 2—14
  type VI, 2
  type V, 2
  type IX, 1
    genes, 18—28
  type X, 41—45, 192
    function of, 1
  type XI, 1
    gene of, 14—18
  and vitamin D deficiency, 188
Collagenase
  action of on collagen helix, 227—228
  increased synthesis of, 229
  inhibitors of, 231—232
  in rheumatoid arthritis, 238, 239
Collagen fibrils
  tensile strength of, 1
  type II, 191
  type IX in, 19—21
Collagen-proteoglycan matrix, 81—82
Complementary DNA clones
  of core protein gene, 29—36
  for link protein genes, 37—39
  hybridization of to mRNAs, 35—36
  for matrix protein genes, 40—41
  type II collagen, 2—4
  type V collagen-related, 15
  type X collagen, 42—43
  type XI collagen, 15—16
Compressibility, of proteoglycans, 70—72
Connective tissue cells, energy source for, 101
Connective tissue polysaccharides, exclusion properties of, 84—85
Core protein gene
  molecular cloning of, 29—36
  regulation of, 36—37
  structure of, 28—29
Core proteins
  in cartilage degradation, 222
  mRNA of, 36—37
Corneal stroma, type IX collagen in, 21
C-propeptide
  in calcification of cartilage, 183
  calcium binding function of, 195
  in hypophosphatasia, 202
  in Kniest dysplasia, 201
  of type II collagen, 194—195, 203
  of type XI collagen, 16
Cross-linked collagen, 216—217
Cystatins, 231
Cysteine proteinases, precursors of, 230
Cysteine residues, of core protein, 34
Cysteinyl residues
  of matrix protein genes, 40
  of type IX collagen gene, 27
  of XI collagen, 17
Cytochrome oxidase
  and calcium accumulation, 117—118
  oxygen and, 103
Cytokines, see also specific types
  in cartilage degradation, 232—236
    inhibition of, 236—238
  growth-inducing functions of, 140—141
  in proteinase synthesis modulation, 228—229
Cytosine residues, methylation of, 11—12
Cytosolic calcium ions, 113

D

Darcy's law, 79
Decalcification, 180

Deer antler cells, granules in, 115
Degenerative cartilage disease, immune component in, 141
Degradation
enzymes in, 222—232
inhibitors of, 216
intracellular, 216
mediators of, 232—241
potential pathways for, 214
relationship between structure and susceptibility to, 213—222
Demineralized bone matrix (DBM), 141—142
Dermatan sulfate
in chondrogenesis, 165
in matrix synthesis, 185
Descemet's membrane, hexagonal lattice structure of, 45
Dextran sulfate, 165
Diacylglycerol (DAG), 132
Differentiation factors, 132
early work with, 141
functions of, 141—142
types of, 142—145
Differentiation memory, 157
Diffusional volume expansion, 82
Diffusion coefficients, 63
of chondroitin sulfate chains, 69
effective quantities for, 64
Disulfide bonds, 222
interchain, 213—214
DNA probes, encoding human type IX collagen chains, 28
Domain structure, 24—27
Donnan distribution, 65—66

E

EGF-like domains, encoding for, 40—41
Elastase, 225
in cartilage breakdown, 87
Embryonic endochondral bone formation, 141
Endochondral bone inducing factors, 143—145
Endochondral ossification, 43, 158—159, 193, 203
factors in, 141—142, 144—145
impairment of cartilage calcification in, 186
Endoplasmic reticulum
calcium release from, 104, 114
in calcium transport, 113
Endoplasmin, in calcium transport, 113
Endothelial cell growth factor (ECGF), 137
Energy metabolism
and calcium transport, 112—119
in cartilage, 97—123
oxygen tension and, 103—107
Energy transduction, mechanisms for in tissues, 97—99
Enzyme/inhibitor systems, 213
Enzymes
in collagen depolymerization, 225—227
regulation of activity of, 228
Epidermal growth factor (EGF), 132, 231—232
effects of, 140—141
mitogenic effect of, 140—141
in proteinase synthesis, 228—229
Epiphyseal cartilage
energy status of, 119—122
maturation and mineralization of, 111—112
Epiphyseal-metaphyseal junction, oxygen supply to, 104—105
Epiphyses
endochondral ossification of, 179
microcirculation of, 105
mineralization of, 120—121
oxygen supply to, 104—105
Exon/intron structure
of core protein gene, 29—32
of type II collagen gene, 4—7
Exon shuffling, 29
Exon structure
of link protein, 38
of type IX collagen genes, 22—24
type X collagen, 42—43
Extracellular matrix
biphasic, 61
in calcification, 202
calcium of in chondrogenesis, 158—160
cellular changes related to hypertrophy and calcification of, 188—202
components of, 60
in chondrogenesis, 166—167
interaction of, 217
macromolecular, 60
compositional analysis of, 61—63
compositional heterogeneity of, 63
degradation of, 193—194
dynamics of, 63—84
factors regulating, 186—188
intracellular link of, 171

maturation of collagen in, 215
mechanical-osmotic pressure balance in, 79—81
molecular transport in, 88—90
multicomponent systems in, 88—90
restricted, 88
molecules of, 164
partitioning of macromolecules at cartilage/fluid interfaces of, 84—88
physicochemical aspects of, 59—92
processing of proteoglycans in, 217—218, 220
synthesis of, 185—188

F

Fatty acids, 102
Fenestrae, 105
Fibrillar collagen family
specialized subclass of, 7
triple-helical domains in, 41
Fibroblast growth factor (FGF), 132
effects of saline and on intrachondral lesions, 139
function and utility of, 137—138
heparin binding property of, 137
in proteinase synthesis modulation, 228—229
and related peptides, 136—138
Fibroblastic cells, 143
Fibronectin, heparin-binding domain of, 167
Fick's Law, 64
Flavoprotein of succinate dehydrogenase ($FADH_2$), reduced, 98
Fractures, factors in repair of, 141
Freeze-drying techniques, 116

G

Gap junctions, communication through, 155—156
G-binding protein, ribosylation of, 110—111
Gelatin, in cartilage degradation, 228
Gel-polymer network, 77—78
Glucose, utilization of, 100—101
Glutamate, metabolism of, 101—102
Glutamine
deamidation of, 101—102
in energy economy, 122
as energy source, 101
Glutamine metabolism pathways, 101
Glycines, in helix structure, 215
Glycolysis
anerobic, 100
normalization of, 101
Glycolytic regulatory system, 119
Glycopeptidase F, 145
Glycoprotein, synthesis of, 36—37
Glycosaminoglycan chains
anionic charge groups along, 91—92
in extracellular matrix, 62
Glycosaminoglycans (GAGs)
in cartilage degradation, 225
in chondrogenesis, 164—165, 167
clusters of, 222
in extracellular matrix, 62
FGF and CDGF effects on, 137
hydrated, 218
increased production of with TGF-beta, 138—140
interaction of with poly-L-lysine, 162
negatively charged, 221
polyanion effect on release and metabolism of, 168
synthesis of, 123
Glycosidase activity, 225
Glycosylation, 221
Gly-Ser-rich domain, 33
Gly-X-Y repeat structure, imperfections in, 26
Ground substance, structural studies of, 61
Growth cartilage
calmodulin mRNA levels in, 113
hypertrophic zone of, 122
NAD and NADH concentrations in, 120
Growth factors, 132, see also specific types
cytokine, 140—141
fibroblast, 136—138
functions of, 132—133
insulin-like, 133—136
as molecular messengers, 155
signal transduction mechanisms of, 132
transforming, 138—140
Growth hormone, 186
Growth plates, 203
angiogenesis and, 193
biosynthetic machinery of, 185—188
calcium transport and, 118
and cartilage matrix hypertrophy and calcification, 188—202
and cellular commitment to hypertrophy, 192—193

cellular/matrix changes and interactions in, 184
growth hormone and insulin-like growth factor-1 in, 186
hypertrophic zone of, 191
oxygen tension in, 104—105
structure and organization of, 179—185
synthesis of, 185—186
vitamin D in, 186—188
GTP/GDP ratio, 122

H

Happel-Brenner model, 75—76
Helical collagen, in cartilage degradation, 227—228
Heparan sulfate proteoglycans
action of, 166
chondrogenesis stimulatory effect of, 164—167
Heparin, 165, 167
Heparinase, 137
Heparitinase, 137
Humoral factors, delivery of, 155
Hyaline articular cartilage, 1
impermeability of, 86
Hyaluronate
in chondrogenesis, 166—167
hydraulic conductivity of, 74—75
sedimentation of, 75
Hyaluronic acid, 2
core protein interaction with, 28
interaction of with proteoglycans, 198, 217
Hybridization analysis, of link protein RNA, 38
Hydraulic conductivities, 68
of chondrosarcoma, 70
influence of, 77
specific, 72—77
Hydrodynamic frictional coefficient, 66—67
Hydrogen bonds, interchain, 214
Hydropathy profiles, of type II procollagen, 4
Hydroxyapatite, 118
formation of, 188—189
precipitation of, 203
synthesis of, 190
Hydroxyproline, 228
3-Hydroxypyridium cross-link, 216
Hypertrophic cartilage, 44—45
Hypertrophic cells, 181, 192
Hypertrophic zone, 181
electron micrograph of, 191, 196
uncalcified, 186
Hypertrophy
cellular changes related to, 188—192
cellular commitment to, 192—193
Hypophosphatasia, 202
Hypoxia, localized, 120—121

I

Imbibition rate, 77—78
Inducing factors, 132, 141—142
Inflammatory cycle, 239
Inositol 1,4,5-trisphosphate (IP_3), 132
In situ hybridization
with cDNA probes for type I collagen mRNA, 12
using type II collagen mRNA probes, 9—10
Insulin, and glucose utilization, 100—101
Insulin-like growth factor-II, 134—136
Insulin-like growth factor-I, 134—136
in matrix synthesis regulation, 186
Insulin-like growth factors (IGFs), 133—136
cellular receptors for, 134
limited availability of, 134—136
source of, 134
Interferon gamma, in inflammation, 239—240
Interglobular domain (IGD), 33
Interleukin-1
in cartilage degradation, 236, 238
in chronic inflammation, 141
in collagen degradation, 236
in osteoarthritis, 241
in proteinase synthesis modulation, 228—229
in proteoglycan degradation, 234—236
Interleukins, 239
Intervertebral disks, type IX collagen in, 19
Intrafibrillar water, 62—63

K

Keel bone, 192
Keratan sulfate
calcium binding to, 200
increased synthesis of, 106
measurement of in serum, 225
Keratan sulfate enriched region, 33

Kininogens, 231
Kniest dysplasia, 201
  and type II collagen defects, 14

L

Lectin homology region, 35
Lectin sequences, vertebrate, 34
Limb bud, development of, 167
Limb chondrogenesis, 153—154
  matrix trapping model of, 170
  three-dimensional developmental programming of, 156
Limb mesenchymal cells
  condensation, 166
  poly-L-lysine effects on, 168—169
Link protein, 28, 37
  amino acid sequencing of, 39
  in cartilage degradation, 221—222, 224—225
  cDNA encoding, 33
  domain A of, 32—33
  domain organization of, 32
  mRNA of, 39—40
  proteoglycan of, 2
  RNA of, 38
Link protein gene, 1
  cloning of, 37—39
  regulation of expression of, 39—40
Lysine-containing heteropolypeptides, 161
Lysosomal enzymes, 216
  increased synthesis of, 229
Lysozyme
  immobilization of, 87
  in matrix, 162—163
Lysyl residues, in fibrillar collagens, 17

M

Macromolecular partitioning, at cartilage/fluid interfaces, 84—88
Macrophage-derived growth factor (MDGF), 137
Marfan syndrome, type II collagen gene linkage to, 14
Matrix, see Extracellular matrix
Matrix protein gene
  cloning of, 40—41
  regulation of, 41
Matrix-to-cell interactions, 155
Matrix trapping, 169—170
Matrix vesicles, 190
Mechanical compression, 80—81
Mechanical loading
  in cartilage degradation, 232—233
  decreased, 232—233
  increased, 233
Mechanical-osmotic pressure balance, 79—81
Membrane-to-membrane interactions, 155
Mesenchymal cells
  differentiation of into chondrocytes, 153
  extracellular signals to, 171
  in *in vitro* chondrogenesis, 154—155
  poly-L-lysine effects on, 168—169
Mesenchymal condensation, 166
Metalloproteinases
  inhibition of, 230—231
  and matrix degradation, 194
  synthetic inhibitors of, 237
  tissue inhibitor of, 228
  transcription of, 228—229
Metaphysis, region of, 182
Methylation studies, of type II collagen genes, 11—12
MHC antigens, class II, 239—240
Mineralization, see also Calcification
  chondrocytes in regulation of, 202—203
  initiation of and energy status of epiphyseal cartilage, 119—122
  matrix vesicles in, 190
  mitochondrial granules and, 114—116
  and oxygen tension, 106—107
Mineralization nodules, 180
Mitochondria
  ATP generation by, 99
  calcium of, 119, 188
  calicum transport by, 116—119
  collapse of membrane potential of, 104
  in energy metabolism, 98
  respiratory activities of, 117
Mitochondrial granules, 114—116
Molecular cloning
  of alpha1(II) collagen gene, 2—4
  of alpha1(X) collagen gene, 42—43
  of alpha1(XI) collagen genes, 15—16
  of core protein gene, 29—36
  of type IX collagen genes, 21—24
Molecular transport, in extracellular matrices, 88—90
Monomer size, relationship of to function, 218—221
Multiplication stimulating activity (MSA), 133
  interactions of with CDFs, 134

N

NAD
biosynthesis pathways of, 108
in growth cartilage, 120
intracellular concentration of, 108
rapid efflux of, 111
in ribosylation reactions, 122—123
tissue levels of, 109—110
utilization of, 107
NAD/NADH ratio, 107, 120
increased, 103
NADP
synthesis of, 100
utilization of, 107
NADP-dependent pentose phosphate shunt pathway, 100
NADP/NADPH ratio, 99
Nanomelia mutation, 36—37
Neahypoxia, 103
Nicotinamide, in epiphyseal cartilage maturation, 111—112
Nicotinamide adenine dinucleotide (NADH)
energy available as, 99
in energy metabolism, 97
generation of, 98
in growth cartilage, 120
Nonfibrillar collagens, type IX collagen as, 27—28
Northern blot hybridization, 38—39
Notochords, chondrogenic effects of, 164
Nucleotide sequence analysis, of type X collagen gene, 42—43
Nutrient requirements, and energy generation, 99—102
Nutrition loss, 181

O

Odontoblasts, granules in, 115
O-linked oligosaccharides, 216—217
Oncogene products, in transduction, 132
Osmotic flow, 77—79
caused by polyelectrolytes, 80
Osmotic imbibition, 81
Osmotic premeability coefficient, 77
Osmotic pressure, 91
balance of with mechanical pressure, 79—81
difference, 77
equation for, 64—65
in extracellular matrix, 62—63
influence of, 77
of macromolecules, 65—66
Osteoarthritis
in cartilage degradation, 240—241
familial, and type II collagen gene, 14
Osteoblasts
cartilaginous scaffolds for, 181
granules in, 115
osteoid secretion by, 183
Osteoclasts, 183
granules in, 115
Osteogenesis imperfecta, lethal, 14
Osteogenin, 144
Osteoid, secretion of, 183
Osteoinductive factors, 144, 146
purified, 144—145
Oxidative phosphorylation, ADP stimulated, 117
Oxygen
as electron acceptor, 103
supply of to cartilage in developing bone, 107
supply of to growth cartilage, 104—105
Oxygen tension
in articular cartilage, 105—106
and energy metabolism, 103—107
and growth of isolated chondrocytes, 106—107
and oxygen supply to cartilage, 104—105
Oxygen vapor pressure, 103

P

Partition coefficient
change in, 88
variation in, 86
Pentose phosphate shunt, 100
Perichondrial cells, IGFs produced by, 136
Permeability coefficient conductivity, 73
Plasma membrane
calcium ATP-ase, 113
calcium transport across, 112—114
Plasmin, 230
Platelet-derived growth factor (PDGF), 132
effects of, 140
Polyadenylation sequences, 35—36
Polyadenylation site, for link protein clones, 38—39
Polyanionic proteoglycan compartment, osmotic pressure of, 65—66
Polyanion-polycation binding mechanism, 87—88

Polyanions
  in chondrogenesis, 163—167
  interaction of with polycations, 167—169
Poly A tract, in link protein cloning, 38—39
Polycations
  in chondrogenesis, 161—163
  interaction of with polyanions, 167—169
Polyionic molecules, in chondrogenesis, 161—171
Poly-L-lysine (PL), 161
  action of on pre-chondrogenic mesenchyme and mature chondrocytes, 162
  in chondrogenesis, 162—163, 167—168
  effects of on limb mesenchymal cells, 168—169
  functional roles of, 167
Prehypertrophic cells, 184
Procollagen
  chondrogenesis stimulatory activity of, 169
  type II, 192
Proenzyme precursors
  activation of, 229—230
  in cartilage degradation, 229—230
Proliferative zone, 179, 183
  cells of, 180
Proline residues, 215
Prostaglandins
  increase in, 239
  interleukin-1 stimulation of, 141
  in rheumatoid arthritis, 238
Proteinase inhibitors, 237—238
  in cartilage degradation, 230—232
Proteinases
  in cartilage degradation, 222—223, 228—229
  modulation of synthesis of, 228—229
  physiological inhibitors of, 230—232
Protein factors, see also Differentiation factors; Growth factors
  classes of, 132
  in growth and maintenance of cartilage phenotype, 131—132
  specificity of signal induced by, 131
Proteins, see also specific types
  cationic, 85—88
  synthesis of and adenine nucleotide balance, 121—122
Proteoglycan-collagen network
  dynamics of, 81—82
Proteoglycan compartment, concentration polarization of, 76—77
Proteoglycans, see also Extracellular matrix
  aggregation of, 203, 218—221
    abnormal, 201—202
    collapsed, 198—200
    composition of, 2
    in internal swelling pressure, 1
    maturation of, 219
  binding component of, 143
  in biphasic matrix, 61
  in calcification of cartilage, 183
  in cartilage degradation, 233—236
  cartilage-specific, 142
  charge-charge interactions in, 87
    with other macromolecules, 85—86
  compressibility of, 70—72
  concentration of, 197—198
  conformation of, 82—84
  core protein structure of, 28—37
    cDNA encoding, 31
    domain structure of, 30
    genes encoding, 1
  degradation products of, 234
  electroneutral kinetic unit of, 79
  flux of, 71
  frictional interaction of, 66—68
  GAGs in, 218
  growth factors in synthesis of, 186
  in growth plate cartilage, 200—201
  high resistance to flow of, 73—74
  hydraulic conductivity of, 72—77
  immobility of, 82
  inhibition of loss of, 81
  interaction of with extracellular matrix components, 143
  interaction of with hyaluronic acid, 198
  in matrix calcification, 197—201
  negative charge of, 163
  organization and composition of, 202
  osmotic pressure of, 62—66, 79—80, 91, 221
  permeability of to albumin, 86
  polyionic interactions with, 168—169
  processing of in cartilage degradation, 217—218, 220
  progressive depletion of in osteoarthritis, 240—241
  proteolysis of, 224
  released from bovine nasal cartilage, 234—236

resistance of to flow, 91
sedimentation coefficient of, 67—68
sedimentation of, 70—71
structural studies of, 61
structure-function of, 90—92
types of, 163—164
vitamin D in synthesis of, 186—188
water binding capacity of, 90—91
Pyridine nucleotide
and chondrogenesis, 107—112
metabolism of and cartilage mineralization, 123
oxidation of and calcium efflux from mitochondria, 119

R

Redox measurements, 119—121
Reserve zone, 179
Reticuloplasmin, in calcium transport, 113
Rheumatoid arthritis, cartilage degradation in, 238—240
Rickets, 186
hypertrophic zone in, 202
Rotational diffusion, 63—64

S

Sedimentation
of hyaluronate, 75
of proteoglycans, 71
Segmental mobility, of chondroitin sulfate chains, 66—69
Ser-Gly-rich domains, 32
boundary between lectin homology and, 35
Serum inhibitors, 231
Signal transduction mechanism, 132
Skeletal growth factor (SGF), 134
Skeletogenesis, 158—159
Sodium chloride, distribution of between femoral head cartilage and sodium chloride solution, 68
Solvent-solute exchange, 78—79
Somatomedins, 133—136
Spondyloepiphyseal dysplasia, type II collagen gene alleles in, 13—14
Steric exclusion, 84—85
Steroids, inhibition of cartilage degradation by, 236—237
Stickler syndrome, linkage of with type II collagen gene, 14
Stromelysin, 194, 225, 228
Sulfation factor, 133

T

12-O-Tetradecanoyl phorbol-13-acetate (TPA), 132
Thermolysin, 225
Tissue hydration, 63
Tissue inhibitor of metalloproteinases (TIMP), 228, 231—232
Trabeculae, 181
calcified, 183
Transforming growth factor-beta (TGF-beta), 132, 231—232
in cartilage differentiation, 9
effects of on organ cultures, 140
in endochondral bone formation, 144—145
family of, 138—140
in osteoinduction, 146
role of in bone formation, 142—143
source of, 142
Translational diffusion, 63, 77
process of, 64
Translational motion, factors governing, 64
Tricarboxylic acid (TCA) cycle, 98, 100
oxidative decarboxylation by, 102
Triple-helical domains
and cartilage degradation, 213—215
of type IX collagen gene, 24—27
of type X collagen genes, 41
of type XI collagen genes, 17
Tumor necrosis factor (TNF), 231
in osteoarthritis, 241
in proteinase synthesis modulation, 228
Type I collagen mRNA, in matrix synthesis, 185
Type II collagen
alpha chains in, 13—14
in calcification of cartilage, 183
in cartilage degradation, 213
C-propetide of, 194—195
fibrils of, 191
in matrix synthesis, 185
mRNA of, 2, 4, 7—8
decreased levels of, 9
pericellular cleavage of, 193—194
protein and gene structure of, 2
triple-helical composition of, 13—14
Type II collagen gene, 2—14
carboxyl-terminal propeptide of, 5
cDNA sequences of in human and rat, 6
chromosomal location on, 13—14

C-terminal protease cleavage site of, 5
encoding alpha3 chain of type XI collagen, 6
and Marfan syndrome, 14
methylation of cytosine residues of, 11—12
N-propeptide triple-helical domain of, 6—7
quantitative analysis of expression of, 8—9
regulation of, 7—12
and spondyloepiphyseal dysplasia, 13—14
Stickler syndrome linkage with, 14
studies of expression of, 8
transfection experiments with, 11
Type II procollagen, 4
Type IX collagen, 18—21
in cartilage collagen fibrils, 19—21
heterotrimeric molecule of, 24
in inflammation, 240
monoclonal antibody against, 21
polypeptide subunits of, 18—19
Type IX collagen genes
chromosomal location of, 28
exon structure of, 23—24
molecular cloning of, 21—24
multiple domain structure of, 24—27
in nonfibrillar collagen family, 27—28
Type X collagen, 41—42
in calcification, 192, 195—197, 202
function of, 1
and matrix degradation, 194
overproduction of in vitamin D deficiency, 188
Type X collagen gene
function of in hypertrophic cartilage, 44—45
molecular cloning of, 42—43
regulation of, 43—44
structure of, 41—42
Type XI collagen
discovery of, 14—15
function of, 15
interaction of with other matrix components, 217
Type XI collagen genes
carboxyl propeptide region of, 17
chromosomal location of, 17—18
molecular cloning of, 15—16
triple-helical domain of, 17

U

Ultramicrotomy techniques, 116

V

Vascularization, in cartilage calcification, 193
Versican, domains in, 29—32
Vitamin D
in alkaline phosphatase synthesis, 189—190
in chondrocytes, 202
deficiency of, 187—188
in matrix synthesis regulation, 186—188
Vitreous, type IX collagen in, 19
Volume expansion concept, 65
Von Willebrand factor, sequence domains in, 40—41

W

Water binding, 90—91